Of Related Interest

Choice and Chance: An Introduction to Inductive Logic

by *Brian Skyrms*

CONTENTS

Derivation and Counterexample: An Introduction to Philosophical Logic

by *Karel Lambert* and *Bas C. van Fraassen*

Derivation and Counterexample: An Introduction to Philosophical Logic is an introduction to the standard body of elementary symbolic logic but is designed specifically for those whose interest is in philosophical aspects and uses of logic.

Contents: Introduction. PART ONE: THE LOGIC OF STATEMENTS. Validity and Logical Truth; Invalidity and Consistency. PART TWO: THE LOGIC OF GENERAL TERMS. Validity and Logical Truth; Invalidity and Consistency. PART THREE: THE LOGIC OF SINGULAR TERMS. Unanalyzed Singular Terms; Analyzed Singular Terms. PART FOUR: FINITE METALOGIC. Metatheory of the Logic of Statements; Metatheory of Free Logic. PART FIVE: PHILOSOPHICAL IMPLICATIONS AND APPLICATIONS OF FREE LOGIC. Philosophical Implications and Applications of Free Logic.

PROBABILITIES, PROBLEMS, AND PARADOXES

Readings in Inductive Logic

PROBABILITIES, PROBLEMS, AND PARADOXES

Readings in Inductive Logic

Sidney A. Luckenbach
San Fernando Valley State College

Dickenson Publishing Company, Inc.
Encino, California, and Belmont, California

Library of Congress Catalog Card Number: 73-163775
ISBN: 0-8221-0010-X

Printed in the United States of America
10 9 8 7 6 5 4 3 2 1

TABLE OF CONTENTS

Introduction

I. TO THE READER

This book of readings is designed primarily for an introductory, though not necessarily elementary, course in inductive logic. While the selections are ideally suited for use with a contemporary text on inductive logic in an undergraduate course (see the Bibliography), they could also be used either to supplement the lectures of an instructor who prefers to dispense with a textbook or to provide the philosophically more mature reader with a relatively brief sampling of some of the basic concepts, problems, and methods of analysis of present-day inductive logic. By and large, the readings, individually and collectively, stand on their own. Some of the papers, especially those in Part I, presuppose some knowledge of the probability calculus; and a few presuppose an understanding of some of the elementary ideas and techniques of deductive logic. In neither case, however, should such presupposed knowledge or understanding be construed as constituting any formal prerequisite.

In selecting the articles reprinted here, one question reigned supreme: What topics, problems, and approaches should be covered in order to prepare a student for further, more advanced work, either on his own or in a second course? The following papers constitute what I believe to be a reasonable answer to this question. Mirroring the historical development of the philosophy of inductive logic, the Prescript by David Hume provides the setting for much of the subsequent discussion and analysis. Part I provides a fairly adequate account of the three dominant theories of *probability,* the key concept in inductive logic, along with three correlative views of induction. Additional accounts of induction are set forth in Part II, centering around the three major problems of inductive logic—*Hume's problem* (the traditional problem of the justification of induction), *Hempel's paradox of confirmation* (the raven paradox), and *Goodman's new riddle of induction* (the grue-bleen paradox)—and their solutions or dissolutions. Several papers in Part II, as well as a few in Part I, also put forth suggestions for further study and development. Finally, the Postscript, consisting of Wilfrid Sellars' "Are There Non-Deductive Logics?," returns us to—if, in point of fact, we were ever entitled to leave—the general philosophical setting of probability and induction.

In addition to the fact that the papers in this volume represent dominant points of view or discuss significant concepts and problems, two other criteria governed their selection. On the one hand, I thought it generally inadvisable to include selections reprinted in other anthologies of similar design and scope. On the other hand, the papers reprinted here are essentially expository and explanatory in nature. (In line with this, I have not thought it advisable to provide a commentary on the selections, and have limited my remarks to this introduction of the topics considered in the readings.) To be sure, there are philosophical sins, of commission and especially omission, in each of the papers reprinted here. But, equally certain, there is a superabundance of mere criticism and polemics these days, especially in philosophy. Yet, what is really called for is positive, concerted, and constructive efforts, not negativistic, disputatious obfuscation. What inductive logic needs most is comprehensive "theories," and the aim of this anthology is to assist the student through the initial steps toward the realization of this goal.

II. LOGIC

Logic might be characterized as the study of the distinction between *correct* (good, rationally justifiable) and *incorrect* (bad, not rationally justifiable) reasoning (see the selections by Hume and Sellars). Needless to say, however, the logician is not concerned with reasoning in the sense of the *process* of drawing inferences or conclusions from known facts or assumptions, but rather with the "completed process," that is, the combination of the (explicitly stated) initial claims or assumptions (the *premises*) and the *conclusion*. Furthermore, the logician, *qua* logician, is solely concerned with what might be called the *logical* or *evidential* relation between premises and conclusion. Is the relation correct or incorrect, good or bad, strong or weak? Or, in other words, what kind and what degree of evidence, support, or grounds do the stated premises provide for the stated conclusion?

The completed inference process is commonly called an *argument*. Still, in its drive for generality, logic abstracts not only from the particular contexts of inferences but also from the actual (or even potential) inference patterns of sapient organisms, and merely requires that an *argument* consist of two parts, a set of statements called its *premises,* and a statement called its *conclusion*. The task peculiar to logic, then, is to explicate and evaluate the evidential link between premises and conclusions of arguments. From a logical point of view, the paradigm of a correct argument, a *valid* argument, is one in which the premises constitute conclusive evidence for the conclusion, while the arguments at the other extreme are those whose premises supply little or no evidential support for the conclusion. (It is to be noted that several papers reprinted here reject this twofold logical abstraction, arguing that it is productive of naught but confusion and paradox, especially with respect to inductive logic. See the papers by Mackie, Grunstra, and Sellars.)

Traditionally, arguments are divided into two *non-overlapping* classes, deductive and inductive, each with its respective criteria of correctness. For various reasons, such a dichotomous classification proves to be of questionable merit. For example, how, on this traditional view, is one to distinguish deductively incorrect arguments from inductive arguments (correct or incorrect)? Certainly no standard (logical) definition of deductive correctness (i.e., validity) will enable one to do this. To demur, at this stage, that no such distinction is needed, is, in effect, to admit that the initial disjoint classification was a pseudo-one. On the other hand, the claim that a deductive argument is one that purports to be deductively valid and an inductive argument is one that purports to be inductively correct—though perhaps a justifiable claim in principle—is, in essence, a rejection of the abstract concept of "argument" as adumbrated above, as well as a rejection of the procedures and principles of much of contemporary logic.

Another reason against this traditional division—a reason that is of major significance for understanding more general aspects of present-day philosophy of inductive logic—is the fact that for some logicians there is no such thing as induction or inductive arguments in the traditional sense of the terms. For them, inductive arguments are simply a (special?) subclass of deductive arguments—for example, those with a peculiar premise like the Principle of Uniformity of Nature, or those with (or relying on) a specific definition of confirmation (see Carnap's "On Inductive Logic"). With such an approach, it is still legitimate to speak of "inductive logic," for this can be defined as the study of certain concepts (probability, confirmation, simplicity, etc.), principles, postulates, rules (the principle of uniformity, the law of causation, the rule of induction, etc.), and problems (the traditional problem of induction, the paradoxes of confirmation, etc.).

Perhaps the best approach is to begin with the notion of argument as spelled out above, and then to explicate *deductive-validity* (or simply *validity*) and *inductive-correctness* (*inductive-strength*). Following this approach, one might propose the following definitions:

> An *argument* is a set of statements, some of which are its premises, and one its conclusion.

> *Logic* is the study, involving both explication and evaluation, of the evidential relation between premises and conclusions of arguments.
>
> An argument is *deductively valid* if, and only if, it is impossible that the conclusion be false and the premises true. (The premises constitute conclusive evidence for the conclusion.)
>
> An argument is *inductively strong* if, and only if, it is not deductively valid and it is *improbable* that the conclusion be false given that the premises are true. (The premises constitute persuasive, though not conclusive, evidence for the conclusion.)

Admittedly, these definitions are rough and require further analysis. Nevertheless, given them, one could characterize deductive logic as essentially, though not merely, the study of deductively valid arguments, and inductive logic as essentially, though not solely, the study of inductively strong arguments. One of the major tasks of deductive logic, then, is to provide a rigorous explication of validity. The analogous task of inductive logic is to supply an adequate analysis of inductively strong arguments or, in short, an adequate analysis of probability. A survey of the major types of arguments commonly classified as inductive—statistical and inductive generalization (induction by enumeration), conversion of deduction, analogical arguments, etc. (see Bibliography reference 36, Arthur Pap)—reveals that, tradition, dictionaries, and speech and English departments notwithstanding, the defining characteristic of inductive arguments is *not* that they go from the specific to the general, but rather that their premises provide at best partial support for, or render probable, their conclusions.

III. PROBABILITY

Probability, according to many philosophers, is our guide in life. To be sure, most of our beliefs, decisions, and actions, in ordinary affairs as well as in science, are *based on* probabilities or probable knowledge rather than certainties. This is true not only for the *making* of decisions and the *acquiring* of beliefs, but for the justification of beliefs, decisions, and actions as well. What, then, is this thing called *probability*?

To provide a complete catalogue of the numerous answers to this question, though of considerable merit, is an extremely time-consuming undertaking and one more suited to a text on probability and induction (see Bibliography reference 1, Henry E. Kyburg). Certainly the uses of the word "probability" are varied, the nuances manifold. Probabilities are applied to events, properties, statements, hypotheses, and theories. The word "probability" is used sometimes in a classificatory sense (the probable and the improbable), sometimes in a comparative sense (more or less probable than), and sometimes in a quantitative sense (probable to degree m/n). On a different level, the word "probably" is often used to lessen the degree of one's commitment to some claim, belief, or promise (see Bibliography reference 44, Stephen Toulmin). For example, to claim, "There will probably be smog in the basin tomorrow" is a way to cautiously or guardedly claim, "There will be smog in the basin tomorrow."

Little is to be gained, however, from any *casual* or *partial* inspection of the role of the word "probability" or its cognates in ordinary and not-so-ordinary reasoning. Furthermore, however useful an extended analysis of "probability" in ordinary language might be, this anthology is designed with other purposes in mind. The three views represented in Part I are the dominant ones in contemporary inductive logic. Initially, they appear to be concerned very little with ordinary usage. In the final analysis, however, the proponents of each of the views believe that something has been said, by way of explication or reconstruction, concerning at least some of the more ordinary uses of the notion. Whether or not this is the case, the reader must decide for himself.

The three views represented below (the *relative frequency, logical,* and *personalistic* theories) have at least one thing in common: they are all primarily concerned with explicating a quantitative concept of probability, and, as a first step, they all presuppose the probability calculus. In other words, each, in its particular fashion, is an "interpretation" of the calculus, a theory of probability partially determined by the probability calculus. As interpretations, the selections stand on their own, but the probability calculus requires some comment. In order to facilitate the presentation of the calculus, as well as to prepare the reader for some of the discussions below, a few words are in order on some basic ideas of the logic of statements and the logic of sets.

Truth-functional statements and connectives. If one thinks of compound statements as those built from "simpler" statements by means of connectives, one is led by a process of abstraction to those statements which are not further decomposable, the basic or *simple* statements. Exactly what these latter statements are is difficult to say and has provoked much controversy in contemporary philosophy. In the elementary logic of statements we are primarily concerned with a special class of compound statements, those which are truth-functional. A *truth-functional compound statement* is a statement the truth value (truth or falsity) of which is uniquely determined by the truth value(s) of its constituent statement(s). The connectives that give rise to truth-functional compound statements are called *truth-functional connectives.*

Five such connectives are required in the sequel: in English they are "not," "and," "or," "if then," and "if and only if" (and their equivalents); their symbolic counterparts are, respectively, "$\sim$," "&," "$\vee$," "$\supset$," and "$\equiv$." Let us use the letters "p," "q," and "r" to stand for statements or sentences that are true or false. The formula "$\sim p$," then, is the negation of "p": if "p" is true, "$\sim p$" is false; if "p" is false, "$\sim p$" is true. The formula "$p \& q$" is the conjunction of "p" and "q": only if both "p" and "q" are true is "$p \& q$" true; otherwise "$p \& q$" is false. The formula "$p \vee q$" is the disjunction of "p" and "q": it is false only if both "p" and "q" are false, otherwise it is true. The formula "$p \supset q$" (read "if p, then q") is a conditional statement, the antecedent of which is "p" and the consequent "q": "$p \supset q$" is false only if "p" is true and "q" is false; otherwise it is true. The formula "$p \equiv q$" (read "p if and only if q" or "if p then q, and if q then p") is a biconditional statement which is true only if "p" and "q" have the same truth value (either both true or both false); otherwise it is false. In each case, the truth value of the compound statement is uniquely determined by the truth value of the constituent statements.

It is important to note that not all compound statements of ordinary discourse, or even of science, are truth-functional. Statements like "Jack believes that God exists," "Jill said that $2+2=5$," and "Jim was arrested after John was murdered," are not truth-functional; though compound, their truth values are not uniquely determined by the truth value of their respective constituent(s). Similarly, statements properly expressed in the subjunctive mood are not truth-functional. While "If the moon *is* made of green cheese, then I am Basil the Bulgar Slayer" is trivially true because of the false antecedent, the statement "If the moon *were* made of soft cheese, a rocket *would* not make a successful landing" is not so. This non-truth-functional character of certain statements is of major concern for inductive logic in the following sense: *If* science requires statements expressing causal or law-like relations, and *if,* as many contend, the latter are properly expressed in the subjunctive and hence not amenable to truth-functional analysis, then a comprehensive theory of probability and induction—or, at the minimum, a comprehensive logic of science—must countenance concepts and methods of analysis not found in standard (purely formal, syntactical, truth-functional) logic.

It is obvious that some statements are true and some are false. What is perhaps less obvious is that some are true (or false) in a different way, or for different reasons, than others. In short, some statements are contingently true (false), that is, true (false) because of the actual constitution of the world, while others are necessarily true (false), true (false) come what may, true (false) in all possible worlds.

A *logical* or *analytic truth* is a statement necessarily true, true in all possible worlds (e.g., "$p \vee \sim p$," "$p \supset$ p," "All black cats are black," etc.). A *logically false* statement (a *self-contradiction*) is a statement necessarily false, false in all possible worlds (e.g., "$p \& \sim p$," "Some black cats are not black," etc.). *Contingent* statements are those which are neither logically true nor logically false. Statements "p" and "q" are said to be *logically equivalent* if they make the same claim (alternatively, if "$p \equiv q$" is logically true). The formula "p" *logically implies* "q" if "$p \supset q$" is logically true. Finally, two statements are said to be *mutually exclusive* if they make conflicting claims (that is, if both cannot be true). (At this stage the reader might consider the following problem: Logical truths, because necessarily true, are assigned the maximum probability value (1), whereas logically false statements, because necessarily false, are assigned the minimum probability value (0). Are there statements other than logical truths which are to be assigned the probability 1? Similarly, are there statements other than self-contradictions which are to be assigned the probability 0?)

Some basic notions of set theory. Similar notions can be introduced for the logic of sets. Intuitively, a *set* is a class or collection, possibly empty, of elements. Let us use the letters "A," "B," "C" for sets. Also, let "W" represent the domain *of discourse,* the class of all elements under consideration, and "ϕ" the *null* or *empty set,* the set that has no members. "$\bar{A}$" designates the *complement* of A, the set containing all those elements in the domain of discourse which are not in A; "$A \cap B$" stands for the *intersection* of A and B, the set of things belonging to both A and B; and "$A \cup B$" represents the *union* of A and B, the set of elements that are members of A or B or both. A and B are said to be *identical* if and only if they have all members in common; A is a *subset* of B if and only if every member of A is also a member of B; and finally, A and B are said to be *mutually exclusive* if and only if they have no elements in common.

The Probability Calculus. The probability calculus is, as its name aptly denotes, a calculus, an (initially) uninterpreted system. In essence, it merely provides a *theoretical* or *formal* or *implicit* definition of "probability"; in other words, it may be regarded as specifying certain minimal conditions that any correct use of (quantitative) probability must satisfy. Furthermore, other than those concepts concerning the set-theoretical or logical structure of the entities to which probabilities are assigned, the calculus utilizes only one *primitive* or *basic* notion: probability.

Three cursory formulations of the probability calculus are presented for the following reasons. As mentioned earlier, probabilities are applied to a number of different kinds of elements—events, sets, statements, etc. To cover this diversity of applications, as well as to prepare the reader for the diverse approaches represented in Part I, two formulations are put forth, one in terms of sets and one in terms of statements. On the other hand, while probability is, for most theorists, a *relational* or *conditional* property of events or sets or statements—that is, "the probability of X" is short for "the probability of X *given* (some) Y"—and not a property of events or sets or statements *simpliciter,* a *non-relational* or *unconditional* approach has, according to some writers, the merit of rendering more intuitive, and thus enhancing the initial understanding of, the problem considered in Part I, that of defining "probability" or interpreting the probability calculus. The first two versions, then, are in terms of *conditional probabilities,* while the third is in terms of *unconditional* or *pure statement probabilities.* (In each formulation there is much redundancy in the set of principles or fundamental properties listed. They are presented in this form merely to help the reader obtain a fuller understanding of the nature and scope of the probability calculus.)

Version I. The Probability Calculus in Terms of Sets

Where $A \neq \phi$, let "$\Pr(B,A)$" represent the conditional probability of B given A:

P1. $\Pr(B,A)$ is a single-valued real function such that $0 \leqslant \Pr(B,A) \leqslant 1$. (For each B and A ($\neq \phi$), there is one and only one probability value of B given A, a real number in the closed interval 0 to 1.)
P2. If A is a subset of B, $\Pr(B,A) = 1$
P3. If $B = C$, $\Pr(B,A) = \Pr(C,A)$
P4. $\Pr(W,A) = \Pr(B \cup \bar{B},A) = 1$
P5. $\Pr(\phi,A) = \Pr(B \cap \bar{B},A) = 0$
P6. $\Pr(B \cap C,A) = \Pr(B,A) \times \Pr(C,A \cap B)$
P7. $\Pr(B \cup C,A) = \Pr(B,A) + \Pr(C,A) - \Pr(B \cap C,A)$
P8. If B and C are mutually exclusive, $\Pr(B \cup C,A) = \Pr(B,A) + \Pr(C,A)$
P9. $\Pr(\bar{B},A) = 1 - \Pr(B,A)$

Version II. The Probability Calculus in Terms of Statements

Where "p" is not logically false, let $\Pr(q,p)$ represent the conditional probability of q given p:

P1. $\Pr(q,p)$ is a single-valued real function such that $0 \leqslant \Pr(q,p) \leqslant 1$
P2. If "p" logically implies "q," $\Pr(q,p) = 1$
P3. If "q" and "r" are logically equivalent, $\Pr(q,p) = \Pr(r,p)$
P4. If "q" is logically true, $\Pr(q,p) = 1$
P5. If "q" is logically false, $\Pr(q,p) = 0$
P6. $\Pr(q \,\&\, r, p) = \Pr(q,p) \times \Pr(r,q \,\&\, p)$
P7. $\Pr(q \vee r,p) = \Pr(q,p) + \Pr(r,p) - \Pr(q \,\&\, r,p)$
P8. If "q" and "r" are mutually exclusive, $\Pr(q \vee r,p) = \Pr(q,p) + \Pr(r,p)$
P9. $\Pr(\sim q,p) = 1 - \Pr(q,p)$

Version III. The Probability Calculus in Terms of Statements

Let $\Pr(p)$ represent the unconditional or pure statement probability of p:

P1. $\Pr(p)$ is a single-valued real function such that $0 \leqslant \Pr(p) \leqslant 1$

P2. If "p" and "q" are logically equivalent, $\Pr(p) = \Pr(q)$

P3. If "p" is logically true, $\Pr(p) = 1$

P4. If "p" is logically false, $\Pr(p) = 0$

DEFINITION: *Where* $Pr(\mathrm{p}) \neq 0$, *the conditional probability of* q *given* p *is that probability such that* $Pr(\mathrm{q}\ given\ \mathrm{p}) = \frac{Pr(\mathrm{p}\ \&\ \mathrm{q})}{Pr(\mathrm{p})}$

DEFINITION: *Statements* "p" *and* "q" *are said to be* stochastically (*or* probabilistically) independent *if and only if* $Pr(\mathrm{q}\ given\ \mathrm{p}) = Pr(\mathrm{q})$ or $Pr(\mathrm{p}) = 0$ *or* $Pr(\mathrm{q}) = 0$

P5. $\Pr(p\ \&\ q) = \Pr(p) \times \Pr(q \text{ given } p)$

P6. If "p" and "q" are independent, $\Pr(p\ \&\ q) = \Pr(p) \times \Pr(q)$

P7. $\Pr(p \vee q) = \Pr(p) + \Pr(q) - \Pr(p\ \&\ q)$

P8. If "p" and "q" are mutually exclusive, $\Pr(p \vee q) = \Pr(p) + \Pr(q)$

P9. $\Pr(\sim p) = 1 - \Pr(p)$

Version I is equivalent to that presented by Hans Reichenbach in *The Theory of Probability* and should be referred to when reading the papers by Reichenbach and Wesley C. Salmon reprinted here. Version II is equivalent to Rudolf Carnap's version in *The Logical Foundations of Probability* and is utilized in his articles in this book. Version III is essentially that presented by Brian Skyrms in *Choice and Chance* and has considerable merit, even though it gives rise to a few problems.

In Versions I and II, *conditional probability*—$\Pr(B,A)$ for sets and $\Pr(q,p)$ for statements—is taken as basic. If desired, an unconditional application of probability to sets or statements can be provided via $\Pr(A) = \Pr(A,W)$ for sets and, where "p" is logically true, $\Pr(q) = \Pr(q,p)$ for statements. Naturally, such a procedure presupposes some independent route to conditional probabilities. Version III, on the other hand, takes *unconditional* or *pure statement probability* to be basic (an analogous approach could be adopted for events, sets, etc.) and then introduces conditional probabilities in terms of it. Of course, this approach presupposes some independent route to pure statement probabilities, especially to the probability of simple statements *and* conjunctions thereof.

What has been called "the philosophical problem of probability" is that of providing an *adequate interpretation,* or possibly more than one adequate interpretation, of the probability calculus. Exactly what constitutes an adequate interpretation differs, often in *ad hoc* fashion, from writer to writer. Criteria of adequacy are discussed in a number of papers, especially in those by Reichenbach, Carnap and Salmon. For now, let us merely say that an *adequate interpretation* of the probability calculus is, at the very minimum, a definition of "$\mathrm{Pr}(p)$" or "$\mathrm{Pr}(q,p)$" or "$\mathrm{Pr}(B,A)$" which (1) satisfies the postulates of the probability calculus, that is, renders them true, and (2) yields a reasonable explication of some, ideally all, of the senses in which probability is utilized in science and ordinary reasoning. Despite the differences among them, the proponents of each of the views represented in Part I believe they have provided an adequate interpretation in this sense. Furthermore, the views represented here all agree on what probabilities *are*—they are simply numbers. The issue that gives rise to such diverse views concerns the manner in which specific probability values *are ascertained*—relative frequentists from relative frequencies, Carnap from the structure of language and conditions of rationality, and personalists from preferences and betting ratios. Whether and to what extent there is genuine, rather than merely verbal, disagreement concerning the meaning of probability or the adequate interpretation of the probability calculus is one of the many important questions the reader must consider for himself.

Reichenbach's treatment of the relative frequency interpretation and its justification is richly and provocatively enhanced by the two papers by Salmon. Carnap's logical interpretation is presented in "On Inductive Logic," and his more recent and somewhat revised views on the personalistic interpretation and on the general nature of inductive logic are found in "The Aim of Inductive Logic." Carnapian themes are also considered in van Fraassen's paper (along with a thought-provoking correlation of the problem of the selection of a probability metric with the traditional philosophical problem of individuation), as well as in the papers by Salmon, Mackie, and Leblanc (though from a less favorable point of view). Finally, there are two papers on the personalistic interpretation: the classic paper by Ramsey and Jeffrey's "Probable Knowledge." In this context one should also consider Carnap's "The Aim of Inductive Logic" and Salmon's "Inquiries Into the Foundations of Science." For other interpretations, see the works of Laplace, Keynes, von Mises, Popper, and Savage listed in the Bibliography.

IV. EMPIRICISM, EPISTEMOLOGY, AND INDUCTION

This anthology might just as well have been titled "Empiricism and Induction," for (1) it is empiricists, by and large, who show the greatest (positive) interest in induction and its problems, and (2) the papers reprinted here, perhaps too one-sidedly, represent this philosophical tradition.

Since it would be at best an understatement to say that most contemporary philosophers and scientists are empiricists, a brief look at the role of induction in empiricist epistemology is in order and might help to provide a setting for many of the discussions below.

A cursory survey of the empiricist tradition from Hume to Reichenbach, Feigl, and Salmon clearly reveals the vital role of induction and probability. In fact, with little, if any, distortion, induction might be viewed as the cornerstone of empiricism: it is central to epistemology, to the theory of concept formation, as well as to the justification of all factual knowledge (not to mention the "discovery" of empirical generalizations and laws of nature). To highlight the centrality of induction and probability in empiricist theories, let us consider a rather fantastic tale, though one propagated by a large number of modern and contemporary philosophers and scientists.

The tale, most suitably called the *Romulus-Remus fantasy,* is the naive empiricist's view regarding epistemology in general and the nature and role of induction in specific. A first formulation of the Romulus-Remus fantasy might go as follows: Man is born into the world with a *tabula rasa* (Locke), in a state of total nescience (Harrod). All knowledge is derived by means of sense experience—*Nihil est in intellectu quod non fuerit in sensu.* This derivation is conceived in such a manner that the acquisition of knowledge is not rendered dependent on any special feature of one's physical or (especially) social environment. After all, Romulus and Remus survived and matured intellectually in their environment! The ultimate or fundamental units of knowledge—the data of immediate experience, the given—are the simple ideas according to Locke, the atomic statements for Russell and Wittgenstein, and the protocol statements of early positivism. The remainder of human knowledge, the non-simple or non-atomic, is *constructed* out of these "atoms" by rather simple and straightforward procedures. Sensation thus "forms" (causes?) the simple concepts, although some elementary inductive procedures might be at play even at this level. (Many empiricists fail to specify what is involved in this "formation" or "derivation" or "abstraction" process, or they refer one to the psychologist with a sleight of hand. Perhaps some empiricists have in mind some simplistic form of conceptual atomism, e.g., that concepts are "picked up" from experience like sea shells from the sea shore.) Other concepts are "formed" by some process of "combining" or "compounding," and finally Mill's methods and/or the straight rule of statistical generalization account for empirical generalizations and laws. With potent but empty minds, given only the data of immediate experience and some simple inductive moves, humans learn all that they know and do all that they do.

If we were to formulate a principle or thesis for this most naive and extreme form of "Romulus-Remus empiricism," it might be something like: All knowledge comes from sense experience. In the minds of most empiricists, however, such a thesis is tantamount to the denial of the synthetic *a priori*—or, positively put, equivalent to the affirmation that all synthetic statements are "inductive" or "probabilistic." However put, the characteristic claim of this extreme form of empiricism might be said to be: *Induction is the indispensable foundation of all factual knowledge.*

The second and perhaps more significant, though restricted, manner in which induction comes into play for empiricists is with respect to generalizations—roughly, that induction is the indispensable foundation of all factual *science*. Traditionally, induction was thought of as inferences or arguments going from the specific to the general. In line with this view, it was (and still is for some) natural to think of *all* empirical generalizations, including, for many theorists, laws, as the end products of inductive reasoning. Surely, one could have argued, theories, laws, and generalizations are not arrived at or discovered deductively. The techniques in this area must be inductive. Did Mill not claim that his methods were adequate for the discovery of causal relations? (A principle or thesis for this variety of empiricism might be something like: *Induction is the indispensable foundation of all factual science.*)

A third and final perspective highlighting the significant role of induction and probability in empiricist theories is obtained by considering the empiricist meaningfulness criterion, often referred to as the thesis or principle of empiricism. The empiricist (positivist, behaviorist, operationalist) attempts to restrict the domain of the cognitively (or empirically) meaningful to those statements which are in principle testable, or confirmable or disconfirmable, on the basis of observational or experimental evidence. In short, his criterion is simply that a statement *S* is cognitively (or empirically) meaningful if, and only if, *S* is analytic or contradictory or confirmable in principle. The cash value of such a proposal depends, of course, on the explication of "testability" or "confirmability," and this requires a theory of confirmation, or, in Carnap's terms, a "theory of probability$_1$." (Concerning these issues, see the papers by Reichenbach, Carnap ("On Inductive Logic"), Hempel ("Studies in the Logic of Confirmation"), and Feigl.)

Perhaps one and all will acknowledge the vital role played by probability and induction in the general empiricist schema. What is perhaps not so obvious is that such emphasis calls for a *comprehensive theory,* not merely bits and pieces and countless promissory notes. This, then, is *the* task confronting the empiricist—to construct a full-fledged system of inductive logic which is consistent (i.e., free of paradoxes and inconsistencies), reasonably complete, and rationally justified. The construction of such a system—and hence the viability of empiricism—hinges, in large part, on the solution of three problems: 1) Hume's Problem: What rationally justifies any inference from the observed to the unobserved, the known to the unknown?

(See the selections by Hume, Reichenbach, Hempel ("Recent Problems of Induction"), Feigl, Black, and Salmon.); 2) Hempel's Paradox: What is confirming evidence (positive instances) for a hypothesis? (See both papers by Hempel and that by Mackie.); and 3) Goodman's new riddle of induction: What hypotheses are confirmed by positive instances? (See Hempel's "Recent Problems of Induction," and the articles by Leblanc and Grunstra.) In principle, until these problems are solved, inductive logic remains in a rather low state; and, consequently, empiricism takes on the appearance of being quite heavily dependent on wishful thinking or dogma concerning probability and induction.

Prescript

SCEPTICAL DOUBTS CONCERNING THE HUMAN UNDERSTANDING

David Hume

All the objects of human reason or inquiry may naturally be divided into two kinds, to wit, "Relations of Ideas," and "Matters of Fact." Of the first kind are the sciences of Geometry, Algebra, and Arithmetic, and, in short, every affirmation which is either intuitively or demonstratively certain. *That the square of the hypotenuse is equal to the square of the two sides* is a proposition which expresses a relation between these figures. *That three times five is equal to the half of thirty* expresses a relation between these numbers. Propositions of this kind are discoverable by the mere operation of thought, without dependence on what is anywhere existent in the universe. Though there never were a circle or triangle in nature, the truths demonstrated by Euclid would forever retain their certainty and evidence.

Matters of fact, which are the second objects of human reason, are not ascertained in the same manner, nor is our evidence of their truth, however great, of a like nature with the foregoing. The contrary of every matter of fact is still possible, because it can never imply a contradiction and is conceived by the mind with the same facility and distinctness as if ever so conformable to reality. *That the sun will not rise tomorrow* is no less intelligible a proposition and implies no more contradiction than the affirmation *that it will rise.* We should in vain, therefore, attempt to demonstrate its falsehood. Were it demonstratively false, it would imply a contradiction and could never be distinctly conceived by the mind.

It may, therefore, be a subject worthy of curiosity to inquire what is the nature of that evidence which assures us of any real existence and matter

---

From Sections IV and V of Hume's *An Enquiry Concerning Human Understanding,* first published in 1748.

of fact beyond the present testimony of our senses or the records of our memory. . . .

All reasonings concerning matter of fact seem to be founded on the relation of *cause* and *effect*. By means of that relation alone we can go beyond the evidence of our memory and senses. If you were to ask a man why he believes any matter of fact which is absent, for instance, that his friend is in the country or in France, he would give you a reason, and this reason would be some other fact: as a letter received from him or the knowledge of his former resolutions and promises. A man finding a watch or any other machine in a desert island would conclude that there had once been men in that island. All our reasonings concerning fact are of the same nature. And here it is constantly supposed that there is a connection between the present fact and that which is inferred from it. Were there nothing to bind them together, the inference would be entirely precarious. The hearing of an articulate voice and rational discourse in the dark assures us of the presence of some person. Why? Because these are the effects of the human make and fabric, and closely connected with it. If we anatomize all the other reasonings of this nature, we shall find that they are founded on the relation of cause and effect. . . .

I shall venture to affirm, as a general proposition which admits of no exception, that the knowledge of this relation is not, in any instance, attained by reasonings *a priori,* but arises entirely from experience, when we find that any particular objects are constantly conjoined with each other. Let an object be presented to a man of ever so strong natural reason and abilities—if that object be entirely new to him, he will not be able, by the most accurate examination of its sensible qualities, to discover any of its causes or effects. Adam, though his rational faculties be supposed, at the very first, entirely perfect, could not have inferred from the fluidity and transparency of water that it would suffocate him, or from the light and warmth of fire that it would consume him. No object ever discovers, by the qualities which appear to the senses, either the causes which produced it or the effects which will arise from it; nor can our reason, unassisted by experience, ever draw any inference concerning real existence and matter of fact. . . .

But to convince us that all the laws of nature and all the operations of bodies without exception are known only by experience, the following reflections may perhaps suffice. Were any object presented to us, and were we required to pronounce concerning the effect which will result from it without consulting past observation, after what manner, I beseech you, must the mind proceed in this operation? It must invent or imagine some event which it ascribes to the object as its effect; and it is plain that this invention must be entirely arbitrary. The mind can never possibly find the effect in the supposed cause by the most accurate scrutiny and examination. For the effect is totally different from the cause, and consequently can never be discovered in it. Motion in the second billiard ball is a quite distinct event from motion in the first, nor is there anything in the one to suggest the smallest hint of the other. A stone or piece of metal raised into the air and left without any support immediately falls. But to consider the matter *a priori,* is there anything we discover in this situation which can beget the idea of a downward rather than an upward or any other motion in the stone or metal?

And as the first imagination or invention of a particular effect in all natural operations is arbitrary where we consult not experience, so must we also esteem the supposed tie or connection between the cause and effect which binds them together and renders it impossible that any other effect could result from the operation of that cause. When I see, for instance, a billiard ball moving in a straight line toward another, even suppose motion in the second ball should by accident be suggested to me as the result of their contact or impulse, may I not conceive that a hundred different events might as well follow from that cause? May not both these balls remain at absolute rest? May not the first ball return in a straight line or leap off from the second in any line or direction? All these suppositions are consistent and conceivable. Why, then, should we give the preference to one which is no more consistent or conceivable than the rest? All our reasonings *a priori* will never be able to show us any foundation for this preference.

In a word, then, every effect is a distinct event from its cause. It could not, therefore, be discovered in the cause, and the first invention or conception of it, *a priori,* must be entirely arbitrary. And even after it is suggested, the conjunction of it with the cause must appear equally arbitrary, since there are always many other effects which, to reason, must seem fully as consistent and natural. In vain, therefore, should we pretend to determine any single event or infer any cause or effect without the assistance of observation and experience. . . .

. . . When it is asked, *What is the nature of all our reasonings concerning matter of fact?* the proper answer seems to be, That they are founded on the relation of cause and effect. When again it is asked, *What is the foundation of all our reasonings and conclusions concerning that relation?* it may be replied in one word, *experience.* But if we still carry on our sifting humor and ask, *What is the foundation of all conclusions from experience?* this implies a new question which may be of more difficult solution and explication. . . .

It must certainly be allowed that nature has kept us at a great distance from all her secrets and has afforded us only the knowledge of a few superficial qualities of objects, while she conceals from us those powers and principles on which the influence of these objects entirely depends. Our senses inform us of the color, weight, and consistency of bread, but neither sense nor reason can ever inform us of those qualities which fit it for the nourishment and support of the human body. Sight or feeling conveys an idea of the actual motion of bodies, but as to that wonderful force or power which would carry on a moving body forever in a continued change of place, and which bodies never lose but by communicating it to others, of this we cannot form the most distant conception. But notwithstanding this ignorance of natural powers and principles, we always presume when we see like sensible qualities that they have like secret powers, and expect that effects similar to those which we have experienced will follow from them. If a body of like color and consistency with that bread which we have formerly eaten be presented to us, we make no scruple of repeating the experiment and foresee with certainty like nourishment and support. Now this is a process of the mind or thought of which I would willingly know the foundation. It is allowed on all hands that there is no known connection between the sensible qualities and the secret powers, and, consequently, that the mind is not led to form such a conclusion concerning their constant and regular conjunction by anything which it knows of their nature. As to past *experience,* it can be allowed to give *direct* and *certain* information of those precise objects only, and that precise period of time which fell under its cognizance: But why this experience should be extended to future times and to other objects which, for aught we know, may be only in appearance similar, this is the main question on which I would insist. The bread which I formerly ate nourished me; that is, a body of such sensible qualities was, at that time, endued with such secret powers. But does it follow that other bread must also nourish me at another time, and that like sensible qualities must always be attended with like secret powers? The consequence seems nowise necessary. At least, it must be acknowledged that there is here a consequence drawn by the mind that there is a certain step taken, a process of thought, and an inference which wants to be explained. These two propositions are far from being the same: *I have found that such an object has always been attended with such an effect,* and *I foresee that other objects which are in appearance similar will be attended with similar effects.* I shall allow, if you please, that the one proposition may justly be inferred from the other: I know, in fact, that it always is inferred. But if you insist that the inference is made by a chain of reasoning, I desire you to produce that reasoning....

All reasonings may be divided into two kinds, namely, demonstrative reasoning, or that concerning relations of ideas, and moral reasoning, or that concerning matter of fact and existence. That there are no demonstrative arguments in the case seems evident, since it implies no contradiction that the course of nature may change and that an object, seemingly like those which we have experienced, may be attended with different or contrary effects. May I not clearly and distinctly conceive that a body, falling from the clouds and which in all other respects resembles snow, has yet the taste of salt or feeling of fire? Is there any more intelligible proposition than to affirm that all the trees will flourish in December and January, and will decay in May and June? Now, whatever is intelligible and can be distinctly conceived implies no contradiction and can never be proved false by any demonstrative argument or abstract reasoning *a priori*.

If we be, therefore, engaged by arguments to put trust in past experience and make it the standard of our future judgment, these arguments must be probable only, or such as regard matter of fact and real existence, according to the division above mentioned. But that there is no argument of this kind must appear if our explication of that species of reasoning be admitted as solid and satisfactory. We have said that all arguments concerning existence are founded on the relation of cause and effect, that our knowledge of that relation is derived entirely from experience, and that all our experimental conclusions proceed upon the supposition that the future will be conformable to the past. To endeavor, therefore, the proof of this last supposition by probable arguments, or arguments regarding existence, must be evidently going in a circle and taking that for granted which is the very point in question.

...When a man says, *I have found, in all past instances, such sensible qualities, conjoined with such secret powers,* and when he says, *similar sensible qualities will always be conjoined with similar secret powers,* he is not guilty of a tautology, nor are these propositions in any respect the same. You say that the one proposition is an inference from the other; but you must confess that the inference is not intuitive, neither is it demonstrative. Of what nature is it then? To say it is experimental is begging the question. For all inferences from experience suppose, as their foundation, that the future will resemble the past and that similar powers will be conjoined with similar sensible qualities. If there be any suspicion that the course of nature may change, and that the past may be no rule for the future, all experience becomes useless and can give rise to no inference or conclusion. It is impossible, therefore, that any arguments from experience can prove this resemblance of the past to the future, since all these arguments are founded on the supposition of that resemblance. Let the course of things be allowed hitherto ever so regular, that alone, without some new argument or inference, proves not that for the future it will continue so. In vain do you pretend to have learned the nature of bodies from your past experience. Their secret nature, and consequently all their effects and influence, may change without any change in their sensible qualities. This happens sometimes, and with regard to some objects. Why may it not happen always, and with regard to all objects? What logic, what process of argument secures you against this supposition? My practice, you say, refutes my doubts. But you mistake the purport of my question. As an agent, I am quite satisfied in the point; but as a philosopher who has some share of curiosity, I will not say skepticism, I want to learn the foundation of this inference....

It is certain that the most ignorant and stupid peasants, nay infants, nay even brute beasts, improve by experience and learn the qualities of natural objects by observing the effects which result from them. When a child has felt the sensation of pain from touching the flame of a candle, he will be careful not to put his hand near any candle, but will expect a similar effect from a cause which is similar in its sensible qualities and appearance. If you assert, therefore, that the understanding of the child is led into this conclusion by any process of argument or ratiocination, I may justly require you to produce that argument, nor have you any pretense to refuse so equitable a demand. You cannot say that the argument is abstruse and may possibly escape your inquiry, since you confess that it is obvious to the capacity of a mere infant. If you hesitate, therefore, a moment or if, after reflection, you produce an intricate or profound argument, you, in a manner, give up the question and confess that it is not reasoning which engages us to suppose the past resembling the future, and to expect similar effects from causes which are to appearance similar. This is the proposition which I intended to enforce in the present section. If I be right, I pretend not to have made any mighty discovery. And if I be wrong, I must acknowledge myself to be indeed a very backward scholar, since I cannot now discover an argument which, it seems, was perfectly familiar to me long before I was out of my cradle.

Suppose a person, though endowed with the strongest faculties of reason and reflection, to be brought on a sudden into this world; he would, indeed, immediately observe a continual succession of objects and one event following another, but he would not be able to discover anything further. He would not at first, by any reasoning, be able to reach the idea of cause and effect, since the particular powers by which all natural operations are performed never appear to the senses; nor is it reasonable to conclude, merely because one event in one instance precedes another, that therefore the one is the cause, the other the effect. The conjunction may be arbitrary and casual. There may be no reason to infer the existence of one from the appearance of the other: and, in a word, such a person without more experience could never employ his conjecture or reasoning concerning any matter of fact or be assured of anything beyond what was immediately present to his memory or senses.

Suppose again that he has acquired more experience and has lived so long in the world as to have observed similar objects or events to be constantly conjoined together—what is the consequence of this experience? He immediately infers the existence of one object from the appearance of the other, yet he has not, by all his experience, acquired any idea or knowledge of the secret power by which the one object produces the other, nor is it by any process of reasoning he is engaged to draw this inference; but still he finds himself determined to draw it, and though he should be convinced that his understanding has no part in the operation, he would nevertheless continue in the same course of thinking. There is some other principle which determines him to form such a conclusion.

This principle is *custom* or *habit*. For wherever the repetition of any particular act or operation produces a propensity to renew the same act or operation without being impelled by any reasoning or process of the understanding, we always say that this propensity is the effect of *custom*. By employing that word we pretend not to have given the ultimate reason of such a propensity. We only point out a principle of human nature which is universally acknowledged, and which is well known by its effects. Perhaps we can push our inquiries no further or pretend to give the cause of this cause, but must rest contented with it as the ultimate principle which we can assign of all our conclusions from experience. It is sufficient satisfaction that we can go so far without repining at the narrowness of our faculties, because they will carry us no further. And it is certain we here advance a very intelligible proposition at least, if not a true one, when we assert that after the constant conjunction of two objects, heat and flame, for instance, weight and solidity, we are determined by custom alone to expect the one from the appearance of the other. This hypothesis seems even the only one which explains the difficulty why we draw from a thousand instances an inference which we are not able to draw from one instance that is in no respect different from them. Reason is incapable of any such variation. The conclusions which it draws from considering one circle are the same which it would form upon surveying all the circles in the universe. But no man, having seen only one body move after being impelled by another, could infer that every other body will move after a like impulse. All inferences from experience, therefore, are effects of custom, not of reasoning.

Custom, then, is the great guide of human life. It is that principle alone which renders our experience useful to us and makes us expect, for the future, a similar train of events with those which have appeared in the past. Without the influence of custom we should be entirely ignorant of every matter of fact beyond what is immediately present to the memory and senses. We should never know how to adjust means to ends or to employ our natural powers in the production of any effect. There would be an end at once of all action as well as of the chief part of speculation.

I

Probability and Induction

1

A FREQUENTIST THEORY OF PROBABILITY AND INDUCTION

Hans Reichenbach

THE PROBABILITY CONCEPT OF THE LANGUAGE OF EVERYDAY LIFE

The word "probable" is frequently used in everyday language; more often, however, the concept is employed without being explicitly expressed. We must restrict to mere probability not only statements of comparatively great uncertainty, like predictions about the weather, where we would cautiously add the word "probable," but also statements of so high a degree of probability that we do not consider it necessary to mention the unavoidable uncertainty, or statements of the probability character of which we are not even conscious. Sometimes we express the uncertainty by a gesture or the accentuation of words. If we expect the plumber to come for some repairs we may communicate the news to the family by a shrug of the shoulders. But in many instances even this symbolic gesture is missing. Thus, when we go to the station to catch a certain train, it does not occur to us that because of an accident the train might for once be late. Even a desultory consideration of the statements of daily life shows clearly that a great number of them owe their character of "certainty" to a confusion of certainty with a high degree of probability. On close inspection, finally, it becomes evident that there are no statements of absolute certainty, if the statements are not to designate empty logical relations but to assert the existence of specific facts.

From Reichenbach's *Theory of Probability,* pp. 3–5, 68–69, 75–76, 337–340, 343–344, 347–348, 350–354, 359, 361, 364–365, 372–382, 429, 431–434, 440, 442–448, 473–475, 482. Originally published by the University of California Press, © 1949; reprinted by permission of The Regents of the University of California.

Incidentally, it would be a mistake to believe that the concept of probability concerns only statements about the future. In many statements about the past, we evidently use the concept of probability. The historian considers it very probable that Nero ordered the burning of Rome; he believes it less probable that Henrietta of England, who lived at the court of Louis XIV as the Duchess of Orleans, was murdered; he regards it as improbable that Bacon was the author of Shakespeare's plays. Even events of the past can only be asserted as probable.

Although we *apply* the concept of probability in daily life as a matter of course, we find it difficult to say what we *mean* by the concept "probable." We know that a probability statement neither asserts nor denies the facts that it designates, but we do not restrict it to an assertion of the mere possibility of an event, since we make distinctions in the degree of probability. For instance, we regard Nero's responsibility for the burning of Rome as more probable than Bacon's authorship of Shakespeare's plays. But we have only a vague notion of the *applicability* of the concept without being able to explain its *meaning*. Since, furthermore, probability statements appear to be grounded in the insufficiency of human knowledge, they seem to be more or less subjective. It is impossible to know with certainty who the Man in the Iron Mask was; his identity, however, is an objective fact, and if those who knew his origin had left a trustworthy account, they would have spared us the uncertainty. To know how the weather will be tomorrow is not possible for us; but we hope that future meteorology will predict the weather of the next day with certainty, or at least with the same certainty that the arrival of trains is predicted today.

What can be the significance, for philosophical investigation, of a concept whose interpretation is vague and whose origin seems to be rooted in the inadequacy of human knowledge? The analysis of the probability concept of everyday language has, indeed, been rather fruitless for philosophical investigation. Its inefficiency has manifested itself in the philosophical critique of the probability concept carried through in traditional philosophy. Philosophers have been satisfied to construe probability as an uncertainty originating in the imperfection of human knowledge, and to connect the concept of probability with that of possibility; this was virtually all that philosophy could discover so long as it restricted its studies to the probability concept of everyday language. Thus the first line of development of a theory of probability—which, incidentally, goes back to Aristotle—did not supply any significant results.

Philosophers even tried to eliminate the concept of probability from science and to restrict it to prescientific language—which was an evasion rather than a solution of the problem. It cannot be admitted that everyday language uses concepts essentially different from those of scientific language. The analysis of science has shown that there is no sharp borderline between scientific and prescientific statements, that the concepts of daily life are absorbed by the language of science, in which they take on a more concise form and a clearer content without being abandoned. Thus it has become evident that criticism of fundamental concepts in their scientific formulation is more fruitful than reflection about their naïve usage; that scientific expression, through its precise wording, leads to a clearer interpretation of concepts and a deeper understanding of their meanings.

It may be recalled that the analysis of the concepts of time and space remained futile so long as the philosophical discussion did not extend beyond the use of these concepts in daily life. Only with the elaboration of the scientific theory of space and time, carried through in non-Euclidean geometry and the theory of relativity, did it become possible to uncover the ultimate nature of space-time concepts and to achieve a more profound understanding of their application to daily life. Thus we now have a better knowledge of what an architect means when he specifies lengths and widths in the plan of a building, and what a watchmaker does when he synchronizes a number of watches. Another example is logic, the fundamentals of which, though continuously applied in everyday language, remained unclarified until mathematicians undertook the analysis and formulation of logical relations. The probability concept, therefore, can be studied successfully only within the realm of its scientific application.

THE FREQUENCY INTERPRETATION

The laws of the calculus of probability are difficult to understand, however, if one does not envisage a definite interpretation. Thus, for didactic reasons, an interpretation of the probability concept must be added, at this point, to the axiomatic construction. But this method will not prejudice later investigations of the problem. The interpretation is employed merely as a means of illustrating the system of formal laws of the probability concept, and it will always be possible to separate the conceptual system from the interpretation, because, for the derivation of theorems, the axioms will be used in the sense of merely formal statements, without reference to the interpretation.

This presentation follows a method applied in the teaching of geometry, where the conceptual formulation of geometrical axioms is always accompanied by spatial imagery. Although logical precision requires that the premises of the inferences be restricted to the meaning given in the conceptual formulation, the interpretation is used as a parallel meaning in order to make the conceptual part easier to understand. The method of teaching thus follows the historical path of the development of geometry, since, historically speaking, the separation of the conceptual system of geometry from its interpretation is a later discovery. The history of the calculus of probability has followed a similar path. The mathematicians who developed the laws of this calculus in the seventeenth and eighteenth centuries always had in mind an interpretation of probability, usually the frequency interpretation, though it was sometimes accompanied by other interpretations.

In order to develop the frequency interpretation, we define probability as the *limit of a frequency* within an infinite sequence. The definition follows a path that was pointed out by S. D. Poisson[1] in 1837. In 1854 it was used by George Boole,[2] and in recent times it was brought to the fore by Richard von Mises,[3] who defended it successfully against critical objections.

The following notation will be used for the formulation of the frequency interpretation. In order to secure sufficient generality for the definition, we shall not yet assume that all elements x_i of the sequence belong to the class A. We assume, therefore, that the sequence is *interspersed* with elements x_i of a different kind. For instance, the sequence of throws of a coin may be interspersed with throws of a second coin. In this case only certain elements x_i will belong to the class A, if the class is defined as representing the throws of one of the coins only. Similarly, only some among the elements y_i will belong to the class B, which may signify the occurrence of tails lying up. It may happen that y_i represents a case of tails up, whereas the corresponding x_i does not belong to the class A, that is, the event of tails lying up is produced by the second coin. When the frequency is counted out in such a sequence pair, the result is expressed by the symbol

$$\overset{n}{\underset{i=1}{N}} (x_i \epsilon A) \tag{1a}$$

which means the number of such x_i between 1 and n that satisfy $x_i \epsilon A$. The symbol is extended correspondingly to apply to different variables and to different classes and also to a pair, a triplet, and so on, of variables. For instance, the expression

$$\overset{n}{\underset{i=1}{N}} (x_i \epsilon A).(y_i \epsilon B) \tag{1b}$$

represents the number of pairs x_i,y_i such that x_i belongs to A and simultaneously y_i belongs to B; it signifies the number of pairs x_i,y_i that are elements of the common class A and B. To abbreviate the notation, the following symbol is introduced:

$$N^n(A) =_{Df} \overset{n}{\underset{i=1}{N}} (x_i \epsilon A) \qquad N^n(A.B) =_{Df} \overset{n}{\underset{i=1}{N}} (x_i \epsilon A).(y_i \epsilon B) \tag{2}$$

[1] *Recherches sur la probabilité des jugements en matière criminelle et en matière civile* ... (Paris, 1837).

[2] *The Laws of Thought* (London, 1854), p. 295.

[3] "Grundlagen der Wahrscheinlichkeitsrechnung," in *Math. Zs.*, vol. V (1919), p. 52, and later publications.

Furthermore, the *relative frequency* $F^n(A,B)$ is defined by

$$F^n(A,B) = \frac{N^n(A.B)}{N^n(A)} \tag{3}$$

In the special case in which all elements x_i belong to the class A, that is, when the sequence x_i is *compact*, the denominator of the fraction is equal to n, whereas in the numerator the expression A may be dropped; then (3) assumes the simpler form

$$F^n(A,B) = \frac{1}{n} \cdot N^n(B) \tag{4}$$

With the help of the concept of relative frequency, the frequency interpretation of the concept of probability may be formulated:

If for a sequence pair x_iy_i the relative frequency $F^n(A,B)$ goes toward a limit p for $n \rightarrow \infty$, the limit p is called the probability from A to B within the sequence pair. In other words, the following coördinative definition is introduced:

$$P(A,B) = \lim_{n\rightarrow\infty} F^n(A,B) \tag{5}$$

No further statement is required concerning the properties of probability sequences. In particular, randomness (see § 30[4]) need not be postulated.

...all the axioms of the probability calculus follow logically from the frequency interpretation. The result holds not only for infinite but also for finite sequences, provided that in this case we regard the limit of the frequency as given by the value of $F^n(A,B)$ taken for the last element. All the axioms are satisfied tautologically, and are strictly, not only approximately, valid even before the transition to the limit.

The given proof guarantees that the frequency interpretation is an admissible interpretation of the theorems derivable from the axiom system....

THE PROBLEM OF APPLICATION

The problem of application may be divided into two groups of questions. The first group includes the question what is meant by the concept of probability as used in applications; we speak here of the *problem of the meaning of probability statements*. The second group deals with the question whether we are justified in applying the rules of probability to physical objects, whether we are entitled to assert probability statements referring to physical objects when the statements are established by means of the rules

[4] [Ed. note: Section and chapter references made throughout this article refer to sections from Reichenbach's *Theory of Probability*.]

of the calculus. This group may be called the *problem of the assertability of probability statements.* The two problems are closely connected, since the assertability will depend on the meaning assumed for the probability statement. I speak of assertability instead of truth because the requirement of truth would be too strong. It will be seen that a certain group of probability statements—those, namely, that state numerical probability values—cannot be proved as true, but are assertable on other grounds.

The problem of meaning can be formulated as the problem of the interpretation of the *P*-symbol. It was explained above that in the formal conception of the calculus of probability the *P*-symbol remains uninterpreted; the axioms then are conceived as a set of implicit definitions restricting the meaning of the *P*-symbol by subjecting it to certain formal conditions, without, however, defining the meaning exhaustively. Although the axioms exclude a class of interpretations as inadmissible, they leave open a *class of admissible interpretations.* The problem of meaning consists in selecting, among the admissible interpretations, the one that is suitable to cover the applications of the *P*-symbol to physical reality.

There remains, of course, the possibility that several interpretations are used, varying with the subject to which the term "probability" is applied. In fact, the term has two different forms of usage, which call for separate treatment. In the first usage the word "probability" refers to *sequences of events* or of other physical objects; in this usage the word must, without question, be interpreted as referring to a relative frequency, though the precise formulation of the interpretation requires further investigation. In the second usage, however, the term "probability" is applied to *single events* or other single physical objects; we must inquire whether this usage necessitates the introduction of a genuinely different interpretation or whether it is reducible to the frequency interpretation.

THE LOGICAL FORM OF LIMIT STATEMENTS

According to the frequency interpretation, introduced in § 16, the probability of a sequence is defined by the limit of the relative frequency. In order to make the meaning of the limit statement clear, I shall formulate it by means of the logical symbolism.

The limit statement may be formulated in two steps, which will be symbolized in the notation for the frequency sequence f^n counted through (see 1, § 48). On the first step we formulate the condition that, throughout the sequence, terms f^n that are as close to the limit p as we wish will always recur. In symbolic notation (see 3, § 89),

$$(\delta)(n_o)(\exists n)(n > n_o).(f^n = p \pm \delta) \qquad (1)$$

Translated into words this means: however small we choose δ and however large we choose n_0, there is an element f^n beyond f^{n_0} that is situated within $p \pm \delta$. In this case I shall call p a *partial limit* of the sequence.

On the second step we set up a stronger condition, adding the requirement: for every δ, however small, there is an n_0 such that, from f^{n_0} on, all f^n remain within $p \pm \delta$. In symbolic notation this reads

$$(\delta)(\exists n_0)(n)[(n > n_0) \supset (f^n = p \pm \delta)] \qquad (2)$$

Only if (2) is satisfied do we call p the *limit* of the sequence. The limit is distinguished from the partial limit by the requirement of a certain permanence in the approximations attained.

I turn now to the problem of how to find out whether a sequence has a limit according to (2), and which value of p constitutes this limit. This analysis requires a study of the form in which the sequence is given.

In principle, a class or a sequence of elements can be given in two ways. First, the elements can be pointed out or enumerated individually, as in a list of students attending a certain class. A class or sequence given in this way is said to be *extensionally given*. Second, the class or sequence can be defined by a rule that does not enumerate the elements individually. Of this sort, for instance, is the class definition, "all men whose height is between $5\frac{1}{2}$ and 6 feet." The phrase states a rule by means of which it can be determined for every individual object whether it belongs to the class defined. Such a class is said to be *intensionally given*. The difference in the way of giving a class or sequence becomes important when the operators "all" or "there is" are applied.

As long as we deal with finite classes the application of the operators has a simple meaning; difficulties result, however, in the application of the operators to infinite classes or sequences. For all-statements or existential statements referring to infinite classes, the problem of verifiability arises—the question how such statements can be evaluated as true or false. With regard to finite classes, the verification can be achieved, in principle, by checking individually every element of the class. But the procedure is not applicable to infinite classes.

The result of the analysis must now be applied to probability sequences. Both ways of giving, the intensional and the extensional, can be carried through for probability sequences. These terms will be referred only to the attribute class; an intensionally or extensionally given probability sequence is a sequence for which the attribute is given, respectively, intensionally or extensionally. The strictly alternating sequence (1, § 26) represents an example of an intensionally given probability sequence.... Sequences of throws of the die, social statistics, and so on, are examples of extensionally given probability sequences....

With respect to intensionally given probability sequences of an infinite length, statements expressing a frequency interpretation have the usual meaning of mathematical all-statements and existential statements; like the latter, they are strictly verifiable....

With regard to extensionally given probability sequences, however, the result is different. We saw that for them the frequency interpretation of probability leads to completely unverifiable statements because the statements include both all-operators and existential operators. An extensionally given sequence can never be realized in its entirety; we know only an initial section of it, and its infinite remainder is a matter of the future. It is obvious, however, that the relative frequency holding for the initial section imposes no restrictions on the limit of the relative frequency existing for the infinite sequence: given a relative frequency $f^n = \frac{m}{n}$ for the initial section, it is possible to imagine a continuation of the sequence of such a kind that the frequency converges toward any value p selected arbitrarily. In other words, let m' and n' be the respective frequencies of elements counted from the place of the n-th element; then the expression $\frac{m + m'}{n + n'}$ for m' and n' going toward infinite values, will converge toward a limit independent of the constant values m and n. With respect to extensionally given sequences of an infinite length, therefore, the limit statement is not verifiable.

This result makes questionable the meaning of the limit statement for infinite sequences of the extensional kind. Assuming the verifiability theory of meaning, we demand that if a statement is to be regarded as meaningful it should be possible to verify it as either true or false (the word "verify" is used here in the neutral sense, denoting a determination as true or as false). Since such verification is not physically possible for extensionally given sequences of an infinite length, we must inquire in what sense the limit statement can be maintained for them, or what other change of interpretation could be envisaged.

THE MEANING OF LIMIT STATEMENTS

We saw the logical difficulties in the fact that the statement about the limit appears meaningless for extensionally given sequences of an infinite length. A limit at a given value p is compatible with every finite beginning of the probability sequence; since we can count the frequency only in a finite initial section, all limit statements must be called nonverifiable and, consequently, meaningless.

The analysis of the problem of meaning with respect to sequences of physical events has suffered from the fact that it has been too closely attached to the mathematical formulation of the calculus of probability. Only in the mathematical version do we find infinite sequences; the sequences of actual statistics, however, are always finite. They are so, not only with respect to the initial section for which the statistics were compiled, but also with respect to the section of the sequence that lies in the future and is not yet observed. In fact, we are interested only in finite sequences because they will exhaust all the possible observations of a human lifetime or the lifetime of the human race. We wish to find sequences that behave, in a finite length of these dimensions, in a way comparable to a mathematical limit, that is, converging sufficiently well within that length and remaining within the interval of convergence. If a sequence of roulette results or of mortality statistics were to show a noticeable convergence only after billions of elements, we could not use it for the application of probability concepts, since its domain of convergence would be inaccessible to human experience. However, should one of the sequences converge "reasonably" within the domain accessible to human observation and diverge for all its infinite rest, such divergence would not disturb us; we should find that such a *semiconvergent* sequence satisfies sufficiently all the rules of probability. I will introduce the term *practical limit* for sequences that, in dimensions accessible to human observation, converge sufficiently and remain within the interval of convergence. Sequences of this kind will include only a subclass of all sequences having a limit and will include also a subclass consisting of semiconvergent sequences. It is with sequences having a practical limit that all actual statistics are concerned....

These considerations will settle the problem of meaning for probability sequences of physical events. To those who insist that every meaningful statement should be strictly verifiable in a finite number of observations we answer that all sequences with which we are concerned are sequences of a practical limit, and thus a *finitization* of the limit statement can be carried through. There can be no doubt that this finitization will satisfy all requirements of the verifiability theory of meaning in its most rigorous form....

THE ASSERTABILITY OF PROBABILITY STATEMENTS IN THE FREQUENCY INTERPRETATION

We turn now to the problem of the assertability of probability statements, still restricting ourselves to the frequency interpretation of such statements. Two questions must be answered. The first concerns the *assertability* of *probability laws,* that is, theorems of the calculus and thus of formulas that do not include assertions of specific numerical values of probability. The second concerns the *assertability of numerical degrees of probability* and therefore may be formulated as the question of the *ascertainment of the degree of probability.*

The question of the assertability of probability laws is easy to answer. It was shown in § 18 that axioms of the calculus of probability follow tautologically from the frequency interpretation. Therefore, if probability is regarded as meaning the limit of a relative frequency, the validity of probability laws is guaranteed by deductive logic. This result is of major importance for the epistemological critique of probability statements. So far as theorems of the calculus of probability are concerned, the application of probability statements to physical objects offers no greater problems than that of theorems of deductive logic. In other words, with respect to theorems of the calculus there is no problem of application that is specific for probability....

The problem of the ascertainment of the degree of probability, in contradistinction to the problem of probability laws, offers serious difficulties. They originate from the difficulties in the verification of the limit statement explained above (§ 64).

In extensionally given sequences—and probability sequences referring to physical reality are of this type—verification of the limit statement is impossible if the sequences are infinite. It has been emphasized that sequences of physical applications are not infinite and that verification by enumeration is possible in principle. Although this result is sufficient to settle the problem of the *meaning* of the limit statement, it does not help so far as the *assertability* of the limit statement is concerned. The reason is that in all practical applications we wish to know the value of the limit before the sequence is completely produced; indeed, all practical use of probability statements consists in the fact that they are applied for the prediction of relative frequencies, and thus cannot be based on counting the total sequence. It is the *predictive nature* of probability statements, therefore, that presents difficulties to a proof that such statements are assertable.

This consideration makes clear that the ascertainment of the degree of probability must be achieved by methods other than counting the relative frequency in the total sequence. In practical applications, two different methods are used. The first is called the *a posteriori* determination of probabilities; the second, the *a priori* determination of probabilities.

The *a posteriori* determination is identical with a procedure known in logic as *induction by enumeration,* if the term is applied in a somewhat wider sense than in traditional logic. It is based on counting the relative frequency in an initial section of the sequence, and consists in the inference that the relative frequency observed will persist approximately for the rest of the sequence; or, in other words, that the observed value represents, within certain limits of exactness, the value of the limit for the whole sequence. This inference is called the *inductive inference*; its formulation is called the *rule of induction.* A strict formulation of the rule will be given later (§ 87). The inductive inference of traditional logic is of a somewhat narrower form; it refers to initial sections in which all elements have the same attribute B and assumes the permanence of B for the rest of the sequence. This inference, which may be called *classical induction,* is a special case of the general form, namely, the case where the observed relative frequency is $= 1$. To distinguish the general rule of induction from the special case of classical induction, it may be called *statistical induction.*

The problematic character of the inductive inference has often been discussed. In § 64 it was pointed out that a relative frequency $\frac{m}{n}$ observed in an initial section is compatible with any value p for the limit of the relative frequency in the whole infinite series. For sequences of a finite length the frequency $\frac{m}{n}$ will impose certain restrictions on the limit; but if the sequence is long enough—and that is true of practical applications—the restrictions are negligible and the value of the limit is virtually independent of the observed value $\frac{m}{n}$. The inductive inference, therefore, is essentially different from a deductive inference; it carries no logical necessity with it. For the present, however, discussion of the logical nature of the inference will be postponed.

It would be a simple way out of the difficulty if it could be shown that other methods than the inductive rule are available for the ascertainment of a degree of probability. Therefore we must now inquire whether an *a priori* determination of a degree of probability is possible.

THE SO-CALLED *A PRIORI* DETERMINATION OF THE DEGREE OF PROBABILITY

The *a priori* determination of the degree of probability occupies a special position in the historical development of the philosophy of probability. The determination of probabilities on the ground of properties of symmetry, without the use of frequencies, has been regarded by some authors as the nucleus of the problem of probability. It has been contended that every other determination of probabilities must be reduced to the same logical schema. In the pursuit of this idea it was believed that all attempts to solve the philosophical problems of the concept of probability should start at this point.

Mechanisms like the die and the roulette wheel are characterized by the existence of a complete and exclusive disjunction the terms of which are equiprobable. The theorem of addition permits us to infer, from the number r of the terms of the disjunction, that the probability is $= \frac{1}{r}$ for every single term. With respect to such disjunctions it is therefore possible to reduce the determination of the degree of probability to the determination of the concept *equiprobable*. On this fact was based the well-known definition of probability as the ratio of the favorable to the possible cases.

. . . even if the degree of probability can be reduced to equiprobability, the problem of probability is only shifted to this concept. All the difficulties of the so-called *a priori* determination of probability, therefore, center around this issue.

In the search for an argument by which the equiprobability can be derived, a principle was established that must be regarded as the foundation of all *a priori* determination of probability: the *principle of indifference,* also called the *principle of no reason to the contrary.* It maintains that events are equiprobable when there is no reason to assume that one should occur rather than another. Thus we have no reason, it is argued, to favor one of the faces of the die; therefore they are equiprobable. Some authors present the argument in a disguise provided by the concept of *equipossibility*: cases that satisfy the principle of "no reason to the contrary" are said to be *equipossible* and therefore equiprobable. This addition certainly does not improve the argument, even if it originates with a mathematician as eminent as Laplace, since it obviously represents a vicious circle. *Equipossible* is equivalent to *equiprobable.* It appears advisable to discuss the principle of indifference in a form that avoids this fallacy, and ask whether the absence of a reason to the contrary can guarantee equal probabilities.

At first sight the principle seems plausible, since we actually use such inferences in the determination of probabilities. The inference is often associated with a feeling of self-evidence that makes the conclusion seem "logical." But the fact that we use and believe in inferences of this type does not prove that they are valid. On the contrary, a brief reflection on the content of the principle shows that it is invalid. The absence of a reason to the contrary is a condition of our knowledge; equiprobability is a condition holding for physical objects. Why should the occurrence of physical events follow the directive of human ignorance? Perhaps we have no reason to prefer one face of the die to the other; but then we have no reason, either, to assume that the faces are equally probable. To transform the absence of a reason into a positive reason represents a feat of oratorical art that is worthy of an attorney for the defense but is not permissible in the court of logic....

THE THREE FORMS OF AN *A POSTERIORI* DETERMINATION OF DEGREES OF PROBABILITY

The analysis presented in the preceding sections can be summed up in the following statement: *there is no a priori ascertainment of a degree of probability; a probability metric can be determined only a posteriori.*

There are three possible ways for *a posteriori* establishment of a probability metric:

1. Degrees of probability can be directly ascertained through induction by enumeration (*statistical probabilities*).

2. A probability metric can be inferred deductively from known probabilities (*deduced probabilities*).

3. A probability metric can be inferred by means of general inductive methods from known observational data (*hypothetical probabilities*)....

Of the three methods, the second and third can be carried through only when certain probabilities are known; the first method alone is applicable without such knowledge. The first can, therefore, be used as a *primary method* of finding probabilities, whereas the other two are *secondary methods*....

I shall use the term *advanced knowledge* to denote a state of knowledge that includes a sufficient number of probabilities; it then follows that the second and third methods of ascertaining probabilities are applicable only in advanced knowledge. For *primitive knowledge*—a state of knowledge that does not include a knowledge of probabilities—the rule of induction is the only instrument for the ascertainment of probabilities. Incidentally, the rule of induction can also be applied in advanced knowledge and then assumes a somewhat different function, which will be studied in § 86; but for primitive knowledge it is the only available instrument. Many controversies about the legitimacy of probability methods arise from a confusion of these two states of knowledge: in advanced knowledge the whole technique of the calculus of probability is at our disposal and can be adduced to justify the methods employed, whereas primitive knowledge requires other means of justifying inductive inferences.

The given analysis shows that only for statistical induction must a justification be required. The other two methods of ascertaining probabilities are covered by the logical system of the calculus of probability. The axiomatic construction of this calculus is proof that the rule of induction is the only nonanalytic principle necessary for the application of the calculus to reality, if the frequency interpretation of probability is assumed. The rule does not enter into the formal calculus, which is a deductive system like all other mathematical disciplines; however, it is required for the applied calculus of probability, which deals with empirical statistics. The assertability of applied probability statements is thus reduced to the problem of a justification of the rule of induction....

THE FREQUENCY INTERPRETATION OF THE PROBABILITY OF THE SINGLE CASE

...Assume that the frequency of an event B in a sequence is $= \frac{5}{6}$. Confronted by the question whether an individual event B will happen, we prefer to answer in the affirmative because, if we do so repeatedly, we shall be right in $\frac{5}{6}$ of the cases. Obviously, we cannot maintain that the assertion about the individual event is true. In what sense, then, can the assertion be made? The answer is given by regarding the statement not as an assertion but as a *posit*. We do not say that B will occur, but we posit B. We do so if

$P(B) > \frac{1}{2}$; otherwise we posit $\bar{B}$. The word "posit" is used here in the same sense as the word "wager" or "bet" in games of chance. When we bet on a horse we do not want to say by such a wager that it is true that the horse will win; but we behave as though it were true by staking money on it. *A posit is a statement with which we deal as true, although the truth value is unknown.* We would not do so, of course, without some reason; we decide for a posit, or a wager, when it seems to be the best we can make. The term "best" occurring here has a meaning that can be numerically interpreted; it refers to the posit that will be the most successful when applied repeatedly.

If we wish to improve a posit, we must make a selection S such that $P(A.S,B) > P(A,B)$.[5] If we now posit B only in the case $A.S$ and omit a posit in the case $A.\bar{S}$, we obtain a relatively greater number of successes than by the original posit. It is even more favorable to construct the selection so that at the same time $P(A.\bar{S},B)$ is $< \frac{1}{2}$. We then always posit $\bar{B}$ in the case $A.\bar{S}$. We can thus make a posit for each element of the original sequence and obtain a greater number of successes. The procedure may be called the *method of the double posit.*

It should be noticed that we cannot improve a posit without knowing a selection S that leads to a greater probability. If we were to posit arbitrarily sometimes B, sometimes $\bar{B}$, we would in general construct a selection S for which $P(A.S,B) = P(A,B) = P(A.\bar{S},B)$, that is, a selection of the domain of invariance. Positing B in the case $A.S$ would then lead to the same relative number of successes as in the main sequence, whereas positing $\bar{B}$ in the case $A.\bar{S}$ would lead to a smaller ratio of successes.

In dealing with exclusive disjunctions $B_1 \vee \ldots \vee B_r$ of more than two terms, if we are compelled to posit only one of the terms (if a posit $B_2 \vee \ldots \vee B_r$ is impossible because of practical reasons), the B_k that carries the greatest probability will be the best posit. In this case, therefore, the probability $\frac{1}{2}$ no longer represents a critical value....

If we are asked to find the probability holding for an individual future event, we must first incorporate the case in a suitable reference class. An individual thing or event may be incorporated in many reference classes, from which different probabilities will result. This ambiguity has been called the *problem of the reference class.* Assume that a case of illness can be characterized by its inclusion in the class of cases of tuberculosis. If additional information is obtained from an X-ray, the same case may be incorporated in the class of serious cases of tuberculosis. Depending on the classification, different probabilities will result for the prospective issue of the illness.

We then proceed by considering *the narrowest class for which reliable sta-*

[5] If we have $P(A.S,B) < P(A,B)$, the selection $\bar{S}$ will have the desired property because it then follows from the rule of elimination that $P(A.\bar{S},B) > P(A,B)$. See the discussion following (11*b*, § 19). [Ed. note: Footnotes have been renumbered for use in this book.]

tistics can be compiled. If we are confronted by two overlapping classes, we shall choose their common class. Thus, if a man is 21 years old and has tuberculosis, we shall regard the class of persons of 21 who have tuberculosis. Classes that are known to be irrelevant for the statistical result may be disregarded. A class C is irrelevant with respect to the reference class A and the attribute class B if the transition to the common class $A.C$ does not change the probability, that is, if $P(A.C,B) = P(A,B)$. For instance, the class of persons having the same initials is irrelevant for the life expectation of a person.

We do not affirm that this method is perfectly unambiguous. Sometimes it may be questioned whether a transition to a narrower class is advisable, because, perhaps, the statistical knowledge regarding the class is incomplete. We are dealing here with a method of technical statistics; the decision for a certain reference class will depend on balancing the importance of the prediction against the reliability available. It is no objection to this interpretation that it makes the probability constructed for the single case dependent on the state of our knowledge. This knowledge may even be of such a kind that it does not determine one class as the best. For instance, we may have reliable statistics concerning a reference class A, and likewise reliable statistics for a reference class C, whereas we have insufficient statistics for the reference class $A.C$. The calculus of probability cannot help in such a case because the probabilities $P(A,B)$ and $P(C,B)$ do not determine the probability $P(A.C,B)$. The logician can only indicate a method by which our knowledge may be improved. This is achieved by the rule: look for a larger number of cases in the narrowest common class at your disposal.

Whereas the probability of a single case is thus made dependent on our state of knowledge, this consequence does not hold for a probability referred to classes. If the reference class is stated, the probability of an attribute is objectively determined, though we may be mistaken in the numerical value we assume for it on the basis of inductions. The probability of death for men 21 years old concerns a frequency that holds for events of nature and has nothing to do with our knowledge about them, nor is it changed by the fact that the death probability is higher in the narrower class of tuberculous men of the same age. The dependence of a single-case probability on our state of knowledge originates from the impossibility of giving this concept an independent interpretation; there exist only substitutes for it, given by class probabilities, and the choice of the substitute depends on our state of knowledge. My thesis that there exists only one concept of probability, which applies both to classes and to single cases, must therefore be given the more precise formulation: there exists only one legitimate concept of probability, which refers to classes, and the pseudoconcept of a probability of a single case must be replaced by a substitute constructed in terms of class probabilities....

Besides choosing a suitable reference class, we must also choose a sequence into which the individual case considered is to be incorporated. This choice is usually less difficult than that of the reference class because the frequency will be the same for most sequences that can be reasonably chosen. We often follow the time order of the events observed or of the observations made. One of the rules to be required is that a knowledge of the attribute of the individual case should not be used for the construction of the order of the sequence. (See the remarks on random sequences in § 30.). . .

This interpretation of probability statements is complicated by the following peculiarity. If we have an implication (A $\supset$ *B*), we can add an arbitrary term in the implicans, that is, we can derive the implication $(A.C \supset B)$. If we have a probability implication $(A \underset{p}{\ni} B)$, however, the addition of a term in the implicans will, in general, lead to a different degree of probability, that is, to a probability implication $(A.C \underset{q}{\ni} B)$ where q is different from p. This is why the choice of the reference class is easily made for a general implication, whereas it is difficult to make it for a probability implication. Once a class A is found such that $(A \supset B)$ holds, we know that if $x_i \epsilon A$ we shall have $y_i \epsilon B$; it does not matter to what other classes the event x_i belongs. For a probability implication there is no such simple relation. We must be aware of the possibility that, if x_i belongs to both A and C, the reference to the common class $A.C$ may lead to a value of the probability different from the one resulting for the reference class A.

Therefore, we can ask only for the best reference class available, the reference class that, on the basis of our present knowledge, will lead to the greatest number of successful predictions, whether they concern hits of bombs, cases of diseases, or political events. If no statistics are available for the common class $A.C$, we shall base our probability calculations on the reference class A, and must renounce the improvement in the success ratio that might result from the use of the reference class $A.C$ in combination with the method of the double posit. Such a procedure seems reasonable if we realize that narrowing the reference class means nothing but increasing the success ratio, and that there is no reference class that permits the prediction of a single case. This goal, which could be reached if we had a knowledge of synthetic logical implications combining past and future events, is unattainable if probability implications are all that we have to connect the past with the future.

We must renounce all remnants of absolutism in order to understand the significance of the frequency interpretation of a probability statement about a single case. But there is no place for absolutism in the theory of probability statements concerning physical reality. Such statements are used as rules of behavior, as rules that determine the most successful behavior attainable in a given state of knowledge. Whoever wants to find more in these statements will eventually discover that he has been pursuing a chimera.

THE LOGICAL INTERPRETATION OF PROBABILITY

In order to construct a logical form for probability statements concerning single cases, it is advisable to introduce a change in the logical classification of probabilities. In the frequency interpretation, a probability is regarded as a property of a sequence of events. Correspondingly, the statement about the probability of a single case is regarded as stating a property, though fictitious, of an individual event. It is possible, however, to go from events to sentences about events, and to regard a probability, not as a property of the event, but as a property of the sentence about the event. Instead of saying, for instance, that the probability of obtaining face 6 with a die is $= \frac{1}{6}$, we can say that the probability of the sentence, "Face 6 will turn up," is $= \frac{1}{6}$. By this transition, probability is made a rating of propositions; and probability statements belong not in the object language but in the metalanguage. ... I shall introduce the term *logical interpretation of probability* for this conception; the conception previously used will be called *object interpretation.* The logical interpretation offers the advantage that the probability attached to the single case assumes the form of a truth value of a proposition or, rather, since the proposition can be maintained only in the sense of a posit, of the truth value of a posit. We shall use the term *weight* for the truth value of the posit; the probability of the single case, therefore, is regarded, in the logical interpretation, as the weight of a posit. A posit the weight of which is known is called an *appraised posit.*

When we say that probability assumes the form of a truth value, we use the latter term in a wider sense than usual. Classical logic is two-valued; it knows only the two truth values *true* and *false.* In regarding probability as a truth value we construct a multivalued logic, differing from other such logics in that it is a logic with a continuous scale of truth values ranging from 0 to 1. The formal construction of the *probability logic* will be carried through in chapter 10. ...

Because of its analogy with the concept of a statement of known truth value, the concept of appraised posit is indispensable for the understanding of language. It defines the logical category under which probability statements concerning individual cases are to be subsumed, and allots to such statements a legitimate place within the body of knowledge. A two-valued logic has no place for unknown truth values; so long as the truth value of a statement is not known, as for statements about future events, classical logic does not allow us to judge the truth of the statement. All it offers is a "wait and see." Our actual behavior, however, does not follow this maxim of passivity. We form judgments about the likelihood of the event and use them as a guide for action. We must do so, because action presupposes judgments about future events; if we had to wait until direct observation informed us about the occurrence or nonoccurrence of an event, it would be too late to plan actions influencing the event.

Probability logic supplies the logical form of a truth evaluation by degrees that is applicable before the occurrence of an event. It allows us to coördinate to the sentence about the individual event a fictitious truth value, derived from the frequency within an appropriate sequence, in such a way that, so far as actions are concerned, the fictitious truth value, or weight, satisfies to a certain extent the requirements that can be asked with respect to a truth value. The logical interpretation repeats the procedure followed in the object-language interpretation of the probability statement about the single case: the metalinguistic conception of the probability of the single case as a weight of a statement, too, is constructed by a transfer of meaning from the general to the particular case. The numerical value of the frequency in the sequence is transferred to the individual statement in the sense of a rating, although the individual statement taken alone exhibits no features that could be measured by the rating.

In spite of the fictitious nature of the rating so constructed, the system of posits endowed with weights can be substituted for a system of statements known as true or false. The essential difference between the two systems consists in the fact that in the substitute system our action is determined, not by a knowledge of the truth value of the statement about the individual event, but by a knowledge of a truth frequency in a sequence. The substitution of this statistical knowledge for unavailable specific knowledge is justified because it offers success in the greatest number of cases. This is why we can act when the truth of the sentence about the individual event remains unknown: the frequency interpretation of probability replaces the unattainable ascertainment of truth by a procedure that accords the best success attainable in the totality of cases. . . .

PROBABILITY MEANING

The interpretation of probability as a truth value permits the introduction of a new category of meaning. The verifiability theory of meaning, in its strict form, makes meaning dependent on verifiability as true or false. The theory can be extended, however, so that a sentence is regarded as meaningful when it is possible to determine a weight for the sentence. By "possible" we understand here "physically possible," that is, "compatible with physical laws." The meaning so defined is called *probability meaning*.

The advantage of the new category of meaning derives from the fact that a determination of weight may be physically possible, whereas a corresponding absolute verification is not physically, but only logically, possible. With respect to simple sentences, for example, sentences concerning the weather of the next day, the distinction is irrelevant; here it is physically possible both to determine a weight in advance and to verify after the occurrence of an event. But this cannot be done with more complicated sentences. Thus, a statement about the temperature of the sun cannot be strictly verified in the sense of a physical possibility of verification, but it is physically possible to determine a weight for it. *Probability meaning,* therefore, represents a wider category than *physical truth meaning,* that is, a meaning defined by the physical possibility of strict verification. But it is a narrower category than *logical meaning*, a meaning defined by the logical possibility of strict verification (see § 66). When the term "verification" is used in a wider sense, to include the determination of a weight, probability meaning represents a *physical* meaning, since it is based on the physical possibility of verification. It can be shown that probability meaning constitutes the very category of meaning that underlies conversational and scientific language, for which physical truth meaning is too narrow and logical meaning too wide.

[EXPLANATORY INDUCTION AND INDUCTION BY ENUMERATION]

The word "induction" is usually employed in a sense more comprehensive than the one so far envisaged in this work. The rule by which we infer that the frequency observed in an initial section will persist for the whole sequence is regarded as a special case of induction, often called *induction by enumeration*....

... there is another form of induction that, at first sight, seems to be of a different nature. It is based on the application of causal explanation and may therefore be called *explanatory induction*. It consists in the inference from certain observational data to a hypothesis, or theory, from which the data are derivable and which, conversely, is regarded as being made probable by the data....

So far as explanatory induction is concerned, a fallacious interpretation was discussed at the end of § 21, in the analysis of the so-called *inference by confirmation*. This inference has also been called the *hypothetico-deductive* method, the term being meant to indicate the deductive relation from the hypothesis to the observational data. The fallacy consists in the belief that an inductive relation holds for the reversed direction, or, more precisely speaking, that the implication $a \supset b$ entitles us to regard a as probable when b is given.

Explanatory induction must not be interpreted in the same sense. Only for a superficial consideration does explanation have the form of the hypothetico-deductive method, or of the inference by confirmation. In deeper analysis it reveals a much more complicated structure. Explanatory induction must be regarded, not as an inference in its own right, but as a combination of probability inferences such as are formulated by the rule of Bayes. The complicated nature of the inference is made clear by the fact that the inference is applied only when much more is known than the occurrence of the consequences of an assumption; without an estimate of the antecedent probabilities the inference is never made.

For scientific theories, estimates of the antecedent probabilities are often given in the form of considerations about the plausibility or "naturalness" of a theory, that is, by arguments that make the theory credible, independent of the observed confirmation.... The misunderstandings of the inference by indirect evidence and its interpretation as an inference by confirmation may perhaps be psychologically explained as an oversight of the role which the antecedent probabilities play in the inference. These probabilities are easily overlooked because they often need not be known otherwise than in the form of crude estimates, while the result of the inference can be very precise (see §§ 62, 70).

The thesis may be generalized: all inductive inferences that do not have the form of induction by enumeration must be construed in terms of the theorems of the calculus of probability. In fact, the calculus of probability contains the key to a theory of induction in advanced knowledge. Philosophers who believe that a philosophical theory of induction is to be developed independent of the statistical methods employed in the sciences make the mistake of overlooking the existing mathematical methodology: all the questions concerning induction in advanced knowledge, or *advanced induction*, are answered in the calculus of probability. While logicians were vainly looking for an inductive logic that could account for scientific method, mathematicians constructed a mathematical system that covers all forms of probability inference and thus of scientific inference—a system that can even be transcribed into a system of logic, as was shown in chapter 10. The logician of our day who is aware of the fallacies of the philosophy of rationalism abandons all attempts at a construction of an inductive logic from pure reason. The inductive method presented by the calculus of probability is a much more powerful instrument than any substitute devised under the name of rational belief; moreover, it admits of an empiricist interpretation that rejects all forms of synthetic self-evidence.

The probability character of explanatory induction is also indicated by the fact that the merging of scientific theories of different domains leads to an increase in reliability. Thus Newton's combination of Galileo's theory of falling bodies and Kepler's laws of planetary motion led to a theory of gravitation that was superior in reliability to either of the theories included in it. Some logicians have regarded the unification of theories as the expression of a tendency to logical elegance, or economy. Such an interpretation, however, seriously misrepresents the nature of scientific method. The unification of theories is an instrument for connecting scientific results in such a way that the combination obtains a higher probability than each of its parts taken separately. The schema of these inferences can be understood when it is interpreted in terms of the theorems of the calculus of probability. Such an analysis makes clear that the theory of advanced induction is identical with the theory of probability.

Since the axiomatic construction of the calculus of probability leads to the result that, when the frequency theory is assumed, all probability inferences are reducible to deductive inferences with the addition of induction by enumeration, it follows that all inductive inferences are reducible to induction by enumeration. This thesis was at the basis of Hume's theory of induction—though he thought only of classical induction—but he had no proof for it. The proof can be given only by the axiomatic construction of the calculus of probability.

The thesis, furthermore, must not be oversimplified to the statement that all inductive inferences can be construed directly as induction by enumeration; the reduction is possible only indirectly through the reducibility of the axioms of probability to the frequency interpretation....

The thesis is further obscured by a confusion between the context of discovery and the context of justification, if I may be allowed to use certain terms that I have introduced elsewhere. The finding of explanation belongs in the context of discovery and can be analyzed only psychologically, not logically; it is a process of intuitive guessing and cannot be portrayed by a rational procedure controlled by logical rules. Rationalization belongs in the context of justification; it can be applied only when given inductive conclusions are to be judged appropriate to given facts. It is in this context that the thesis belongs. Testing the relations between given observational data and given inductive conclusions is a procedure expressible in terms of theorems of the calculus of probability; and the inductive inferences of the test procedure are, therefore, ultimately reducible to induction by enumeration.

I do not maintain, by this thesis, that the finding of inductive explanation could be achieved by enumerating observations and simple generalization, such as Bacon hoped to attain in his tables; nor do I claim to have better methods. I refuse to answer the challenge of setting up rules of a logic of discovery. There are no such rules. Philosophers who believe that induction could become a sort of philosopher's stone, supplying methods that automatically transform facts into theories, misunderstand the task of logical analysis and burden the theory of induction with an unsolvable problem. Like deductive logic, the logic of induction concerns, not the psychological process of finding solutions, but the critical process of testing given solutions; it applies to the rational reconstruction of knowledge and thus belongs in the context of justification, not in the context of discovery.

THE PROBABILITY OF HYPOTHESES

The thesis that explanatory induction can be construed in terms of the theorems of the calculus of probability and is therefore reducible to induction by enumeration is attacked by the argument that the probability of hypotheses is not interpretable as a frequency.... [However] the probability of a hypothesis or a scientific theory can be defined in terms of frequencies. Applied to the individual hypothesis, the probability assumes the character of a weight; all that was said about the use of a weight for statements of single cases holds likewise for the weight of hypotheses. In fact, speaking of the probability of an individual hypothesis offers no more logical difficulties than speaking of the probability of an individual event, say, the death of a certain person.

It is sometimes argued that in cases of the latter kind the choice of the reference class is easily made—that, for example, the reference class "all persons in the same condition of health" offers itself quite naturally. But critics of the frequency interpretation of the probability of theories forget how much experience and inductive theory is invested in the choice of the reference class of the probability of death. Should we some day reach a stage in which we have as many statistics on theories as we have today on cases of disease and subsequent death, we could select a reference class that satisfies the condition of homogeneity (see § 86), and the choice of the reference class for the probability of theories would seem as natural as that of the reference class for the probability of death. In some domains we have actually been witnesses of such a development. For instance, we know how to find a reference class for the probability of good weather tomorrow; but before the evolution of a scientific meteorology this reference class seemed as ambiguous as that of a scientific theory may seem today. The selection of a suitable reference class is always a problem of advanced knowledge.

INDUCTION BY ENUMERATION IN ADVANCED KNOWLEDGE

The theory of induction by enumeration in advanced knowledge was presented in § 62. The inductive inference was shown there to refer to a probability lattice and to be reducible to an application of the rule of Bayes. Assuming that the antecedent probabilities are known and that the sequences are normal sequences, the following questions can be answered.

1. Given an initial section of a sequence with the relative frequency f^n, what value $p \pm \delta$ is the best posit for the limit of the frequency?

2. What is the second-level probability v_n that a limit at $p \pm \delta$ will be reached?

If the antecedent probabilities are not known but the sequences at least are known to be of the Bernoulli type, we can prove the following theorem:

3. The greater n, the greater the second-level probability v_n that the limit will be at $p = f^n \pm \delta$; and v_n converges to 1 with $n \rightarrow \infty$.

We see that the theory of induction by enumeration, in advanced knowledge, is as complete as can be required; the theory tells us what value to posit, how good our posit is, and that it will become better and better with larger numbers.

In discussions of induction the question has been asked, What is a large number? In fact, the conception of a large number varies greatly with the field in which the induction is applied. For the test of a new medicine a few hundred cases may be sufficient; insurance companies count millions of cases; and physicists dealing with problems of the kinetic theory of gases do not speak of large numbers unless they are compelled to use a mode of writing in terms of powers of the number ten. It is obvious that the definition of a large number belongs in advanced knowledge: it is a number so large that the probability of the second level for a sufficient convergence from that number on is high enough. A "sufficient convergence" and a "high-enough probability" are matters of definition and depend on what is attainable; the large number is a function of these definitions, computable or appraisable only within advanced knowledge (see 12, § 62).

A large number is not always necessary for an inductive inference. Sometimes the inference can be based on only one instance. Such inferences occur when previous inductions have created a situation in which one experiment, called a *crucial experiment*, is sufficient for an induction. Instances of crucial experiments are found in what a physician calls *differential diagnosis.* One Wassermann test may be regarded as sufficient evidence of a case of syphilis, because previous inductions have established a relation between the test and the disease and have shown that repeated applications of the same test to the same person usually lead to the same result. The existence of crucial experiments has been misconstrued as evidence against an inductive interpretation of scientific methods, but it can be incorporated without difficulty in an interpretation for which all inductive inferences are reducible to induction by enumeration.

A further condition that can be satisfied only in advanced knowledge is the condition of a *homogeneous reference class.* A class of tuberculosis cases is a homogeneous class. But it would seem unwise to compile death statistics in a class of persons with different diseases or in a class including both human beings and animals. The definition of the predicate "homogeneous" depends on the state of our knowledge. An inhomogeneous class can be defined as a class for which we know methods by means of which the class can be so subdivided, without the use of the attribute considered (see § 30), that subclasses of different frequencies for the attribute result. The subdivision will sometimes be achievable by reference to other attributes, such as are given in the example of different diseases, or biological species. However, it can be achieved also by dividing the total sequence, or ordered class, into consecutive sections. The latter method amounts to a determination of the dispersion (see § 52). For an inhomogeneous class the probability of the second level concerning the persistence of the observed frequency is lower than for a homogeneous class. It is obvious that such considerations are restricted to advanced knowledge....

In advanced knowledge the inference of induction by enumeration is justified by the theorems of the calculus of probability. Statements about the limit of the frequency and about the reliability of the inductive conclusion can be given a probability meaning (§ 74); in other words, within the frame of advanced knowledge the inductive inference is an appropriate instrument for operations controlled by a logic of probability. There is no question of the legitimacy of induction in advanced knowledge.

It is in advanced knowledge, too, that such formulas as the rule of succession (22, § 62) find their places. The equality of antecedent probabilities on which the applicability of the rule depends must be known before the rule can be used. The argument that equality may be assumed in the absence of knowledge to the contrary makes use of the principle of indifference, but it was shown (§ 68) that the principle is untenable. Logic cannot supply a probability metric; only experience and observation can inform us about degrees of probability or about the equality of such degrees. But even experience and observation can supply such knowledge only if the observational results are linked and carried on by inductive inferences. There is no circularity involved in such a procedure if it is used in establishing special forms of inductive inferences; and the rule of succession, therefore, occupies a legitimate position in advanced knowledge. But it would be circular to base the general use of the inductive inference on formulas that presuppose the knowledge of a probability metric. The ultimate justification of induction must be given by other means than formulas that are derivable in the calculus of probability; the problem falls entirely within the province of primitive knowledge.

THE RULE OF INDUCTION

We turn now to the consideration of induction in primitive knowledge. So long as no probabilities have been established, the inductive rule cannot be based on theorems of the calculus of probability; therefore we cannot prove that the inductive rule leads to the posit of greatest weight, nor do we know how probable it is that the limit posited will be reached. We cannot even prove that the posit becomes better with a greater number of observed instances. In spite of our ignorance, however, we must use the inductive rule, since otherwise we could not establish any probability values and could never proceed to the advanced state of knowledge in which the theorems of probability take over the functions of a guide in inductive method.

To facilitate the discussion of induction in primitive knowledge, or *primitive induction*, we shall proceed by steps. We shall not begin with the analysis of a state in which nothing is known about the progress of sequences, but shall leave the discussion of that question to a later inquiry (§ 91). Rather, we shall introduce the assumption that the sequences under consideration have a limit of the frequency, although the limit is unknown. Let us see to what extent this assumption can help in the solution of the inductive problem.

Again the concept of *posit* will be used for the interpretation of the statements to be considered. The statement that the observed frequency will persist can be maintained only in the sense of a posit, since it is obvious that we cannot prove it to be true. But it is not an appraised posit, since we have no weight for it. In what sense, then, can the inductive posit be justified if we have no proof that the posit will lead to the greatest number of successes?

To answer the question, we must analyze the way in which the rule of induction is used. The inductive posit is not meant to be a final posit. We have the possibility of correcting a first posit, of replacing it by a new one when new observations have led to different results. From this point of view, the following analysis of the inductive procedure can be made. If the sequence has a limit of the frequency, there must exist an n such that from there on the frequency $f^i (i > n)$ will remain within the interval $f^n \pm \delta$, where δ is a quantity that we can choose as small as we like, but that, once chosen, is kept constant. Now if we posit that the frequency f^i will remain within the interval $f^n \pm \delta$, and if we correct this posit for greater n by the same rule, we must finally come to the correct result. The inductive procedure, therefore, represents a *method of anticipation*; in applying the inductive rule we anticipate a result that for iterated procedure must finally be reached in a finite number of steps. We thus speak here of an *anticipative posit.* In contradistinction to an *appraised posit*, the weight of which is known, it may also be called a *blind posit*, since it is used without a knowledge of how good it is; the term "blind" is meant to express the fact that it is a posit without a rating. . . .

The method of the anticipative posit may be formulated as follows:

> RULE OF INDUCTION. *If an initial section of n elements of a sequence* x_i *is given, resulting in the frequency* f^n*, and if, furthermore, nothing is known about the probability of the second level for the occurrence of a certain limit p, we posit that the frequency* $f^i (i > n)$ *will approach a limit p within* $f^n \pm \delta$ *when the sequence is continued.*

The distinction between appraised and anticipative posits leads to two different kinds of posit. It is a common feature of both that their use is not justified for the individual case, but only in repeated applications. With respect to the grounds of their use, however, the two posits must be distinguished. The appraised posit is justified by the *principle of the greatest number of successes*. This kind of posit, therefore, can be used only when the corresponding weight is known. The anticipative posit cannot be justified by a maximum principle. It involves another form of justification based on the *principle of finite attainability*. If the sequence has a limit, the anticipative posit is justified because, in repeated applications, it leads to any desired approximation of the value of the limit in a finite number of steps.

This argument may be called an *asymptotic justification*. It includes an explanation why the value f^n found for the last observed element of the sequence is preferable to any earlier value. If after 100 elements we find $f^n = \frac{1}{2}$, after 200 elements, $f^n = \frac{2}{3}$, we do not claim that $\frac{2}{3}$ is a better value than $\frac{1}{2}$ in the sense that it is more probable. Such a proof is impossible in primitive knowledge and can be given only in advanced knowledge (see § 86). But if the procedure of going through all elements successively can be justified, we know, at least, that in selecting the f^n of the later element we are closer to the end of the procedure. The choice of the last f^n is therefore a matter of economy.

The posit f^n is not the only form of anticipative posit. We could also use a posit of the form

$$f^n + c_n \tag{1}$$

where c_n is an arbitrary function, which is so chosen that it converges to 0 with n increasing to infinite values. All posits of this form will converge asymptotically toward the same value, though they will differ for small n. We shall prefer the inductive posit f^n, for which $c_n = 0$. To do so we can, however, adduce only grounds of descriptive simplicity;[6] that is, the inductive posit is simpler to handle. . . .

When we use the logical conception of probability, the rule of induction must be regarded as a *rule of derivation*, belonging in the metalanguage. The rule enables us to go from given statements about frequencies in observed initial sections to statements about the limit of the frequency for the whole sequence. It is comparable to the rule of inference of deductive logic (see § 5), but differs from it in that the conclusion is not tautologically implied by the premises. The inductive inference, therefore, leads to something new; it is not empty, like the deductive inference, but supplies an addition to the content of knowledge. It is a consequence of the synthetic nature of inductive inference that the conclusion cannot be asserted as true, but can be asserted only in the sense of a posit. Since we could show that all inductive inferences are reducible to induction by enumeration, in the wider sense of statistical induction (§§ 67, 84), the rule of induction is the only rule of derivation that distinguishes *inductive logic* from *deductive logic*. In other words, inductive logic contains all the rules of derivation of deductive logic with the addition of the inductive rule. . . .

[6] Descriptive simplicity is a property of a description that has no bearing upon its truth. It must be distinguished from inductive simplicity, which classifies descriptions leading to different predictions. See *EP,* § 42. In the same book, on p. 355, I tried to give other reasons for preferring the posit f^n. Dr. Norman Dalkey has since convinced me that they are invalid. It is, however, sufficient for the theory of induction that the posit f^n is descriptively simpler.

THE JUSTIFICATION OF INDUCTION

We used the assumption of the existence of a limit of the frequency in order to prove that, if no probabilities are known, the anticipative posit is the best posit because it leads to success in a finite number of steps. With respect to the individual act of positing, however, the limit assumption does not supply any sort of information. The posit may be wrong, and we can only say that if it turns out to be wrong we are willing to correct it and to try again. But if the limit assumption is dispensable for every individual posit, it can be omitted for the method of positing as a whole. The omission is required because we have no proof for the assumption. But the absence of proof does not mean that *we know that there is no limit*; it means only that *we do not know whether there is a limit*. In that case we have as much reason to try a posit as in the case that the existence of a limit is known; for, if a limit of the frequency exists, we shall find it by the inductive method if only the acts of positing are continued sufficiently. Inductive positing in the sense of a trial-and-error method is justified so long as it is not known that the attempt is hopeless, that there is no limit of the frequency. Should we have no success, the positing was useless; but why not take our chance?

The phrase "take our chance" is not meant here to state that there is a certain probability of success; it means only that there is a possibility of success in the sense that there is no proof that success is excluded. Furthermore, the suggestion to try predictions by means of the inductive method is not an advice of a trial at random, of trying one's luck, so to speak; it is the proposal of a systematic method of trial so devised that if success is attainable the method will find it....

If there is a limit of the frequency, the use of the rule of induction will be a sufficient condition to find the limit to a desired degree of approximation. There may be other methods, but this one, at least, is sufficient. Consequently, when we do not know whether there is a limit, we can say, if there is any way to find a limit, the rule of induction will be such a way. It is, therefore, a necessary condition for the existence of a limit, and thus for the existence of a method to find it, that the aim be attainable by means of the rule of induction.

To clarify these logical relations, we shall formulate them in the logical symbolism. We abbreviate by a the statement, "There exists a limit of the frequency"; by b the statement, "I use the rule of induction in a repeated procedure"; by c the statement, "I shall find the limit of the frequency." We then have the relation[7]

$$a \supset (b \supset c) \qquad (1)$$

[7] The implications occurring here must be regarded as nomological operations: the first as a tautological implication, the second as a relative nomological implication. See *ESL*, § 63.

This means, $b \supset c$ is the *necessary* condition of a, or, in other words, the attainability of the aim by the use of the rule of induction is a necessary condition of the existence of a limit. Furthermore, if a is true, b is a *sufficient* condition of c. This means, if there is a limit of the frequency, the use of the rule of induction is a sufficient instrument to find it.

It is in this relation that I find the justification of the rule of induction. Scientific method pursues the aim of predicting the future; in order to construct a precise formulation for this aim we interpret it as meaning that scientific method is intended to find limits of the frequency. Classical induction and predictions of individual events are included in the general formulation as the special case that the relative frequency is = 1. It has been shown that if the aim of scientific method is attainable it will be reached by the inductive method. This result eliminates the last assumption we had to use for the justification of induction. The assumption that there is a limit of the frequency must be true if the inductive procedure is to be successful. But we need not know whether it is true when we merely ask whether the inductive procedure is justified. It is justified as an attempt at finding the limit. Since we do not know a sufficient condition to be used for finding a limit, we shall at least make use of a necessary condition. In positing according to the rule of induction, always correcting the posit when additional observation shows different results, we prepare everything so that if there is a limit of the frequency we shall find it. If there is none, we shall certainly not find one—but then all other methods will break down also.

The answer to Hume's question is thus found. Hume was right in asserting that the conclusion of the inductive inference cannot be proved to be true; and we may add that it cannot even be proved to be probable. But Hume was wrong in stating that the inductive procedure is unjustifiable. It can be justified as an instrument that realizes the necessary conditions of prediction, to which we resort because sufficient conditions of prediction are beyond our reach. The justification of induction can be summarized as follows:

> *Thesis θ. The rule of induction is justified as an instrument of positing because it is a method of which we know that if it is possible to make statements about the future we shall find them by means of this method....*

A blind man who has lost his way in the mountains feels a trail with his stick. He does not know where the path will lead him, or whether it may take him so close to the edge of a precipice that he will be plunged into the abyss. Yet he follows the path, groping his way step by step; for if there is any possibility of getting out of the wilderness, it is by feeling his way along the path. As blind men we face the future; but we feel a path. And we know: if we can find a way through the future it is by feeling our way along this path.

2

ON INDUCTIVE LOGIC

Rudolf Carnap

§ 1. INDUCTIVE LOGIC

Among the various meanings in which the word 'probability' is used in everyday language, in the discussion of scientists, and in the theories of probability, there are especially two which must be clearly distinguished. We shall use for them the terms 'probability$_1$' and 'probability$_2$'. Probability$_1$ is a logical concept, a certain logical relation between two sentences (or, alternatively, between two propositions); it is the same as the concept of degree of confirmation. I shall write briefly "$\mathfrak{c}$" for "degree of confirmation," and "$\mathfrak{c}(h, e)$" for "the degree of confirmation of the hypothesis *h* on the evidence *e*"; the evidence is usually a report on the results of our observations. On the other hand, probability$_2$ is an empirical concept; it is the relative frequency in the long run of one property with respect to another. The controversy between the so-called logical conception of probability, as represented e.g. by Keynes,[1] and Jeffreys,[2] and others, and the frequency conception, maintained e.g. by v. Mises[3] and Reichenbach,[4] seems to me futile. These two theories deal with two different probability concepts which are both of great importance for science. Therefore, the theories are not incompatible, but rather supplement each other.[5]

[1] J. M. Keynes, *A Treatise on Probability,* 1921.

[2] H. Jeffreys, *Theory of Probability*, 1939.

[3] R. v. Mises, *Probability, Statistics, and Truth,* (orig. 1928) 1939.

[4] H. Reichenbach, *Wahrscheinlichkeitslehre,* 1935.

[5] The distinction briefly indicated here, is discussed more in detail in my paper "The Two Concepts of Probability," . . . in *Philos. and Phenom. Research,* 1945.

In a certain sense we might regard deductive logic as the theory of L-implication (logical implication, entailment). And inductive logic may be construed as the theory of degree of confirmation, which is, so to speak, partial L-implication. "*e* L-implies *h*" says that *h* is implicitly given with *e*, in other words, that the whole logical content of *h* is contained in *e*. On the other hand, "$\mathfrak{c}(h, e) = \frac{3}{4}$" says that *h* is not entirely given with *e* but that the assumption of *h* is supported to the degree $\frac{3}{4}$ by the observational evidence expressed in *e*.

In the course of the last years, I have constructed a new system of inductive logic by laying down a definition for degree of confirmation and developing a theory based on this definition. A book containing this theory is in preparation. The purpose of the present paper is to indicate briefly and informally the definition and a few of the results found; for lack of space, the reasons for the choice of this definition and the proofs for the results cannot be given here. The book will, of course, provide a better basis than the present informal summary for a critical evaluation of the theory and of the fundamental conception on which it is based.[6]

§ 2. SOME SEMANTICAL CONCEPTS

Inductive logic is, like deductive logic, in my conception a branch of semantics. However, I shall try to formulate the present outline in such a way that it does not presuppose knowledge of semantics.

Let us begin with explanations of some semantical concepts which are important both for deductive logic and for inductive logic.[7]

The system of inductive logic to be outlined applies to an infinite sequence of finite language systems L_N ($N = 1, 2, 3$, etc.) and an infinite language system L_∞. L_∞ refers to an infinite universe of individuals, designated by the individual constants 'a_1', 'a_2', etc. (or 'a', 'b', etc.), while L_N refers to a finite universe containing only N individuals designated by 'a_1', 'a_2', $\cdots$ 'a_N'. Individual variables 'x_1', 'x_2', etc. (or 'x', 'y', etc.) are the only variables occurring in these languages. The languages contain a finite number of predicates of any degree (number of arguments), designating properties of the individuals or relations between them. There are, furthermore, the customary connectives of negation ('$\sim$', corresponding to "not"), disjunction ('$\vee$', "or"), conjunction '$\cdot$', "and"); universal and existential quantifiers ("for every *x*," "there is

[6] In an article by C. G. Hempel and Paul Oppenheim in the present issue of this journal [*Philosophy of Science*, vol. 12, 1945], a new concept of degree of confirmation is proposed, which was developed by the two authors and Olaf Helmer in research independent of my own.

[7] For more detailed explanations of some of these concepts see my *Introduction to Semantics*, 1942.

an x"); the sign of identity between individuals '=', and 't' as an abbreviation for an arbitrarily chosen tautological sentence. (Thus the languages are certain forms of what is technically known as the lower functional logic with identity.) (The connectives will be used in this paper in three ways, as is customary: (1) between sentences, (2) between predicates (§ 8), (3) between names (or variables) of sentences (so that, if 'i' and 'j' refer to two sentences, '$i \vee j$' is meant to refer to their disjunction).)

A sentence consisting of a predicate of degree n with n individual constants is called an *atomic sentence* (e.g. 'Pa_1', i.e. 'a_1 has the property P', or 'Ra_3a_5', i.e. 'the relation R holds between a_3 and a_5'). The conjunction of all atomic sentences in a finite language L_N describes one of the possible states of the domain of the N individuals with respect to the properties and relations expressible in the language L_N. If we replace in this conjunction some of the atomic sentences by their negations, we obtain the description of another possible state. All the conjunctions which we can form in this way, including the original one, are called *state-descriptions* in L_N. Analogously, a state-description in L_∞ is a class containing some atomic sentences and the negations of the remaining atomic sentences; since this class is infinite, it cannot be transformed into a conjunction.

In the actual construction of the language systems, which cannot be given here, semantical rules are laid down determining for any given sentence j and any state-description i whether j holds in i, that is to say whether j would be true if i described the actual state among all possible states. The class of those state-descriptions in a language system L (either one of the systems L_N or L_∞) in which j holds is called the *range* of j in L.

The concept of range is fundamental both for deductive and for inductive logic; this has already been pointed out by Wittgenstein. If the range of a sentence j in the language system L is universal, i.e. if j holds in every state-description (in L), j must necessarily be true independently of the facts; therefore we call j (in L) in this case *L-true* (logically true, analytic). (The prefix 'L-' stands for "logical"; it is not meant to refer to the system L.) Analogously, if the range of j is null, we call j *L-false* (logically false, self-contradictory). If j is neither L-true nor L-false, we call it *factual* (synthetic, contingent). Suppose that the range of e is included in that of h. Then in every possible case in which e would be true, h would likewise be true. Therefore we say in this case that e *L-implies* (logically implies, entails) h. If two sentences have the same range, we call them *L-equivalent*; in this case, they are merely different formulations for the same content.

The L-concepts just explained are fundamental for deductive logic and therefore also for inductive logic. Inductive logic is constructed out of deductive logic by the introduction of the concept of degree of confirmation. This introduction will here be carried out in three steps: (1) the definition of regular $\mathfrak{c}$-functions (§ 3), (2) the definition of symmetrical $\mathfrak{c}$-functions (§ 5), (3) the definition of the degree of confirmation $\mathfrak{c}^*$ (§ 6).

§ 3. REGULAR $\mathfrak{c}$-FUNCTIONS

A numerical function $\mathfrak{m}$ ascribing real numbers of the interval 0 to 1 to the sentences of a finite language L_N is called a regular $\mathfrak{m}$-function if it is constructed according to the following rules:

(1) We assign to the state-descriptions in L_N as values of $\mathfrak{m}$ any positive real numbers whose sum is 1.

(2) For every other sentence j in L_N, the value $\mathfrak{m}(j)$ is determined as follows:

(a) If j is not L-false, $\mathfrak{m}(j)$ is the sum of the $\mathfrak{m}$-values of those state-descriptions which belong to the range of j.

(b) If j is L-false and hence its range is null, $\mathfrak{m}(j) = 0$.

(The choice of the rule (2)(a) is motivated by the fact that j is L-equivalent to the disjunction of those state-descriptions which belong to the range of j and that these state-descriptions logically exclude each other.)

If any regular $\mathfrak{m}$-function $\mathfrak{m}$ is given, we define a corresponding function $\mathfrak{c}$ as follows:

(3) For any pair of sentences e, h in L_N, where e is not L-false, $\mathfrak{c}(h,e) = \dfrac{\mathfrak{m}(e \cdot h)}{\mathfrak{m}(e)}$.

$\mathfrak{m}(j)$ may be regarded as a measure ascribed to the range of j; thus the function $\mathfrak{m}$ constitutes a metric for the ranges. Since the range of the conjunction $e \cdot h$ is the common part of the ranges of e and of h, the quotient in (3) indicates, so to speak, how large a part of the range of e is included in the range of h. The numerical value of this ratio, however, depends on what particular $\mathfrak{m}$-function has been chosen. We saw earlier that a statement in deductive logic of the form "e L-implies h" says that the range of e is entirely included in that of h. Now we see that a statement in inductive logic of the form "$\mathfrak{c}(h,e) = \frac{3}{4}$" says that a certain part—in the example, three fourths—of the range of e is included in the range of h.[8] Here, in order to express the partial inclusion numerically, it is necessary to choose a regular $\mathfrak{m}$-function for measuring the ranges. Any $\mathfrak{m}$ chosen leads to a particular $\mathfrak{c}$ as defined above. All functions $\mathfrak{c}$ obtained in this way are called *regular* $\mathfrak{c}$*-functions.*

One might perhaps have the feeling that the metric $\mathfrak{m}$ should not be chosen once for all but should rather be changed according to the accumulating experiences.[9] This feeling is correct in a certain sense. However, it is to

[8] See F. Waismann, "Logische Analyse des Wahrscheinlichkeitsbegriffs," *Erkenntnis,* vol. 1, 1930, pp. 228–248.

[9] See Waismann, *op. cit.,* p. 242.

be satisfied not by the function $\mathfrak{m}$ used in the definition (3) but by another function $\mathfrak{m}_e$ dependent upon e and leading to an alternative definition (5) for the corresponding $\mathfrak{c}$. If a regular $\mathfrak{m}$ is chosen according to (1) and (2), then a corresponding function $\mathfrak{m}_e$ is defined for the state-descriptions in L_N as follows:

(4) Let i be a state-description in L_N, and e a non-L-false sentence in L_N.
(a) If e does not hold in i, $\mathfrak{m}_e(i) = 0$.
(b) If e holds in i, $\mathfrak{m}_e(i) = \dfrac{\mathfrak{m}(i)}{\mathfrak{m}(e)}$.

Thus $\mathfrak{m}_e$ represents a metric for the state-descriptions which changes with the changing evidence e. Now $\mathfrak{m}_e(j)$ for any other sentence j in L_N is defined in analogy to (2)(a) and (b). Then we define the function $\mathfrak{c}$ corresponding to $\mathfrak{m}$ as follows:

(5) For any pair of sentences e, h in L_N, where e is not L-false, $\mathfrak{c}(h, e) = \mathfrak{m}_e(h)$.

It can easily be shown that this alternative definition (5) yields the same values as the original definition (3).

Suppose that a sequence of regular $\mathfrak{m}$-functions is given, one for each of the finite languages L_N ($N = 1, 2$, etc.). Then we define a corresponding $\mathfrak{m}$-function for the infinite language as follows:

(6) $\mathfrak{m}(j)$ in L_∞ is the limit of the values $\mathfrak{m}(j)$ in L_N for $N \rightarrow \infty$.

$\mathfrak{c}$-functions for the finite languages are based on the given $\mathfrak{m}$-functions according to (3). We define a corresponding $\mathfrak{c}$-function for the infinite language as follows:

(7) $\mathfrak{c}(h, e)$ in L_∞ is the limit of the values $\mathfrak{c}(h, e)$ in L_N for $N \rightarrow \infty$.

The definitions (6) and (7) are applicable only in those cases where the specified limits exist.

We shall later see how to select a particular sub-class of regular $\mathfrak{c}$-functions (§ 5) and finally one particular $\mathfrak{c}$-function $\mathfrak{c}^*$ as the basis of a complete system of inductive logic (§ 6). For the moment, let us pause at our first step, the definition of regular $\mathfrak{c}$-functions just given, in order to see what results this definition alone can yield, before we add further definitions. The theory of regular $\mathfrak{c}$-functions, i.e. the totality of those theorems which are founded on the definition stated, is the first fundamental part of inductive logic. It turns out that we find here many of the fundamental theorems of the classical theory of probability, e.g. those known as the theorem (or principle) of multiplication, the general and the special theorems of addition, the theorem of division and, based upon it, Bayes' theorem.

One of the cornerstones of the classical theory of probability is the principle of indifference (or principle of insufficient reason). It says that, if our evidence *e* does not give us any sufficient reason for regarding one of two hypotheses *h* and *h′* as more probable than the other, then we must take their probabilities$_1$ as equal: $\mathfrak{c}(h, e) = \mathfrak{c}(h', e)$. Modern authors, especially Keynes, have correctly pointed out that this principle has often been used beyond the limits of its original meaning and has then led to quite absurd results. Moreover, it can easily be shown that, even in its original meaning, the principle is by far too general and leads to contradictions. Therefore the principle must be abandoned. If it is and we consider only those theorems of the classical theory which are provable without the help of this principle, then we find that these theorems hold for all regular $\mathfrak{c}$-functions. The same is true for those modern theories of probability$_1$ (e.g. that by Jeffreys, *op. cit.*) which make use of the principle of indifference. Most authors of modern axiom systems of probability$_1$ (e.g. Keynes, *op. cit.*, Waismann, *op. cit.*, Mazurkiewicz,[10] Hosiasson,[11] v. Wright[12]) are cautious enough not to accept that principle. An examination of these systems shows that their axioms and hence their theorems hold for all regular $\mathfrak{c}$-functions. Thus these systems restrict themselves to the first part of inductive logic, which, although fundamental and important, constitutes only a very small and weak section of the whole of inductive logic. The weakness of this part shows itself in the fact that it does not determine the value of $\mathfrak{c}$ for any pair *h, e* except in some special cases where the value is 0 or 1. The theorems of this part tell us merely how to calculate further values of $\mathfrak{c}$ if some values are given. Thus it is clear that this part alone is quite useless for application and must be supplemented by additional rules. (It may be remarked incidentally, that this point marks a fundamental difference between the theories of probability$_1$ and of probability$_2$ which otherwise are analogous in many respects. The theorems concerning probability$_2$ which are analogous to the theorems concerning regular $\mathfrak{c}$-functions constitute not only the first part but the whole of the logico-mathematical theory of probability$_2$. The task of determining the value of probability$_2$ for a given case is—in contradistinction to the corresponding task for probability$_1$—an empirical one and hence lies outside the scope of the logical theory of probability$_2$.)

§ 4. THE COMPARATIVE CONCEPT OF CONFIRMATION

Some authors believe that a metrical (or quantitative) concept of degree of confirmation, that is, one with numerical values, can be applied, if at

[10] St. Mazurkiewicz, "Zur Axiomatik der Wahrscheinlichkeitsrechnung," *C. R. Soc. Science Varsovie,* Cl. III, vol. 25, 1932, pp. 1–4.

[11] Janina Hosiasson-Lindenbaum, "On Confirmation," *Journal of Symbolic Logic,* vol. 5, 1940, pp. 133–148.

[12] G. H. von Wright, *The Logical Problem of Induction* (Acta Phil. Fennica, 1941, Fasc. III). See also C. D. Broad, *Mind,* vol. 53, 1944.

all, only in certain cases of a special kind and that in general we can make only a comparison in terms of higher or lower confirmation without ascribing numerical values. Whether these authors are right or not, the introduction of a merely comparative (or topological) concept of confirmation not presupposing a metrical concept is, in any case, of interest. We shall now discuss a way of defining a concept of this kind.

For technical reasons, we do not take the concept "more confirmed" but "more or equally confirmed." The following discussion refers to the sentences of any finite language L_N. We write, for brevity, "$MC(h, e, h', e')$" for "h is confirmed on the evidence e more highly or just as highly as h' on the evidence e'."

Although the definition of the comparative concept MC at which we aim will not make use of any metrical concept of degree of confirmation, let us now consider, for heuristic purposes, the relation between MC and the metrical concepts, i.e. the regular $\mathfrak{c}$-functions. Suppose we have chosen some concept of degree of confirmation, in other words, a regular $\mathfrak{c}$-function $\mathfrak{c}$, and further a comparative relation MC; then we shall say that MC is in accord with $\mathfrak{c}$ if the following holds:

(1) For any sentences h, e, h', e', if $MC(h, e, h', e')$ *then* $\mathfrak{c}(h, e) \geqslant \mathfrak{c}(h', e')$.

However, we shall not proceed by selecting one $\mathfrak{c}$-function and then choosing a relation MC which is in accord with it. This would not fulfill our intention. Our aim is to find a comparative relation MC which grasps those logical relations between sentences which are, so to speak, prior to the introduction of any particular $\mathfrak{m}$-metric for the ranges and of any particular $\mathfrak{c}$-function; in other words, those logical relations with respect to which all the various regular $\mathfrak{c}$-functions agree. Therefore we lay down the following requirement:

(2) The relation MC is to be defined in such a way that it is in accord with *all* regular $\mathfrak{c}$-functions; in other words, if $MC(h, e, h', e')$, then for every regular $\mathfrak{c}$, $\mathfrak{c}(h, e) \geqslant \mathfrak{c}(h', e')$.

It is not difficult to find relations which fulfill this requirement (2). First let us see whether we can find quadruples of sentences h, e, h', e' which satisfy the following condition occurring in (2):

(3) For every regular $\mathfrak{c}$, $\mathfrak{c}(h, e) \geqslant \mathfrak{c}(h', e')$.

It is easy to find various kinds of such quadruples. (For instance, if e and e' are any non-L-false sentences, then the condition (3) is satisfied in all cases where e L-implies h, because here $\mathfrak{c}(h, e) = 1$; further in all cases where e' L-implies $\sim h'$, because here $\mathfrak{c}(h', e') = 0$; and in many other cases.) We could, of course, define a relation MC by taking some cases where we know that the condition (3) is satisfied and restricting the relation to these cases. Then the relation would fulfill the requirement (2); however, as long as there are cases which satisfy the condition (3) but which we have not included in the relation, the relation is unnecessarily restricted. Therefore we lay down the following as a second requirement for *MC*:

(4) *MC* is to be defined in such a way that it holds in all cases which satisfy the condition (3); in such a way, in other words, that it is the most comprehensive relation which fulfills the first requirement (2).

These two requirements (2) and (4) together stipulate that $MC(h, e, h' e')$ is to hold if and only if the condition (3) is satisfied; thus the requirements determine uniquely one relation *MC*. However, because they refer to the c-functions, we do not take these requirements as a definition for *MC*, for we intend to give a purely comparative definition for *MC*, a definition which does not make use of any metrical concepts but which leads nevertheless to a relation *MC* which fulfills the requirements (2) and (4) referring to c-functions. This aim is reached by the following definition (where '$=_{Df}$' is used as sign of definition).

(5) $MC(h, e, h', e') =_{Df}$ the sentences h, e, h', e' (in L_N) are such that e and e' are not L-false and at least one of the following three conditions is fulfilled:

(a) e L-implies h,
(b) e' L-implies $\sim h'$,
(c) $e' \cdot h'$ L-implies $e \cdot h$ and simultaneously e L-implies $h \vee e'$.

((a) and (b) are the two kinds of rather trivial cases earlier mentioned; (c) comprehends the interesting cases; an explanation and discussion of them cannot be given here.)

The following theorem can then be proved concerning the relation *MC* defined by (5). It shows that this relation fulfills the two requirements (2) and (4).

(6) For any sentences h, e, h', e' in L_N the following holds:
(a) If $MC(h, e, h', e')$, then, for every regular c, $c(h, e) \geqslant c(h', e')$.
(b) If, for every regular c, $c(h, e) \geqslant c(h', e')$, then $MC(h, e, h', e')$.

(With respect to L_∞, the analogue of (6)(a) holds for all sentences, and that of (6)(b) for all sentences without variables.)

§ 5. SYMMETRICAL c-FUNCTIONS

The next step in the construction of our system of inductive logic consists in selecting a narrow sub-class of the comprehensive class of all regular c-functions. The guiding idea for this step will be the principle that inductive logic should treat all individuals on a par. The same principle holds for deductive logic; for instance, if '$\cdot\cdot a \cdot\cdot b \cdot\cdot$' L-implies '$\text{--}b\text{--}c\text{--}$' (where the first expression in quotation marks is meant to indicate some sentence containing '*a*' and '*b*', and the second another sentence containing '*b*' and '*c*'), then L-implication holds likewise between corresponding sentences with other individual constants, e.g. between '··*d*··*c*··' and '--*c*--*a*--'. Now we require that this should hold also for inductive logic, e.g. that $\mathfrak{c}$('--*b*--*c*--', '··*a*··*b*··') = $\mathfrak{c}$('--*c*--*a*--', '··*d*··*c*··'). It seems that all authors on probability$_1$ have assumed this principle—although it has seldom, if ever, been stated explicitly—by formulating theorems in the following or similar terms: "On the basis of observations of *s* things of which s_1 were found to have the property *M* and s_2 not to have this property, the probability that another thing has this property is such and such." The fact that these theorems refer only to the number of things observed and do not mention particular things shows implicitly that it does not matter which things are involved; thus it is assumed, e.g., that $\mathfrak{c}$('*Pd*', '*Pa*·*Pb*· ~ *Pc*') = $\mathfrak{c}$('*Pc*', '*Pa*·*Pd*· ~ *Pb*').

The principle could also be formulated as follows. Inductive logic should, like deductive logic, make no discrimination among individuals. In other words, the value of $\mathfrak{c}$ should be influenced only by those differences between individuals which are expressed in the two sentences involved; no differences between particular individuals should be stipulated by the rules of either deductive or inductive logic.

It can be shown that this principle of non-discrimination is fulfilled if $\mathfrak{c}$ belongs to the class of symmetrical $\mathfrak{c}$-functions which will now be defined. Two state-descriptions in a language L_N are said to be *isomorphic* or to have the same structure if one is formed from the other by replacements of the following kind: we take any one-one relation *R* such that both its domain and its converse domain is the class of all individual constants in L_N, and then replace every individual constant in the given state-description by the one correlated with it by *R*. If a regular $\mathfrak{m}$-function (for L_N) assigns to any two isomorphic state-descriptions (in L_N) equal values, it is called a symmetrical $\mathfrak{m}$-function; and a $\mathfrak{c}$-function based upon such an $\mathfrak{m}$-function in the way explained earlier (see (3) in § 3) is then called a *symmetrical* *$\mathfrak{c}$-function.*

§ 6. THE DEGREE OF CONFIRMATION $\mathfrak{c}^*$

Let *i* be a state-description in L_N. Suppose there are n_i state-descriptions in L_N isomorphic to *i* (including *i* itself), say *i*, *i'*, *i''*, etc. These n_i state-descriptions exhibit one and the same structure of the universe of L_N with respect to all the properties and relations designated by the primitive predicates in L_N. This concept of structure is an extension of the concept of structure or relation-number (Russell) usually applied to one dyadic relation. The common structure of the isomorphic state-descriptions *i*, *i'*, *i''*, etc. can be described by their disjunction $i \vee i' \vee i'' \vee \cdots$. Therefore we call this disjunction, say j, a ***structure-description*** in L_N. It can be shown that the range of j contains only the isomorphic state-descriptions i, i', i'', etc. Therefore (see (2)(a) in § 3) $\mathfrak{m}(j)$ is the sum of the $\mathfrak{m}$-values for these state-descriptions. If $\mathfrak{m}$ is symmetrical, then these values are equal, and hence

$$\mathfrak{m}(j) = n_i \times \mathfrak{m}(i) \tag{1}$$

And, conversely, if $\mathfrak{m}(j)$ is known to be q, then

$$\mathfrak{m}(i) = \mathfrak{m}(i') = \mathfrak{m}(i'') = \cdots = q/n_i \tag{2}$$

This shows that what remains to be decided, is merely the distribution of $\mathfrak{m}$-values among the structure-descriptions in L_N. We decide to give them equal $\mathfrak{m}$-values. This decision constitutes the third step in the construction of our inductive logic. This step leads to one particular $\mathfrak{m}$-function $\mathfrak{m}^*$ and to the $\mathfrak{c}$-function $\mathfrak{c}^*$ based upon $\mathfrak{m}^*$. According to the preceding discussion, $\mathfrak{m}^*$ is characterized by the following two stipulations:

(a) $\mathfrak{m}^*$ is a symmetrical $\mathfrak{m}$-function; (3)
(b) $\mathfrak{m}^*$ has the same value for all structure-descriptions (in L_N).

We shall see that these two stipulations characterize just one function. Every state-description (in L_N) belongs to the range of just one structure-description. Therefore, the sum of the $\mathfrak{m}^*$-values for all structure-descriptions in L_N must be the same as for all state-descriptions, hence 1 (according to (1) in § 3). Thus, if the number of structure-descriptions in L_N is m, then, according to (3)(b),

$$\text{for every structure-description } j \text{ in } L_N,\ \mathfrak{m}^*(j) = \frac{1}{m} \tag{4}$$

Therefore, if i is any state-description in L_N and n_i is the number of state-descriptions isomorphic to i, then, according to (3)(a) and (2),

$$\mathfrak{m}^*(i) = \frac{1}{mn_i} \tag{5}$$

(5) constitutes a definition of $\mathfrak{m}^*$ as applied to the state-descriptions in L_N. On this basis, further definitions are laid down as explained above (see (2) and (3) in § 3): first a definition of $\mathfrak{m}^*$ as applied to all sentences in L_N, and then a definition of $\mathfrak{c}^*$ on the basis of $\mathfrak{m}^*$. Our inductive logic is the theory of this particular function $\mathfrak{c}^*$ as our concept of degree of confirmation.

It seems to me that there are good and even compelling reasons for the stipulation (3)(a), i.e. the choice of a symmetrical function. The proposal of any non-symmetrical $\mathfrak{c}$-function as degree of confirmation could hardly be regarded as acceptable. The same can not be said, however, for the stipulation (3)(b). No doubt, to the way of thinking which was customary in the classical period of the theory of probability, (3)(b) would appear as validated, like (3)(a), by the principle of indifference. However, to modern, more critical thought, this mode of reasoning appears as invalid because the structure-descriptions (in contradistinction to the individual constants) are by no means alike in their logical features but show very conspicuous differences. The definition of $\mathfrak{c}^*$ shows a great simplicity in comparison with other concepts which may be taken into consideration. Although this fact may influence our decision to choose $\mathfrak{c}^*$, it cannot, of course, be regarded as a sufficient reason for this choice. It seems to me that the choice of $\mathfrak{c}^*$ cannot be justified by any features of the definition which are immediately recognizable, but only by the consequences to which the definition leads.

There is another $\mathfrak{c}$-function $\mathfrak{c}_W$ which at the first glance appears not less plausible than $\mathfrak{c}^*$. The choice of this function may be suggested by the following consideration. Prior to experience, there seems to be no reason to regard one state-description as less probable than another. Accordingly, it might seem natural to assign equal $\mathfrak{m}$-values to all state-descriptions. Hence, if the number of the state-descriptions in L_N is n, we define for any state-description i

$$\mathfrak{m}_W(i) = 1/n \tag{6}$$

This definition (6) for $\mathfrak{m}_W$ is even simpler than the definition (5) for $\mathfrak{m}^*$. The measure ascribed to the ranges is here simply taken as proportional to the cardinal numbers of the ranges. On the basis of the $\mathfrak{m}_W$-values for the state-descriptions defined by (6), the values for the sentences are determined as before (see (2) in § 3), and then $\mathfrak{c}_W$ is defined on the basis of $\mathfrak{m}_W$ (see (3) in § 3).[13]

In spite of its apparent plausibility, the function $\mathfrak{c}_W$ can easily be seen to be entirely inadequate as a concept of degree of confirmation. As an example, consider the language L_{101} with '*P*' as the only primitive predicate. Let the number of state-descriptions in this language be n (it is 2^{101}). Then for any state-description, $\mathfrak{m}_W = 1/n$. Let e be the conjunction $Pa_1 \cdot Pa_2 \cdot Pa_3 \cdots Pa_{100}$

[13] It seems that Wittgenstein meant this function $\mathfrak{c}_W$ in his definition of probability, which he indicates briefly without examining its consequences. In his *Tractatus Logico-Philosophicus,* he says: "A proposition is the expression of agreement and disagreement with the truth-possibilities of the elementary [i.e. atomic] propositions" (*4.4); "The world is completely described by the specification of all elementary propositions plus the specification, which of them are true and which false" (*4.26). The truth-possibilities specified in this way correspond to our state-descriptions. Those truth-possibilities which verify a given proposition (in our terminology, those state-descriptions in which a given sentence holds) are called the truth-grounds of that proposition (*5.101). "If T_r is the number of the truth-grounds of the proposition "r," T_{rs} the number of those truth-grounds of the proposition "s" which are at the same time truth-grounds of "r," then we call the ratio T_{rs}:T_r the measure of the *probability* which the proposition "r" gives to the proposition "s" " (*5.15). It seems that the concept of probability thus defined coincides with the function $\mathfrak{c}_W$.

and let h be 'Pa_{101}'. Then $e \cdot h$ is a state-description and hence $\mathfrak{m}_w(e \cdot h) = 1/n$. e holds only in the two state-descriptions $e \cdot h$ and $e \cdot \sim h$; hence $\mathfrak{m}_w(e) = 2/n$. Therefore $\mathfrak{c}_w(h, e) = 1/2$. If e' is formed from e by replacing some or even all of the atomic sentences with their negations, we obtain likewise $\mathfrak{c}_w(h, e') = 1/2$. Thus the $\mathfrak{c}_w$-value for the prediction that a_{101} is P is always the same, no matter whether among the hundred observed individuals the number of those which we have found to be P is 100 or 50 or 0 or any other number. Thus the choice of $\mathfrak{c}_w$ as the degree of confirmation would be tantamount to the principle never to let our past experiences influence our expectations for the future. This would obviously be in striking contradiction to the basic principle of all inductive reasoning.

§ 7. LANGUAGES WITH ONE-PLACE PREDICATES ONLY

The discussions in the rest of this paper concern only those language systems whose primitive predicates are one-place predicates and hence designate properties, not relations. It seems that all theories of probability constructed so far have restricted themselves, or at least all of their important theorems, to properties. Although the definition of $\mathfrak{c}^*$ in the preceding section has been stated in a general way so as to apply also to languages with relations, the greater part of our inductive logic will be restricted to properties. An extension of this part of inductive logic to relations would require certain results in the deductive logic of relations, results which this discipline, although widely developed in other respects, has not yet reached (e.g. an answer to the apparently simple question as to the number of structures in a given finite language system).

Let L_N^p be a language containing N individual constants 'a_1', $\cdots$ 'a_N', and p one-place primitive predicates 'P_1', $\cdots$ 'P_p'. Let us consider the following expressions (sentential matrices). We start with '$P_1x \cdot P_2x \cdots P_px$'; from this expression we form others by negating some of the conjunctive components, until we come to '$\sim P_1x \cdot \sim P_2x \cdots \sim P_px$', where all components are negated. The number of these expressions is $k = 2^p$; we abbreviate them by 'Q_1x', $\cdots$ 'Q_kx'. We call the k properties expressed by those k expressions in conjunctive form and now designated by the k new Q-predicates the *Q-properties* with respect to the given language L_N^p. We see easily that these Q-properties are the strongest properties expressible in this language (except for the L-empty, i.e., logically self-contradictory, property); and further, that they constitute an exhaustive and non-overlapping classification, that is to say, every individual has one and only one of the Q-properties. Thus, if we state for each individual which of the Q-properties it has, then we have described the individuals completely. Every state-description can be brought into the form of such a statement, i.e. a conjunction of N Q-sentences, one for each of the N individuals. Suppose that in a given state-description i the number of individuals having the property Q_1 is N_1, the number for Q_2 is N_2, $\cdots$ that for Q_k is N_k. Then we call the numbers $N_1, N_2, \cdots N_k$ the *Q-numbers* of the state description *i*; their sum is N. Two state-descriptions are isomorphic if and only if they have the same *Q*-numbers. Thus here a structure-description is a statistical description giving the *Q*-numbers N_1, N_2, etc., without specifying which individuals have the properties Q_1, Q_2, etc.

Here—in contradistinction to languages with relations—it is easy to find an explicit function for the number *m* of structure-descriptions and, for any given state-description *i* with the *Q*-numbers $N_1 \cdots N_k$, an explicit function for the number n_i of state-descriptions isomorphic to *i*, and hence also a function for $\mathfrak{m}^*(i)$.[14]

Let *j* be a non-general sentence (i.e. one without variables) in $L_N{}^p$. Since there are effective procedures (that is, sets of fixed rules furnishing results in a finite number of steps) for constructing all state-descriptions in which *j* holds and for computing $\mathfrak{m}^*$ for any given state-description, these procedures together yield an effective procedure for computing $\mathfrak{m}^*(j)$ (according to (2) in § 3). However, the number of state-descriptions becomes very large even for small language systems (it is k^N, hence, e.g., in $L_7{}^3$ it is more than two million.) Therefore, while the procedure indicated for the computation of $\mathfrak{m}^*(j)$ is effective, nevertheless in most ordinary cases it is impracticable; that is to say, the number of steps to be taken, although finite, is so large that nobody will have the time to carry them out to the end. I have developed another procedure for the computation of $\mathfrak{m}^*(j)$ which is not only effective but also practicable if the number of individual constants occurring in *j* is not too large.

The value of $\mathfrak{m}^*$ for a sentence *j* in the infinite language has been defined (see (6) in § 3) as the limit of its values for the same sentence *j* in the finite languages. The question arises whether and under what conditions this limit exists. Here we have to distinguish two cases. (i) Suppose that *j* contains no variable. Here the situation is simple; it can be shown that in this case $\mathfrak{m}^*(j)$ is the same in all finite languages in which *j* occurs; hence it has the same value also in the infinite language. (ii) Let *j* be general, i.e., contain variables.

[14] The results are as follows.

$$m = \frac{(N+k-1)!}{N!(k-1)!} \tag{1}$$

$$n_i = \frac{N!}{N_1!N_2!\cdots N_k!} \tag{2}$$

Therefore (according to (5) in § 6):

$$\mathfrak{m}^*(i) = \frac{N_1!N_2!\cdots N_k!(k-1)!}{(N+k-1)!} \tag{3}$$

Here the situation is quite different. For a given finite language with $\mathcal{N}$ individuals, j can of course easily be transformed into an L-equivalent sentence j'_N without variables, because in this language a universal sentence is L-equivalent to a conjunction of $\mathcal{N}$ components. The values of $\mathfrak{m}^*(j'_N)$ are in general different for each $\mathcal{N}$; and although the simplified procedure mentioned above is available for the computation of these values, this procedure becomes impracticable even for moderate $\mathcal{N}$. Thus for general sentences the problem of the existence and the practical computability of the limit becomes serious. It can be shown that for every general sentence the limit exists; hence $\mathfrak{m}^*$ has a value for all sentences in the infinite language. Moreover, an effective procedure for the computation of $\mathfrak{m}^*(j)$ for any sentence j in the infinite language has been constructed. This is based on a procedure for transforming any given general sentence j into a non-general sentence j' such that j and j', although not necessarily L-equivalent, have the same $\mathfrak{m}^*$-value in the infinite language and j' does not contain more individual constants than j; this procedure is not only effective but also practicable for sentences of customary length. Thus, the computation of $\mathfrak{m}^*(j)$ for a general sentence j is in fact much simpler for the infinite language than for a finite language with a large $\mathcal{N}$.

With the help of the procedure mentioned, the following theorem is obtained:

If j is a purely general sentence (i.e. one without individual constants) in the infinite language, then $\mathfrak{m}^*(j)$ is either 0 or 1.

§ 8. INDUCTIVE INFERENCES

One of the chief tasks of inductive logic is to furnish general theorems concerning inductive inferences. We keep the traditional term "inference"; however, we do not mean by it merely a transition from one sentence to another (viz. from the evidence or premiss e to the hypothesis or conclusion h) but the determination of the degree of confirmation $\mathfrak{c}(h, e)$. In deductive logic it is sufficient to state that h follows with necessity from e; in inductive logic, on the other hand, it would not be sufficient to state that h follows—not with necessity but to some degree or other—from e. It must be specified to what degree h follows from e; in other words, the value of $\mathfrak{c}(h, e)$ must be given. We shall now indicate some results with respect to the most important kinds of inductive inference. These inferences are of special importance when the evidence or the hypothesis or both give statistical information, e.g. concerning the absolute or relative frequencies of given properties.

If a property can be expressed by primitive predicates together with the ordinary connectives of negation, disjunction, and conjunction (without the use of individual constants, quantifiers, or the identity sign), it is called an *elementary property*. We shall use 'M', 'M'', 'M_1', 'M_2', etc. for elementary properties. If a property is empty by logical necessity (e.g. the property designated by '$P \cdot \sim P$') we call it L-empty; if it is universal by logical necessity (e.g. '$P \vee \sim P$'), we call it L-universal. If it is neither L-empty nor L-universal (e.g. 'P_1', '$P_1 \cdot \sim P_2$'), we call it a *factual property*; in this case it may still happen to be universal or empty, but if so, then contingently, not necessarily. It can be shown that every elementary property which is not L-empty is uniquely analyzable into a disjunction (i.e. or-connection) of Q-properties. If M is a disjunction of n Q-properties ($n \geqslant 1$), we say that the (logical) *width* of M is n; to an L-empty property we ascribe the width 0. If the width of M is w ($\geqslant 0$), we call w/k its *relative width* (k is the number of Q-properties).

The concepts of width and relative width are very important for inductive logic. Their neglect seems to me one of the decisive defects in the classical theory of probability which formulates its theorems "for any property" without qualification. For instance, Laplace takes the probability a priori that a given thing has a given property, no matter of what kind, to be 1/2. However, it seems clear that this probability cannot be the same for a very strong property (e.g. '$P_1 \cdot P_2 \cdot P_3$') and for a very weak property (e.g. '$P_1 \vee P_2 \vee P_3$'). According to our definition, the first of the two properties just mentioned has the relative width 1/8, and the second 7/8. In this and in many other cases the probability or degree of confirmation must depend upon the widths of the properties involved. This will be seen in some of the theorems to be mentioned later.

§ 9. THE DIRECT INFERENCE

Inductive inferences often concern a situation where we investigate a whole population (of persons, things, atoms, or whatever else) and one or several samples picked out of the population. An inductive inference from the whole population to a sample is called a direct inductive inference. For the sake of simplicity, we shall discuss here and in most of the subsequent sections only the case of one property M, hence a classification of all individuals into M and $\sim M$. The theorems for classifications with more properties are analogous but more complicated. In the present case, the evidence e says that in a whole population of n individuals there are n_1 with the property M and $n_2 = n - n_1$ with $\sim M$; hence the relative frequency of M is $r = n_1/n$. The hypothesis h says that a sample of s individuals taken from the whole population will contain s_1 individuals with the property M and $s_2 = s - s_1$ with $\sim M$. Our theory yields in this case the same values as the classical theory.[15]

[15] The general theorem is as follows: $\mathfrak{c}^*(h, e) = \dfrac{\binom{n_1}{s_1}\binom{n_2}{s_1}}{\binom{n}{s}}$

If we vary s_1, then $\mathfrak{c}^*$ has its maximum in the case where the relative frequency s_1/s in the sample is equal or close to that in the whole population.

If the sample consists of only one individual c, and h says that c is M, then $\mathfrak{c}^*(h, e) = r$.

As an approximation in the case that n is very large in relation to s, Newton's theorem holds.[16] If furthermore the sample is sufficiently large, we obtain as an approximation Bernoulli's theorem in its various forms.

It is worthwhile to note two characteristics which distinguish the direct inductive inference from the other inductive inferences and make it, in a sense, more closely related to deductive inferences:

(i) The results just mentioned hold not only for $\mathfrak{c}^*$ but likewise for all symmetrical $\mathfrak{c}$-functions; in other words, the results are independent of the particular $\mathfrak{m}$-metric chosen provided only that it takes all individuals on a par.
(ii) The results are independent of the width of M. This is the reason for the agreement between our theory and the classical theory at this point.

§ 10. THE PREDICTIVE INFERENCE

We call the inference from one sample to another the predictive inference. In this case, the evidence e says that in a first sample of s individuals, there are s_1 with the property M, and $s_2 = s - s_1$ with $\sim M$. The hypothesis h says that in a second sample of s' other individuals, there will be s'_1 with M, and $s'_2 = s' - s'_1$ with $\sim M$. Let the width of M be w_1; hence the width of $\sim M$ is $w_2 = k - w_1$.[17]

The most important special case is that where h refers to one individual c only and says that c is M. In this case,

$$\mathfrak{c}^*(h, e) = \frac{s_1 + w_1}{s + k} \tag{1}$$

Laplace's much debated rule of succession gives in this case simply the value $\frac{s_1 + 1}{s + 2}$ for any property whatever; this, however, if applied to different prop-

[16] $\mathfrak{c}^*(h, e) = \binom{s}{s_1} r^{s_1} (1 - r)^{s_2}$

[17] The general theorem is as follows:

$$\mathfrak{c}^*(h, e) = \frac{\binom{s_1 + s'_1 + w_1 - 1}{s'_1}\binom{s_2 + s'_2 + w_2 - 1}{s'_2}}{\binom{s + s' + k - 1}{s'}}$$

erties, leads to contradictions. Other authors state the value s_1/s, that is, they take simply the observed relative frequency as the probability for the prediction that an unobserved individual has the property in question. This rule, however, leads to quite implausible results. If $s_1 = s$, e.g., if three individuals have been observed and all of them have been found to be *M*, the last-mentioned rule gives the probability for the next individual being *M* as 1, which seems hardly acceptable. According to (1), $\mathfrak{c}^*$ is influenced by the following two factors (though not uniquely determined by them):

(i) w_1/k, the relative width of *M*;
(ii) s_1/s, the relative frequency of *M* in the observed sample.

The factor (i) is purely logical; it is determined by the semantical rules. (ii) is empirical; it is determined by observing and counting the individuals in the sample. The value of $\mathfrak{c}^*$ always lies between those of (i) and (ii). Before any individual has been observed, $\mathfrak{c}^*$ is equal to the logical factor (i). As we first begin to observe a sample, $\mathfrak{c}^*$ is influenced more by this factor than by (ii). As the sample is increased by observing more and more individuals (but not including the one mentioned in *h*), the empirical factor (ii) gains more and more influence upon $\mathfrak{c}^*$ which approaches closer and closer to (ii); and when the sample is sufficiently large, $\mathfrak{c}^*$ is practically equal to the relative frequency (ii). These results seem quite plausible.[18]

The predictive inference is the most important inductive inference. The kinds of inference discussed in the subsequent sections may be construed as special cases of the predictive inference.

§ 11. THE INFERENCE BY ANALOGY

The inference by analogy applies to the following situation. The evidence known to us is the fact that individuals *b* and *c* agree in certain properties and, in addition, that *b* has a further property; thereupon we consider the

[18] Another theorem may be mentioned which deals with the case where, in distinction to the case just discussed, the evidence already gives some information about the individual *c* mentioned in *h*. Let M_1 be a factual elementary property with the width w_1 ($w_1 \geqslant 2$); thus M_1 is a disjunction of w_1 *Q*-properties. Let M_2 be the disjunction of w_2 among those w_1 *Q*-properties ($1 \leqslant w_2 < w_1$); hence M_2 L-implies M_1 and has the width w_2. *e* specifies first how the *s* individuals of an observed sample are distributed among certain properties, and, in particular, it says that s_1 of them have the property M_1 and s_2 of these s_1 individuals have also the property M_2; in addition, *e* says that *c* is M_1; and *h* says that *c* is also M_2. Then,

$$\mathfrak{c}^*(h,e) = \frac{s_2 + w_2}{s_1 + w_1}$$

This is analogous to (1); but in the place of the whole sample we have here that part of it which shows the property M_1.

hypothesis that c too has this property. Logicians have always felt that a peculiar difficulty is here involved. It seems plausible to assume that the probability of the hypothesis is the higher the more properties b and c are known to have in common; on the other hand, it is felt that these common properties should not simply be counted but weighed in some way. This becomes possible with the help of the concept of width. Let M_1 be the conjunction of all properties which b and c are known to have in common. The known similarity between b and c is the greater the stronger the property M_1, hence the smaller its width. Let M_2 be the conjunction of all properties which b is known to have. Let the width of M_1 be w_1, and that of M_2, w_2. According to the above description of the situation, we presuppose that M_2 L-implies M_1 but is not L-equivalent to M_1; hence $w_1 > w_2$. Now we take as evidence the conjunction $e \cdot j$; e says that b is M_2, and j says that c is M_1. The hypothesis h says that c has not only the properties ascribed to it in the evidence but also the one (or several) ascribed in the evidence to b only, in other words, that c has all known properties of b, or briefly that c is M_2. Then

$$\mathfrak{c}^*(h, e \cdot j) = \frac{w_2 + 1}{w_1 + 1} \tag{1}$$

j and h speak only about c; e introduces the other individual b which serves to connect the known properties of c expressed by j with its unknown properties expressed by h. The chief question is whether the degree of confirmation of h is increased by the analogy between c and b, in other words, by the addition of e to our knowledge. A theorem[19] is found which gives an affirmative answer to this question. However, the increase of $\mathfrak{c}^*$ is under ordinary conditions rather small; this is in agreement with the general conception according to which reasoning by analogy, although admissible, can usually yield only rather weak results.

Hosiasson[20] has raised the question mentioned above and discussed it in detail. She says that an affirmative answer, a proof for the increase of the degree of confirmation in the situation described, would justify the universally accepted reasoning by analogy. However, she finally admits that she does not find such a proof on the basis of her axioms. I think it is not astonishing that neither the classical theory nor modern theories of probability have been able to give a satisfactory account of and justification for the inference by analogy.

19

$$\frac{\mathfrak{c}^*(h, e \cdot j)}{\mathfrak{c}^*(h, j)} = 1 + \frac{w_1 - w_2}{w_2(w_1 + 1)}$$

This theorem shows that the ratio of the increase of $\mathfrak{c}^*$ is greater than 1, since $w_1 > w_2$.

20 Janina Lindenbaum-Hosiasson, "Induction et analogie: Comparaison de leur fondement," *Mind,* vol. 50, 1941, pp. 351–365; see especially pp. 361–365.

For, as the theorems mentioned show, the degree of confirmation and its increase depend here not on relative frequencies but entirely on the logical widths of the properties involved, thus on magnitudes neglected by both classical and modern theories.

The case discussed above is that of simple analogy. For the case of multiple analogy, based on the similarity of *c* not only with one other individual but with a number *n* of them, similar theorems hold. They show that $\mathfrak{c}^*$ increases with increasing *n* and approaches 1 asymptotically. Thus, multiple analogy is shown to be much more effective than simple analogy, as seems plausible.

§ 12. THE INVERSE INFERENCE

The inference from a sample to the whole population is called the inverse inductive inference. This inference can be regarded as a special case of the predictive inference with the second sample covering the whole remainder of the population. This inference is of much greater importance for practical statistical work than the direct inference, because we usually have statistical information only for some samples and not for the whole population.

Let the evidence *e* say that in an observed sample of *s* individuals there are s_1 individuals with the property *M* and $s_2 = s - s_1$ with $\sim M$. The hypothesis *h* says that in the whole population of *n* individuals, of which the sample is a part, there are n_1 individuals with *M* and n_2 with $\sim M$ ($n_1 \geqslant s_1$, $n_2 \geqslant s_2$). Let the width of *M* be w_1, and that of $\sim M$ be $w_2 = k - w_1$. Here, in distinction to the direct inference, $\mathfrak{c}^*(h, e)$ is dependent not only upon the frequencies but also upon the widths of the two properties.[21]

§ 13. THE UNIVERSAL INFERENCE

The universal inductive inference is the inference from a report on an observed sample to a hypothesis of universal form. Sometimes the term 'induction' has been applied to this kind of inference alone, while we use it in a much wider sense for all non-deductive kinds of inference. The universal inference is not even the most important one; it seems to me now that the role

[21] The general theorem is as follows:

$$\mathfrak{c}^*(h,e) = \frac{\binom{n_1 + w_1 - 1}{s_1 + w_1 - 1}\binom{n_2 + w_2 - 1}{s_2 + w_2 - 1}}{\binom{n + k - 1}{n - s}}$$

Other theorems, which cannot be stated here, concern the case where more than two properties are involved, or give approximations for the frequent case where the whole population is very large in relation to the sample.

of universal sentences in the inductive procedures of science has generally been overestimated. This will be explained in the next section.

Let us consider a simple law l, i.e. a factual universal sentence of the form "all M are M'" or, more exactly, "for every x, if x is M, then x is M'," where M and M' are elementary properties. As an example, take "all swans are white." Let us abbreviate '$M \cdot \sim M'$' ("non-white swan") by 'M_1' and let the width of M_1 be w_1. Then l can be formulated thus: "M_1 is empty," i.e. "there is no individual (in the domain of individuals of the language in question) with the property M_1" ("there are no non-white swans"). Since l is a factual sentence, M_1 is a factual property; hence $w_1 > 0$. To take an example, let w_1 be 3; hence M_1 is a disjunction of three Q-properties, say $Q \vee Q' \vee Q''$. Therefore, l can be transformed into: "Q is empty, and Q' is empty, and Q'' is empty." The weakest factual laws in a language are those which say that a certain Q-property is empty; we call them Q-laws. Thus we see that l can be transformed into a conjunction of w_1 Q-laws. Obviously l asserts more if w_1 is larger; therefore we say that the law l has the strength w_1.

Let the evidence e be a report about an observed sample of s individuals such that we see from e that none of these s individuals violates the law l; that is to say, e ascribes to each of the s individuals either simply the property $\sim M_1$ or some other property L-implying $\sim M_1$. Let l, as above, be a simple law which says that M_1 is empty, and w_1 be the width of M_1; hence the width of $\sim M_1$ is $w_2 = k - w_1$. For finite languages with N individuals, $c^*(l, e)$ is found to decrease with increasing N, as seems plausible.[22] If N is very large, c^* becomes very small; and for an infinite universe it becomes 0. The latter result may seem astonishing at first sight; it seems not in accor-

[22] The general theorem is as follows:

$$c^*(l,e) = \frac{\binom{s+k-1}{w_1}}{\binom{N+k-1}{w_1}} \qquad (1)$$

In the special case of a language containing 'M_1' as the only primitive predicate, we have $w_1 = 1$ and $k = 2$, and hence $c^*(l,e) = \frac{s+1}{N+1}$. The latter value is given by some authors as holding generally (see Jeffreys, op. cit., p. 106 (16)). However, it seems plausible that the degree of confirmation must be smaller for a stronger law and hence depend upon w_1.

If s, and hence N, too, is very large in relation to k, the following holds as an approximation:

$$c^*(l,e) = \left(\frac{s}{N}\right)^{w_1} \!\!\Big/\, k \qquad (2)$$

For the infinite language L_∞ we obtain, according to definition (7) in § 3:

$$c^*(l,e) = 0 \qquad (3)$$

dance with the fact that scientists often speak of "well-confirmed" laws. The problem involved here will be discussed later.

So far we have considered the case in which only positive instances of the law l have been observed. Inductive logic must, however, deal also with the case of negative instances. Therefore let us now examine another evidence e' which says that in the observed sample of s individuals there are s_1 which have the property M_1 (non-white swans) and hence violate the law l, and that $s_2 = s - s_2$ have $\sim M_1$ and hence satisfy the law l. Obviously, in this case there is no point in taking as hypothesis the law l in its original form, because l is logically incompatible with the present evidence e', and hence $\mathfrak{c}^*(l, e') = 0$. That all individuals satisfy l is excluded by e'; the question remains whether at least all unobserved individuals satisfy l. Therefore we take here as hypothesis the restricted law l' corresponding to the original unrestricted law l; l' says that all individuals not belonging to the sample of s individuals described in e' have the property $\sim M_1$. w_1 and w_2 are, as previously, the widths of M_1 and $\sim M_1$ respectively. It is found that $\mathfrak{c}^*(l', e')$ decreases with an increase of N and even more with an increase in the number s_1 of violating cases.[23] It can be shown that, under ordinary circumstances with large N, $\mathfrak{c}^*$ increases moderately when a new individual is observed which satisfies the original law l. On the other hand, if the new individual violates l, $\mathfrak{c}^*$ decreases very much, its value becoming a small fraction of its previous value. This seems in good agreement with the general conception.

For the infinite universe, $\mathfrak{c}^*$ is again 0, as in the previous case. This result will be discussed in the next section.

§ 14. THE INSTANCE CONFIRMATION OF A LAW

Suppose we ask an engineer who is building a bridge why he has chosen the building materials he is using, the arrangement and dimensions of the supports, etc. He will refer to certain physical laws, among them some general laws of mechanics and some specific laws concerning the strength of the materials. On further inquiry as to his confidence in these laws he may apply to them phrases like "very reliable," "well founded," "amply confirmed by numerous experiences." What do these phrases mean? It is clear that they are intended to say something about probability$_1$ or degree of confirmation. Hence, what is meant could be formulated more explicitly in a statement of the form "$\mathfrak{c}(h, e)$ is high" or the like. Here the evidence e is obviously the

[23] The theorem is as follows:

$$\mathfrak{c}^*(l', e') = \frac{\binom{s+k-1}{s_1+w_1}}{\binom{N+k-1}{s_1+w_1}}$$

relevant observational knowledge of the engineer or of all physicists together at the present time. But what is to serve as the hypothesis *h*? One might perhaps think at first that *h* is the law in question, hence a universal sentence *l* of the form: "For every spacetime point *x*, if such and such conditions are fulfilled at *x*, then such and such is the case at *x*." I think, however, that the engineer is chiefly interested not in this sentence *l*, which speaks about an immense number, perhaps an infinite number, of instances dispersed through all time and space, but rather in one instance of *l* or a relatively small number of instances. When he says that the law is very reliable, he does not mean to say that he is willing to bet that among the billion of billions, or an infinite number, of instances to which the law applies there is not one counter-instance, but merely that this bridge will not be a counter-instance, or that among all bridges which he will construct during his lifetime, or among those which all engineers will construct during the next one thousand years, there will be no counter-instance. Thus *h* is not the law *l* itself but only a prediction concerning one instance or a relatively small number of instances. Therefore, what is vaguely called the reliability of a law is measured not by the degree of confirmation of the law itself but by that of one or several instances. This suggests the subsequent definitions. They refer, for the sake of simplicity, to just one instance; the case of several, say one hundred, instances can then easily be judged likewise. Let *e* be any non-L-false sentence without variables. Let *l* be a simple law of the form earlier described (§ 13). Then we understand by the *instance confirmation* of *l* on the evidence *e*, in symbols "$\mathfrak{c}^*_i(l, e)$," the degree of confirmation, on the evidence *e*, of the hypothesis that a new individual not mentioned in *e* fulfills the law *l*.[24]

The second concept, now to be defined, seems in many cases to represent still more accurately what is vaguely meant by the reliability of a law *l*. We suppose here that *l* has the frequently used conditional form mentioned earlier: "For every *x*, if *x* is *M*, then *x* is *M'*" (e.g. "all swans are white"). By the *qualified-instance confirmation* of the law that all swans are white we mean the degree of confirmation for the hypothesis *h'* that the next swan to be observed will likewise be white. The difference between the hypothesis *h* used previously for the instance confirmation and the hypothesis *h'* just described consists in the fact that the latter concerns an individual which is already qualified as fulfilling the condition *M*. That is the reason why we speak here of the qualified-instance confirmation, in symbols "$\mathfrak{c}^*_{qi}$."[25] The results obtained concerning instance confirmation and qualified-instance

[24] In technical terms, the definition is as follows: $\mathfrak{c}^*_i(l, e) =_{Df} \mathfrak{c}^*(h, e)$, where *h* is an instance of *l* formed by the substitution of an individual constant not occurring in *e*.

[25] The technical definition will be given here. Let *l* be 'for every *x*, if *x* is *M*, then *x* is *M'*'. Let *l* be non-L-false and without variables. Let '*c*' be any individual constant not occurring in *e*; let *j* say that *c* is *M*, and *h'* that *c* is *M'*. Then the qualified-instance confirmation of *l* with respect to '*M*' and '*M'*' on the evidence *e* is defined as follows: $\mathfrak{c}_{qi}(\text{'}M\text{'}, \text{'}M'\text{'}, e) =_{Df} \mathfrak{c}^*(h', e \cdot j)$.

confirmation[26] show that the values of these two functions are independent of $\mathcal{N}$ and hence hold for all finite and infinite universes. It has been found that, if the number s_1 of observed counter-instances is a fixed small number, then, with the increase of the sample s, both $\mathfrak{c}^*_i$ and $\mathfrak{c}^*_{qi}$ grow close to 1, in contradistinction to $\mathfrak{c}^*$ for the law itself. This justifies the customary manner of speaking of "very reliable" or "well-founded" or "well confirmed" laws, provided we interpret these phrases as referring to a high value of either of our two concepts just introduced. Understood in this sense, the phrases are not in contradiction to our previous results that the degree of confirmation of a law is very small in a large domain of individuals and 0 in the infinite domain (§ 13).

These concepts will also be of help in situations of the following kind. Suppose a scientist has observed certain events, which are not sufficiently explained by the known physical laws. Therefore he looks for a new law as an explanation. Suppose he finds two incompatible laws l and l', each of which would explain the observed events satisfactorily. Which of them should he prefer? If the domain of individuals in question is finite, he may take the law with the higher degree of confirmation. In the infinite domain, however, this method of comparison fails, because the degree of confirmation is 0 for either law. Here the concept of instance confirmation (or that of qualified-instance confirmation) will help. If it has a higher value for one of the two laws, then this law will be preferable, if no reasons of another nature are against it.

It is clear that for any deliberate activity predictions are needed, and that these predictions must be "founded upon" or "(inductively) inferred from" past experiences, in some sense of those phrases. Let us examine the situation with the help of the following simplified schema. Suppose a man X wants to make a plan for his actions and, therefore, is interested in the prediction h that c is M'. Suppose further, X has observed (1) that many other things were M and that all of them were also M', let this be formulated in the sentence e; (2) that c is M, let this be j. Thus he knows e and j by observa-

[26] Some of the theorems may here be given. Let the law l say, as above, that all M are M'. Let 'M_1' be defined, as earlier, by '$M \cdot \sim M'$' ("non-white swan") and 'M_2' by '$M \cdot M'$' ("white swan"). Let the widths of M_1 and M_2 be w_1 and w_2 respectively. Let e be a report about s observed individuals saying that s_1 of them are M_1 and s_2 are M_2, while the remaining ones are $\sim M$ and hence neither M_1 nor M_2. Then the following holds:

$$\mathfrak{c}^*_i(l, e) = 1 - \frac{s_1 + w_1}{s + k} \quad (1)$$

$$\mathfrak{c}^*_{qi}('M', 'M'', e) = 1 - \frac{s_1 + w_1}{s_1 + w_1 + s_2 + w_2} \quad (2)$$

The values of $\mathfrak{c}^*_i$ and $\mathfrak{c}^*_{qi}$ for the case that the observed sample does not contain any individuals violating the law l can easily be obtained from the values stated in (1) and (2) by taking $s_1 = 0$.

tion. The problem is, how does he go from these premisses to the desired conclusion *h*? It is clear that this cannot be done by deduction; an inductive procedure must be applied. What is this inductive procedure? It is usually explained in the following way. From the evidence *e, X* infers inductively the law *l* which says that all *M* are *M′*; this inference is supposed to be inductively valid because *e* contains many positive and no negative instances of the law *l*; then he infers *h* ("*c* is white") from *l* ("all swans are white") and *j* ("*c* is a swan") deductively. Now let us see what the procedure looks like from the point of view of our inductive logic. One might perhaps be tempted to transcribe the usual description of the procedure just given into technical terms as follows. *X* infers *l* from *e* inductively because $\mathfrak{c}^*(l, e)$ is high; since $l \cdot j$ L-implies *h*, $\mathfrak{c}^*(h, e \cdot j)$ is likewise high; thus *h* may be inferred inductively from $e \cdot j$. However, this way of reasoning would not be correct, because, under ordinary conditions, $\mathfrak{c}^*(l, e)$ is not high but very low, and even 0 if the domain of individuals is infinite. The difficulty disappears when we realize on the basis of our previous discussions that *X* does not need a high $\mathfrak{c}^*$ for *l* in order to obtain the desired high $\mathfrak{c}^*$ for *h*; all he needs is a high $\mathfrak{c}^*_{qi}$ for *l*; and this he has by knowing *e* and *j*. To put it in another way, *X* need not take the roundabout way through the law *l* at all, as is usually believed; he can instead go from his observational knowledge $e \cdot j$ directly to the prediction *h*. That is to say, our inductive logic makes it possible to determine $\mathfrak{c}^*(h, e \cdot j)$ directly and to find that it has a high value, without making use of any law. Customary thinking in every-day life likewise often takes this short-cut, which is now justified by inductive logic. For instance, suppose somebody asks Mr. *X* what color he expects the next swan he will see to have. Then *X* may reason like this: he has seen many white swans and no non-white swans; therefore he presumes, admittedly not with certainty, that the next swan will likewise be white; and he is willing to bet on it. He does perhaps not even consider the question whether all swans in the universe without a single exception are white; and if he did, he would not be willing to bet on the affirmative answer.

We see that the use of laws is not indispensable for making predictions. Nevertheless it is expedient of course to state universal laws in books on physics, biology, psychology, etc. Although these laws stated by scientists do not have a high degree of confirmation, they have a high qualified-instance confirmation and thus serve us as efficient instruments for finding those highly confirmed singular predictions which we need for guiding our actions.

§ 15. THE VARIETY OF INSTANCES

A generally accepted and applied rule of scientific method says that for testing a given law we should choose a variety of specimens as great as possible. For instance, in order to test the law that all metals expand by heat, we should examine not only specimens of iron, but of many different metals. It seems clear that a greater variety of instances allows a more effective examination of the law. Suppose three physicists examine the law mentioned; each of them makes one hundred experiments by heating one hundred metal pieces and observing their expansion; the first physicist neglects the rule of variety and takes only pieces of iron; the second follows the rule to a small extent by examining iron and copper pieces; the third satisfies the rule more thoroughly by taking his one hundred specimens from six different metals. Then we should say that the third physicist has confirmed the law by a more thoroughgoing examination than the two other physicists; therefore he has better reasons to declare the law well-founded and to expect that future instances will likewise be found to be in accordance with the law; and in the same way the second physicist has more reasons than the first. Accordingly, if there is at all an adequate concept of degree of confirmation with numerical values, then its value for the law, or for the prediction that a certain number of future instances will fulfill the law, should be higher on the evidence of the report of the third physicist about the positive results of his experiments than for the second physicist, and higher for the second than for the first. Generally speaking, the degree of confirmation of a law on the evidence of a number of confirming experiments should depend not only on the total number of (positive) instances found but also on their variety, i.e. on the way they are distributed among various kinds.

Ernest Nagel[27] has discussed this problem in detail. He explains the difficulties involved in finding a quantitative concept of degree of confirmation that would satisfy the requirement we have just discussed, and he therefore expresses his doubt whether such a concept can be found at all. He says (pp. 69f): "It follows, however, that the degree of confirmation for a theory seems to be a function not only of the absolute number of positive instances but also of the kinds of instances and of the relative number in each kind. It is not in general possible, therefore, to order degrees of confirmation in a linear order, because the evidence for theories may not be comparable in accordance with a simple linear schema; and a fortiori degrees of confirmation cannot, in general, be quantized." He illustrates his point by a numerical example. A theory $\mathcal{T}$ is examined by a number E of experiments all of which yield positive instances; the specimens tested are taken from two non-overlapping kinds $\mathcal{K}_1$ tnd $\mathcal{K}_2$. Nine possibilities $P_1, \cdots P_9$ are discussed with different numbers of instances in $\mathcal{K}_1$ and in $\mathcal{K}_2$. The total number E increases from 50 in P_1 to 200 in P_9. In P_1, 50 instances are taken from $\mathcal{K}_1$ and none from $\mathcal{K}_2$; in P_9, 198 from $\mathcal{K}_1$ and 2 from $\mathcal{K}_2$. It does indeed seem difficult to find a concept of degree of confirmation that takes into account

[27] E. Nagel, *Principles of the Theory of Probability*. Int. Encycl. of Unified Science, vol. I, no. 6, 1939; see pp. 68–71.

in an adequate way not only the absolute number E of instances but also their distribution among the two kinds in the different cases. And I agree with Nagel that this requirement is important. However, I do not think it impossible to satisfy the requirement; in fact, it is satisfied by our concept $\mathfrak{c}^*$.

This is shown by a theorem in our system of inductive logic, which states the ratio in which the $\mathfrak{c}^*$ of a law l is increased if s new positive instances of one or several different kinds are added by new observations to some former positive instances. The theorem, which is too complicated to be given here, shows that $\mathfrak{c}^*$ is greater under the following conditions: (1) if the total number s of the new instances is greater, *ceteris paribus;* (2) if, with equal numbers s, the number of different kinds from which the instances are taken is greater; (3) if the instances are distributed more evenly among the kinds. Suppose a physicist has made experiments for testing the law l with specimens of various kinds and he wishes to make one more experiment with a new specimen. Then it follows from (2), that the new specimen is best taken from one of those kinds from which so far no specimen has been examined; if there are no such kinds, then we see from (3) that the new specimen should best be taken from one of those kinds which contain the minimum number of instances tested so far. This seems in good agreement with scientific practice. [The above formulations of (2) and (3) hold in the case where all the kinds considered have equal width; in the general and more exact formulation, the increase of $\mathfrak{c}^*$ is shown to be dependent also upon the various widths of the kinds of instances.] The theorem shows further that $\mathfrak{c}^*$ is much more influenced by (2) and (3) than by (1); that is to say, it is much more important to improve the variety of instances than to increase merely their number.

The situation is best illustrated by a numerical example. The computation of the increase of $\mathfrak{c}^*$, for the nine possible cases discussed by Nagel, under certain plausible assumptions concerning the form of the law l and the widths of the properties involved, leads to the following results. If we arrange the nine possibilities in the order of ascending values of $\mathfrak{c}^*$, we obtain this: P_1, P_3, P_7, P_9; P_2, P_4, P_5, P_6, P_8. In this order we find first the four possibilities with a bad distribution among the two kinds, i.e. those where none or only very few (two) of the instances are taken from one of the two kinds, and these four possibilities occur in the order in which they are listed by Nagel; then the five possibilities with a good or fairly good distribution follow, again in the same order as Nagel's. Even for the smallest sample with a good distribution (viz., P_2, with 100 instances, 50 from each of the two kinds) $\mathfrak{c}^*$ is considerably higher—under the assumptions made, more than four times as high—than for the largest sample with a bad distribution (viz. P_9, with 200 instances, divided into 198 and 2). This shows that a good distribution of the instances is much more important than a mere increase in the total number of instances. This is in accordance with Nagel's remark (p. 69): "A large increase in the number of positive instances of one kind may therefore count for less, in the judgment of skilled experimenters, than a small increase in the number of positive instances of another kind."

Thus we see that the concept c* is in satisfactory accordance with the principle of the variety of instances.

§ 16. THE PROBLEM OF THE JUSTIFICATION OF INDUCTION

Suppose that a theory is offered as a more exact formulation—sometimes called a "rational reconstruction"—of a body of generally accepted but more or less vague beliefs. Then the demand for a justification of this theory may be understood in two different ways. (1) The first, more modest task is to validate the claim that the new theory is a satisfactory reconstruction of the beliefs in question. It must be shown that the statements of the theory are in sufficient agreement with those beliefs; this comparison is possible only on those points where the beliefs are sufficiently precise. The question whether the given beliefs are true or false is here not even raised. (2) The second task is to show the validity of the new theory and thereby of the given beliefs. This is a much deeper going and often much more difficult problem.

For example, Euclid's axiom system of geometry was a rational reconstruction of the beliefs concerning spatial relations which were generally held, based on experience and intuition, and applied in the practices of measuring, surveying, building, etc. Euclid's axiom system was accepted because it was in sufficient agreement with those beliefs and gave a more exact and consistent formulation for them. A critical investigation of the validity, the factual truth, of the axioms and the beliefs was only made more than two thousand years later by Gauss.

Our system of inductive logic, that is, the theory of c* based on the definition of this concept, is intended as a rational reconstruction, restricted to a simple language form, of inductive thinking as customarily applied in everyday life and in science. Since the implicit rules of customary inductive thinking are rather vague, any rational reconstruction contains statements which are neither supported nor rejected by the ways of customary thinking. Therefore, a comparison is possible only on those points where the procedures of customary inductive thinking are precise enough. It seems to me, that on these points sufficient agreement is found to show that our theory is an adequate reconstruction; this agreement is seen in many theorems, of which a few have been mentioned in this paper.

An entirely different question is the problem of the validity of our or any other proposed system of inductive logic, and thereby of the customary methods of inductive thinking. This is the genuinely philosophical problem of induction. The construction of a systematic inductive logic is an important step towards the solution of the problem, but still only a preliminary step. It is important because without an exact formulation of rules of induction, i.e. theorems on degree of confirmation, it is not clear what exactly is meant by "inductive procedures," and therefore the problem of the validity of these procedures cannot even be raised in precise terms. On the other hand, a construction of inductive logic, although it prepares the way towards a solution of the problem of induction, still does not by itself give a solution.

Older attempts at a justification of induction tried to transform it into a kind of deduction, by adding to the premisses a general assumption of universal form, e.g. the principle of the uniformity of nature. I think there is fairly general agreement today among scientists and philosophers that neither this nor any other way of reducing induction to deduction with the help of a general principle is possible. It is generally acknowledged that induction is fundamentally different from deduction, and that any prediction of a future event reached inductively on the basis of observed events can never have the certainty of a deductive conclusion; and, conversely, the fact that a prediction reached by certain inductive procedures turns out to be false does not show that those inductive procedures were incorrect.

The situation just described has sometimes been characterized by saying that a theoretical justification of induction is not possible, and hence, that there is no problem of induction. However, it would be better to say merely that a justification in the old sense is not possible. Reichenbach[28] was the first to raise the problem of the justification of induction in a new sense and to take the first step towards a positive solution. Although I do not agree with certain other features of Reichenbach's theory of induction, I think it has the merit of having first emphasized these important points with respect to the problem of justification: (1) the decisive justification of an inductive procedure does not consist in its plausibility, i.e., its accordance with customary ways of inductive reasoning, but must refer to its success in some sense; (2) the fact that the truth of the predictions reached by induction cannot be guaranteed does not preclude a justification in a weaker sense; (3) it can be proved (as a purely logical result) that induction leads in the long run to success in a certain sense, provided the world is "predictable" at all, i.e. such that success in that respect is possible. Reichenbach shows that his rule of induction *R* leads to success in the following sense: *R* yields in the long run an approximate estimate of the relative frequency in the whole of any given property. Thus suppose that we observe the relative frequencies of a property *M* in an increasing series of samples, and that we determine on the basis of each sample with the help of the rule *R* the probability *q* that an unobserved thing has the property *M*, then the values *q* thus found approach in the long run the relative frequency of *M* in the whole. (This is, of course

[28] Hans Reichenbach, *Experience and Prediction*, 1938, §§ 38 ff., and earlier publications.

merely a logical consequence of Reichenbach's definition or rule of induction, not a factual feature of the world.)

I think that the way in which Reichenbach examines and justifies his rule of induction is an important step in the right direction, but only a first step. What remains to be done is to find a procedure for the examination of any given rule of induction in a more thoroughgoing way. To be more specific, Reichenbach is right in the assertion that any procedure which does not possess the characteristic described above (viz. approximation to the relative frequency in the whole) is inferior to his rule of induction. However, his rule, which he calls "the" rule of induction, is far from being the only one possessing that characteristic. The same holds for an infinite number of other rules of induction, e.g., for Laplace's rule of succession (see above, § 10; here restricted in a suitable way so as to avoid contradictions), and likewise for the corresponding rule of our theory of $\mathfrak{c}^*$ (as formulated in theorem (1), § 10). Thus our inductive logic is justified to the same extent as Reichenbach's rule of induction, as far as the only criterion of justification so far developed goes. (In other respects, our inductive logic covers a much more extensive field than Reichenbach's rule; this can be seen by the theorems on various kinds of inductive inference mentioned in this paper.) However, Reichenbach's rule and the other two rules mentioned yield different numerical values for the probability under discussion, although these values converge for an increasing sample towards the same limit. Therefore we need a more general and stronger method for examining and comparing any two given rules of induction in order to find out which of them has more chance of success. I think we have to measure the success of any given rule of induction by the total balance with respect to a comprehensive system of wagers made according to the given rule. For this task, here formulated in vague terms, there is so far not even an exact formulation; and much further investigation will be needed before a solution can be found.

3

DEGREES OF BELIEF

Frank P. Ramsey

The subject of our inquiry is the logic of partial belief, and I do not think we can carry it far unless we have at least an approximate notion of what partial belief is, and how, if at all, it can be measured. It will not be very enlightening to be told that in such circumstances it would be rational to believe a proposition to the extent of $\frac{2}{3}$, unless we know what sort of a belief in it that means. We must therefore try to develop a purely psychological method of measuring belief. It is not enough to measure probability; in order to apportion correctly our belief to the probability we must also be able to measure our belief.

It is a common view that belief and other psychological variables are not measurable, and if this is true our inquiry will be vain; and so will the whole theory of probability conceived as a logic of partial belief; for if the phrase 'a belief two-thirds of certainty' is meaningless, a calculus whose sole object is to enjoin such beliefs will be meaningless also. Therefore unless we are prepared to give up the whole thing as a bad job we are bound to hold that beliefs can to some extent be measured. If we were to follow the analogy of Mr. Keynes' treatment of probabilities we should say that some beliefs were measurable and some not; but this does not seem to me likely to be a correct account of the matter: I do not see how we can sharply divide beliefs into those which have a position in the numerical scale and those which have not. But I think beliefs do differ in measurability in the following two ways. First, some beliefs can be measured more accurately than others; and, secondly, the measurement of beliefs is almost certainly an ambiguous process leading to a variable answer depending on how exactly the measurement is conducted. The degree of a belief is in this respect like the time interval between two events; before Einstein it was supposed that all the ordinary

---

ways of measuring a time interval would lead to the same result if properly performed. Einstein showed that this was not the case; and time interval can no longer be regarded as an exact notion, but must be discarded in all precise investigations. Nevertheless, time interval and the Newtonian system are sufficiently accurate for many purposes and easier to apply.

I shall try to argue later that the degree of a belief is just like a time interval; it has no precise meaning unless we specify more exactly how it is to be measured. But for many purposes we can assume that the alternative ways of measuring it lead to the same result, although this is only approximately true. The resulting discrepancies are more glaring in connection with some beliefs than with others, and these therefore appear less measurable. Both these types of deficiency in measurability, due respectively to the difficulty in getting an exact enough measurement and to an important ambiguity in the definition of the measurement process, occur also in physics and so are not difficulties peculiar to our problem; what is peculiar is that it is difficult to form any idea of how the measurement is to be conducted, how a unit is to be obtained, and so on.

Let us then consider what is implied in the measurement of beliefs. A satisfactory system must in the first place assign to any belief a magnitude or degree having a definite position in an order of magnitudes; beliefs which are of the same degree as the same belief must be of the same degree as one another, and so on. Of course, this cannot be accomplished without introducing a certain amount of hypothesis or fiction. Even in physics we cannot maintain that things that are equal to the same thing are equal to one another unless we take 'equal' not as meaning 'sensibly equal' but a fictitious or hypothetical relation. I do not want to discuss the metaphysics or epistemology of this process, but merely to remark that if it is allowable in physics it is allowable in psychology also. The logical simplicity characteristic of the relations dealt with in a science is never attained by nature alone without any admixture of fiction.

But to construct such an ordered series of degrees is not the whole of our task; we have also to assign numbers to these degrees in some intelligible manner. We can of course easily explain that we denote full belief by 1, full belief in the contradictory by 0, and equal beliefs in the proposition and its contradictory by $\frac{1}{2}$. But it is not so easy to say what is meant by a belief $\frac{2}{3}$ of certainty, or a belief in the proposition being twice as strong as that in its contradictory. This is the harder part of the task, but it is absolutely necessary; for we do calculate numerical probabilities, and if they are to correspond to degrees of belief we must discover some definite way of attaching numbers to degrees of belief. In physics we often attach numbers by discovering a physical process of addition[1]: the measure-numbers of lengths are not assigned arbitrarily subject only to the proviso that the greater length shall

[1] See N. Campbell, *Physics The Elements* (1920), p. 277. [Ed. note: Footnotes have been renumbered for use in this book.]

have the greater measure; we determine them further by deciding on a physical meaning for addition; the length got by putting together two given lengths must have for its measure the sum of their measures. A system of measurement in which there is nothing corresponding to this is immediately recognized as arbitrary, for instance Mohs' scale of hardness[2] in which 10 is arbitrarily assigned to diamond, the hardest known material, 9 to the next hardest, and so on. We have therefore to find a process of addition for degrees of belief, or some substitute for this which will be equally adequate to determine a numerical scale.

Such is our problem; how are we to solve it? There are, I think, two ways in which we can begin. We can, in the first place, suppose that the degree of a belief is something perceptible by its owner; for instance that beliefs differ in the intensity of a feeling by which they are accompanied, which might be called a belief-feeling or feeling of conviction, and that by the degree of belief we mean the intensity of this feeling. This view would be very inconvenient, for it is not easy to ascribe numbers to the intensities of feelings; but apart from this it seems to me observably false, for the beliefs which we hold most strongly are often accompanied by practically no feeling at all; no one feels strongly about things he takes for granted.

We are driven therefore to the second supposition that the degree of a belief is a causal property of it, which we can express vaguely as the extent to which we are prepared to act on it. This is a generalization of the well-known view, that the differentia of belief lies in its causal efficacy, which is discussed by Mr. Russell in his *Analysis of Mind*. He there dismisses it for two reasons, one of which seems entirely to miss the point. He argues that in the course of trains of thought we believe many things which do not lead to action. This objection is however beside the mark, because it is not asserted that a belief is an idea which does actually lead to action, but one which would lead to action in suitable circumstances; just as a lump of arsenic is called poisonous not because it actually has killed or will kill anyone, but because it would kill anyone if he ate it. Mr. Russell's second argument is, however, more formidable. He points out that it is not possible to suppose that beliefs differ from other ideas only in their effects, for if they were otherwise identical their effects would be identical also. This is perfectly true, but it may still remain the case that the nature of the difference between the causes is entirely unknown or very vaguely known, and that what we want to talk about is the difference between the effects, which is readily observable and important.

As soon as we regard belief quantitatively, this seems to me the only view we can take of it. It could well be held that the difference between believing and not believing lies in the presence or absence of introspectible feelings. But when we seek to know what is the difference between believing more firmly

[2] Ibid., p. 271.

and believing less firmly, we can no longer regard it as consisting in having more or less of certain observable feelings; at least I personally cannot recognize any such feelings. The difference seems to me to lie in how far we should act on these beliefs: this may depend on the degree of some feeling or feelings, but I do not know exactly what feelings and I do not see that it is indispensable that we should know. Just the same thing is found in physics; men found that a wire connecting plates of zinc and copper standing in acid deflected a magnetic needle in its neighbourhood. Accordingly as the needle was more or less deflected the wire was said to carry a larger or a smaller current. The nature of this 'current' could only be conjectured: what were observed and measured were simply its effects.

It will no doubt be objected that we know how strongly we believe things, and that we can only know this if we can measure our belief by introspection. This does not seem to me necessarily true; in many cases, I think, our judgment about the strength of our belief is really about how we should act in hypothetical circumstances. It will be answered that we can only tell how we should act by observing the present belief-feeling which determines how we should act; but again I doubt the cogency of the argument. It is possible that what determines how we should act determines us also directly or indirectly to have a correct opinion as to how we should act, without its ever coming into consciousness.

Suppose, however, I am wrong about this and that we can decide by introspection the nature of belief, and measure its degree; still, I shall argue, the kind of measurement of belief with which probability is concerned is not this kind but is a measurement of belief *qua* basis of action. This can I think be shown in two ways. First, by considering the scale of probabilities between 0 and 1, and the sort of way we use it, we shall find that it is very appropriate to the measurement of belief as a basis of action, but in no way related to the measurement of an introspected feeling. For the units in terms of which such feelings or sensations are measured are always, I think, differences which are just perceptible: there is no other way of obtaining units. But I see no ground for supposing that the interval between a belief of degree $\frac{1}{3}$ and one of degree $\frac{1}{2}$ consists of as many just perceptible changes as does that between one of $\frac{2}{3}$ and one of $\frac{5}{6}$, or that a scale based on just perceptible differences would have any simple relation to the theory of probability. On the other hand the probability of $\frac{1}{3}$ is clearly related to the kind of belief which would lead to a bet of 2 to 1, and it will be shown below how to generalize this relation so as to apply to action in general. Secondly, the quantitative aspects of beliefs as the basis of action are evidently more important than the intensities of belief-feelings. The latter are no doubt interesting, but may be very variable from individual to individual, and their practical interest is entirely due to their position as the hypothetical causes of beliefs *qua* bases of action.

It is possible that some one will say that the extent to which we should act on a belief in suitable circumstances is a hypothetical thing, and therefore not capable of measurement. But to say this is merely to reveal ignorance of the physical sciences which constantly deal with and measure hypothetical quantities; for instance, the electric intensity at a given point is the force which would act on a unit charge if it were placed at the point.

Let us now try to find a method of measuring beliefs as bases of possible actions. It is clear that we are concerned with dispositional rather than with actualized beliefs; that is to say, not with beliefs at the moment when we are thinking of them, but with beliefs like my belief that the earth is round, which I rarely think of, but which would guide my action in any case to which it was relevant.

The old-established way of measuring a person's belief is to propose a bet, and see what are the lowest odds which he will accept. This method I regard as fundamentally sound; but it suffers from being insufficiently general, and from being necessarily inexact. It is inexact partly because of the diminishing marginal utility of money, partly because the person may have a special eagerness or reluctance to bet, because he either enjoys or dislikes excitement or for any other reason, e.g. to make a book. The difficulty is like that of separating two different co-operating forces. Besides, the proposal of a bet may inevitably alter his state of opinion; just as we could not always measure electric intensity by actually introducing a charge and seeing what force it was subject to, because the introduction of the charge would change the distribution to be measured.

In order therefore to construct a theory of quantities of belief which shall be both general and more exact, I propose to take as a basis a general psychological theory, which is now universally discarded, but nevertheless comes, I think, fairly close to the truth in the sort of cases with which we are most concerned. I mean the theory that we act in the way we think most likely to realize the objects of our desires, so that a person's actions are completely determined by his desires and opinions. This theory cannot be made adequate to all the facts, but it seems to me a useful approximation to the truth particularly in the case of our self-conscious or professional life, and it is presupposed in a great deal of our thought. It is a simple theory and one which many psychologists would obviously like to preserve by introducing unconscious desires and unconscious opinions in order to bring it more into harmony with the facts. How far such fictions can achieve the required result I do not attempt to judge: I only claim for what follows approximate truth, or truth in relation to this artificial system of psychology, which like Newtonian mechanics can, I think, still be profitably used even though it is known to be false.

It must be observed that this theory is not to be identified with the psychology of the Utilitarians, in which pleasure had a dominating position. The theory I propose to adopt is that we seek things which we want, which may be our own or other people's pleasure, or anything else whatever, and our actions are such as we think most likely to realize these goods. But this is not a precise statement, for a precise statement of the theory can only be made after we have introduced the notion of quantity of belief.

Let us call the things a person ultimately desires 'goods', and let us at first assume that they are numerically measurable and additive. That is to say that if he prefers for its own sake an hour's swimming to an hour's reading, he will prefer two hours' swimming to one hour's swimming and one hour's reading. This is of course absurd in the given case but this may only be because swimming and reading are not ultimate goods, and because we cannot imagine a second hour's swimming precisely similar to the first, owing to fatigue, etc.

Let us begin by supposing that our subject has no doubts about anything, but certain opinions about all propositions. Then we can say that he will always choose the course of action which will lead in his opinion to the greatest sum of good.

It should be emphasized that in this essay good and bad are never to be understood in any ethical sense but simply as denoting that to which a given person feels desire and aversion.

The question then arises how we are to modify this simple system to take account of varying degrees of certainty in his beliefs. I suggest that we introduce as a law of psychology that his behaviour is governed by what is called the mathematical expectation; that is to say that, if p is a proposition about which he is doubtful, any goods or bads for whose realization p is in his view a necessary and sufficient condition enter into his calculations multiplied by the same fraction, which is called the 'degree of his belief in p'. We thus define degree of belief in a way which presupposes the use of the mathematical expectation.

We can put this in a different way. Suppose his degree of belief in p is $\frac{m}{n}$; then his action is such as he would choose it to be if he had to repeat it exactly n times, in m of which p was true, and in the others false. [Here it may be necessary to suppose that in each of the n times he had no memory of the previous ones.]

This can also be taken as a definition of the degree of belief, and can easily be seen to be equivalent to the previous definition. Let us give an instance of the sort of case which might occur. I am at a cross-roads and do not know the way; but I rather think one of the two ways is right. I propose therefore to go that way but keep my eyes open for someone to ask; if now I see someone half a mile away over the fields, whether I turn aside to ask him will depend on the relative inconvenience of going out of my way to cross the fields or of continuing on the wrong road if it is the wrong road. But it will also depend on how confident I am that I am right; and clearly the more confident I am of this the less distance I should be willing to go from the road to check my opinion. I propose therefore to use the distance I would be prepared to go to ask, as a measure of the confidence of my opinion; and what I have said above explains how this is to be done. We can set it out as follows: suppose the disadvantage of going x yards to ask is $f(x)$, the advantage of arriving at the right destination is r, that of arriving at the wrong one w. Then if I should just be willing to go a distance d to ask, the degree of my belief that I am on the right road is given by $p = 1 - \frac{f(d)}{r - w}$.

For such an action is one it would just pay me to take, if I had to act in the same way n times, in np of which I was on the right way but in the others not.

For the total good resulting from not asking each time

$$= npr + n(1-p)w$$
$$= nw + np(r-w),$$

that resulting from asking at distance x each time

$$= nr - nf(x) \qquad \text{[I now always go right.]}$$

This is greater than the preceding expression, provided

$$f(x) < (r-w)(1-p),$$

$\therefore$ the critical distance d is connected with p, the degree of belief, by the relation $f(d) = (r-w)(1-p)$

$$\text{or } p = 1 - \frac{f(d)}{r-w} \qquad \text{as asserted above.}$$

It is easy to see that this way of measuring beliefs gives results agreeing with ordinary ideas; at any rate to the extent that full belief is denoted by 1, full belief in the contradictory by 0, and equal belief in the two by $\frac{1}{2}$. Further, it allows validity to betting as means of measuring beliefs. By proposing a bet on p we give the subject a possible course of action from which so much extra good will result to him if p is true and so much extra bad if p is false. Supposing the bet to be in goods and bads instead of in money, he will take a bet at any better odds than those corresponding to his state of belief; in fact his state of belief is measured by the odds he will just take; but this is vitiated, as already explained, by love or hatred of excitement, and by the fact that the bet is in money and not in goods and bads. Since it is universally agreed that money has a diminishing marginal utility, if money bets are to be used, it is evident that they should be for as small stakes as possible. But then again the measurement is spoiled by introducing the new factor of reluctance to bother about trifles.

Let us now discard the assumption that goods are additive and immediately measurable, and try to work out a system with as few assumptions as possible. To begin with we shall suppose, as before, that our subject has certain beliefs about everything; then he will act so that what he believes to be the total consequences of his action will be the best possible. If then we had the power of the Almighty, and could persuade our subject of our power, we could, by offering him options, discover how he placed in order of merit all possible courses of the world. In this way all possible worlds would be put in an order of value, but we should have no definite way of representing them by numbers. There would be no meaning in the assertion that the difference in value between α and β was equal to that between γ and δ. [Here and elsewhere we use Greek letters to represent the different possible totalities of events between which our subject chooses—the ultimate organic unities.]

Suppose next that the subject is capable of doubt; then we could test his degree of belief in different propositions by making him offers of the following kind. Would you rather have world α in any event; or world β if p is true, and world γ if p is false? If, then, he were certain that p was true, he would simply compare α and β and choose between them as if no conditions were attached; but if he were doubtful his choice would not be decided so simply. I propose to lay down axioms and definitions concerning the principles governing choices of this kind. This is, of course, a very schematic version of the situation in real life, but it is, I think, easier to consider it in this form.

There is first a difficulty which must be dealt with; the propositions like p in the above case which are used as conditions in the options offered may be such that their truth or falsity is an object of desire to the subject. This will be found to complicate the problem, and we have to assume that there are propositions for which this is not the case, which we shall call ethically neutral. More precisely an atomic proposition p is called ethically neutral if two possible worlds differing only in regard to the truth of p are always of equal value; and a non-atomic proposition p is called ethically neutral if all its atomic truth-arguments[3] are ethically neutral.

We begin by defining belief of degree $\frac{1}{2}$ in an ethically neutral proposition. The subject is said to have belief of degree $\frac{1}{2}$ in such a proposition p if he has no preference between the options (1) α if p is true, β if p is false, and (2) α if p is false, β if p is true, but has a preference between α and β simply. We suppose by an axiom that if this is true of any one pair α, β it is true of all such pairs.[4] This comes roughly to defining belief of degree $\frac{1}{2}$ as such a degree of belief as leads to indifference between betting one way and betting the other for the same stakes.

Belief of degree $\frac{1}{2}$ as thus defined can be used to measure values numerically in the following way. We have to explain what is meant by the difference in

[3] I assume here Wittgenstein's theory of propositions; it would probably be possible to give an equivalent definition in terms of any other theory.

[4] α and β must be supposed so far undefined as to be compatible with both p and not-p.

value between α and β being equal to that between γ and δ; and we define this to mean that, if p is an ethically neutral proposition believed to degree $\frac{1}{2}$, the subject has no preference between the options (1) α if p is true, δ if p is false, and (2) β if p is true, γ if p is false.

This definition can form the basis of a system of measuring values in the following way:—

Let us call any set of all worlds equally preferable to a given world a value: we suppose that if world α is preferable to β any world with the same value as α is preferable to any world with the same value as β and shall say that the value of α is greater than that of β. This relation 'greater than' orders values in a series. We shall use α henceforth both for the world and its value.

Axioms

(1) There is an ethically neutral proposition p believed to degree $\frac{1}{2}$.

(2) If p, q are such propositions and the option

α if p, δ if not-p is equivalent to β if p, γ if not-p

then α if q, δ if not-q is equivalent to β if q, γ if not-q.

Def. In the above case we say $\alpha\beta = \gamma\delta$.

Theorems. If $\alpha\beta = \gamma\delta$,
then $\beta\alpha = \delta\gamma$, $\alpha\gamma = \beta\delta$, $\gamma\alpha = \delta\beta$.

(2*a*) If $\alpha\beta = \gamma\delta$, then $\alpha > \beta$ is equivalent to $\gamma > \delta$
and $\alpha = \beta$ is equivalent to $\gamma = \delta$.

(3) If option A is equivalent to option B and B to C then A to C.

Theorem. If $\alpha\beta = \gamma\delta$ and $\beta\eta = \zeta\gamma$,
then $\alpha\eta = \zeta\delta$.

(4) If $\alpha\beta = \gamma\delta$, $\gamma\delta = \eta\zeta$, then $\alpha\beta = \eta\zeta$

(5) $(\alpha, \beta, \gamma) . E! (\imath x) (\alpha x = \beta\gamma)$

(6) $(\alpha, \beta) . E! (\imath x) (\alpha x = x\beta)$

(7) Axiom of continuity:—Any progression has a limit (ordinal).

(8) Axiom of Archimedes.

These axioms enable the values to be correlated one-one with real numbers so that if α^1 corresponds to α, etc.

$$\alpha\beta = \gamma\delta . \equiv . \alpha^1 - \beta^1 = \gamma^1 - \delta^1$$

Henceforth we use α for the correlated real number α^1 also.

Having thus defined a way of measuring value we can now derive a way of measuring belief in general. If the option of α for certain is indifferent with that of β if p is true and γ if p is false,[5] we can define the subject's degree of belief in p as the ratio of the difference between α and γ to that between β and γ; which we must suppose the same for all α's, β's and γ's that satisfy the conditions. This amounts roughly to defining the degree of belief in p by the odds at which the subject would bet on p, the best being conducted in terms of differences of value as defined. The definition only applies to partial belief and does not include certain beliefs; for belief of degree 1 in p, α for certain is indifferent with α if p and any β if not-p.

We are also able to define a very useful new idea—'the degree of belief in p given q'. This does not mean the degree of belief in 'If p then q', or that in 'p entails q', or that which the subject would have in p if he knew q, or that which he ought to have. It roughly expresses the odds at which he would now bet on p, the bet only to be valid if q is true. Such conditional bets were often made in the eighteenth century.

The degree of belief in p given q is measured thus. Suppose the subject indifferent between the options (1) α if q true, β if q false, (2) γ if p true and q true, δ if p false and q true, β if q false. Then the degree of his belief in p given q is the ratio of the difference between α and δ to that between γ and δ, which we must suppose the same for any α, β, γ, δ which satisfy the given conditions. This is not the same as the degree to which he would believe p, if he believed q for certain; for knowledge of q might for psychological reasons profoundly alter his whole system of beliefs.

Each of our definitions has been accompanied by an axiom of consistency, and in so far as this is false, the notion of the corresponding degree of belief becomes invalid. This bears some analogy to the situation in regard to simultaneity discussed above.

I have not worked out the mathematical logic of this in detail, because this would, I think, be rather like working out to seven places of decimals a result only valid to two. My logic cannot be regarded as giving more than the sort of way it might work.

From these definitions and axioms it is possible to prove the fundamental laws of probable belief (degrees of belief lie between 0 and 1):

(1) Degree of belief in p + degree of belief in $\bar{p} = 1$.

(2) Degree of belief in p given q + degree of belief in $\bar{p}$ given $q = 1$.

(3) Degree of belief in (p and q) = degree of belief in p × degree of belief in q given p.

(4) Degree of belief in (p and q) + degree of belief in (p and $\bar{q}$) = degree of belief in p.

[5] Here β must include the truth of p, γ its falsity; p need no longer be ethically neutral. But we have to assume that there is a world with any assigned value in which p is true, and one in which p is false.

The first two are immediate. (3) is proved as follows.

Let degree of belief in $p = x$, that in q given $p = y$.

Then ξ for certain $\equiv \xi + (1 - x)\, t$ if p true, $\xi - xt$ if p false, for any t.

$$\xi + (1 - x)\, t \text{ if } p \text{ true} \equiv$$
$$\begin{cases} \xi + (1 - x)\, t + (1 - y)\, u \text{ if `}p \text{ and } q\text{' true,} \\ \xi + (1 - x)\, t - yu \text{ if } p \text{ true } q \text{ false;} \qquad \text{for any } u. \end{cases}$$

Choose u so that $\xi + (1 - x)\, t - yu = \xi - xt$,

i.e. let $u = t/y \; (y \neq 0)$

Then ξ for certain $\equiv$

$$\begin{cases} \xi + (1 - x)\, t + (1 - y)\, t/y \text{ if } p \text{ and } q \text{ true} \\ \xi - xt \text{ otherwise,} \end{cases}$$

$$\therefore \text{degree of belief in `}p \text{ and } q\text{'} = \frac{xt}{t + (1 - y)\, t/y} = xy. \; (t \neq 0)$$

If $y = 0$, take $t = 0$.

Then ξ for certain $\equiv \xi$ if p true, ξ if p false
$\equiv \xi + u$ if p true, q true; ξ if p false, q false; ξ if p false
$\equiv \xi + u$, pq true; ξ, pq false

$\therefore$ degree of belief in $pq = 0$.

(4) follows from (2), (3) as follows:—

Degree of belief in pq = *that in* $p \times$ that in q given p, by (3). Similarly degree of belief in $p\bar{q}$ = that in $p \times$ that in $\bar{q}$ given p $\therefore$ sum = degree of belief in p, by (2).

These are the laws of probability, which we have proved to be necessarily true of any consistent set of degrees of belief. Any definite set of degrees of belief which broke them would be inconsistent in the sense that it violated the laws of preference between options, such as that preferability is a transitive asymmetrical relation, and that if α is preferable to β, β for certain cannot be preferable to α if p, β if not-p. If anyone's mental condition violated these laws, his choice would depend on the precise form in which the options were offered him, which would be absurd. He could have a book made against him by a cunning better and would then stand to lose in any event.

We find, therefore, that a precise account of the nature of partial belief reveals that the laws of probability are laws of consistency, an extension to partial beliefs of formal logic, the logic of consistency. They do not depend for their meaning on any degree of belief in a proposition being uniquely determined as the rational one; they merely distinguish those sets of beliefs which obey them as consistent ones.

Having any definite degree of belief implies a certain measure of consistency, namely willingness to bet on a given proposition at the same odds for any stake, the stakes being measured in terms of ultimate values. Having degrees of belief obeying the laws of probability implies a further measure of consistency, namely such a consistency between the odds acceptable on different propositions as shall prevent a book being made against you.

Some concluding remarks on this section may not be out of place. First, it is based fundamentally on betting, but this will not seem unreasonable when it is seen that all our lives we are in a sense betting. Whenever we go to the station we are betting that a train will really run, and if we had not a sufficient degree of belief in this we should decline the bet and stay at home. The options God gives us are always conditional on our guessing whether a certain proposition is true. Secondly, it is based throughout on the idea of mathematical expectation; the dissatisfaction often felt with this idea is due mainly to the inaccurate measurement of goods. Clearly mathematical expectations in terms of money are not proper guides to conduct. It should be remembered, in judging my system, that in it value is actually defined by means of mathematical expectation in the case of beliefs of degree $\frac{1}{2}$, and so may be expected to be scaled suitably for the valid application of the mathematical expectation in the case of other degrees of belief also.

Thirdly, nothing has been said about degrees of belief when the number of alternatives is infinite. About this I have nothing useful to say, except that I doubt if the mind is capable of contemplating more than a finite number of alternatives. It can consider questions to which an infinite number of answers are possible, but in order to consider the answers it must lump them into a finite number of groups. The difficulty becomes practically relevant when discussing induction, but even then there seems to me no need to introduce it. We can discuss whether past experience gives a high probability to the sun's rising to-morrow without bothering about what probability it gives to the sun's rising each morning for evermore. For this reason I cannot but feel that Mr. Ritchie's discussion of the problem[6] is unsatisfactory; it is true that we can agree that inductive generalizations need have no finite probability, but particular expectations entertained on inductive grounds undoubtedly do have a high numerical probability in the minds of all of us. We all are more certain that the sun will rise to-morrow than that I shall not throw 12 with two dice first time, i.e. we have a belief of higher degree than $35/36$ in it. If induction ever needs a logical justification it is in connection with the probability of an event like this.

[6] A. D. Ritchie, "Induction and Probability," *Mind*, 1926, p. 318. 'The conclusion of the foregoing discussion may be simply put. If the problem of induction be stated to be "How can inductive generalizations acquire a large numerical probability?" then this is a pseudo-problem, because the answer is "They cannot." This answer is not, however, a denial of the validity of induction but is a direct consequence of the nature of probability. It still leaves untouched the real problem of induction which is "How can the probability of an induction be increased?" and it leaves standing the whole of Keynes' discussion on this point.'

4

PROBABLE KNOWLEDGE

Richard C. Jeffrey

The central problem of epistemology is often taken to be that of explaining how we can know what we do, but the content of this problem changes from age to age with the scope of what we take ourselves to know; and philosophers who are impressed with this flux sometimes set themselves the problem of explaining how we can get along, knowing as little as we do. For knowledge is sure, and there seems to be little we can be sure of outside logic and mathematics and truths related immediately to experience. It is as if there were some propositions—that this paper is white, that two and two are four—on which we have a firm grip, while the rest, including most of the theses of science, are slippery or insubstantial or somehow inaccessible to us. Outside the realm of what we are sure of lies the puzzling region of probable knowledge—puzzling in part because the sense of the noun seems to be cancelled by that of the adjective.

The obvious move is to deny that the notion of knowledge has the importance generally attributed to it, and to try to make the concept of belief do the work that philosophers have generally assigned the grander concept. I shall argue that this is the right move.

1. A PRAGMATIC ANALYSIS OF BELIEF

To begin, we must get clear about the relevant sense of 'belief'. Here I follow Ramsey: 'the kind of measurement of belief with which probability is concerned is . . . a measurement of belief *qua* basis of action'.[1]

[1] Frank P. Ramsey, "Truth and probability," in *The Foundations of Mathematics and Other Logical Essays*, R. B. Braithwaite, ed., London and New York, 1931, p. 171. [Ed. note: Footnotes have been renumbered for use in this book.]

From *The Problem of Inductive Logic*, Imre Lakatos (ed.), pp. 166–180. Copyright 1968, North-Holland Publishing Company. Reprinted by permission of the publisher.

Ramsey's basic idea was that the desirability of a gamble G is a weighted average of the desirabilities of winning and of losing in which the weights are the probabilities of winning and of losing. If the proposition gambled upon is A, if the prize for winning is the truth of a proposition W, and if the penalty for losing is the truth of a proposition L, we then have

$$prob\ A = \frac{des\ G - des\ L}{des\ W - des\ L} \tag{1}$$

Thus, if the desirabilities of losing and of winning happen to be 0 and 1, we have $prob\ A = des\ G$, as illustrated in Figure 1, for the case in which the probability of winning is thought to be $\frac{3}{4}$.

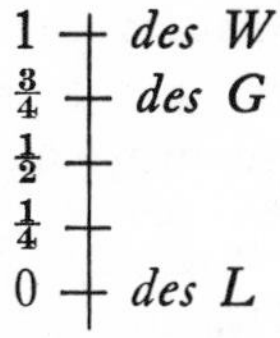

Figure 1

On this basis, Ramsey[2] is able to give rules for deriving the gambler's subjective probability *and* desirability functions from his preference ranking of gambles, provided the preference ranking satisfies certain conditions of consistency. The probability function obtained in this way is a probability measure in the technical sense that, given any finite set of pairwise incompatible propositions which together exhaust all possibilities, their probabilities are non-negative real numbers that add up to 1. And in an obvious sense, probability so construed is a measure of the subject's willingness to act on his beliefs in propositions: it is a measure of degree of belief.

I propose to use what I take to be an improvement of Ramsey's scheme, in which the work that Ramsey does with the operation of forming gambles is done with the usual truth-functional operations on propositions.[3] The basic move is to restrict attention to certain 'natural' gambles, in which the prize for winning is the truth of the proposition gambled upon, and the penalty for losing is the falsity of that proposition. In general, the situation in which the gambler takes himself to be gambling on A with prize W and loss L is one in which he believes the proposition

$$G = AW \vee \overline{A}L$$

[2] "Truth and probability," F. P. Ramsey, *op. cit.*

[3] See Richard C. Jeffrey, *The Logic of Decision*, McGraw-Hill, 1965, the mathematical basis for which can be found in Ethan Bolker, *Functions Resembling Quotients of Measures*, Ph.D. Dissertation, Harvard University, 1965, and *Trans. Am. Math. Soc.*, 124, 1966, pp. 293–312.

If G is a natural gamble we have $W = A$ and $L = \overline{A}$, so that G is the necessary proposition, $T = A \vee \overline{A}$:

$$G = AA \vee \overline{A}\overline{A} = T$$

Now if A is a proposition which the subject thinks good (or bad) in the sense that he places it above T (or below T) in his preference ranking, we have

$$prob\ A = \frac{des\ T - des\ \overline{A}}{des\ A - des\ \overline{A}}, \tag{2}$$

corresponding to Ramsey's formula (1).

Here the basic idea is that if $A_1, A_2, \ldots, A_n$ are an exhaustive set of incompatible ways in which the proposition A can come true, the desirability of A must be a weighted average of the desirabilities of the ways in which it can come true:

$$des\ A = w_1\ des\ A_1 + w_2\ des\ A_2 + \ldots + w_n\ des\ A_n, \tag{3}$$

where the weights are the conditional probabilities,

$$w_i = prob\ A_i / prob\ A \tag{4}$$

Let us call a function *des* which attributes real numbers to propositions a *Bayesian desirability function* if there is a probability measure *prob* relative to which (3) holds for all suitable $A, A_1, A_2, \ldots, A_n$. And let us call a preference ranking of propositions *coherent* if there is a Bayesian desirability function which ranks those propositions in order of magnitude exactly as they are ranked in order of preference. One can show[4] that if certain weak conditions are met by a coherent preference ranking, the underlying desirability function is determined up to a fractional linear transformation, i.e., if *des* and *DES* both rank propositions in order of magnitude exactly as they are ranked in order of preference, there must be real numbers a, b, c, d such that for any proposition A in the ranking we have

$$DES\ A = \frac{a\ des\ A + b}{c\ des\ A + b} \tag{5}$$

The probability measure *prob* is then determined by (2) up to a certain quantization. In particular, if *des* is Bayesian relative to *prob*, then *DES* will be Bayesian relative to *PROB*, where

$$PROB\ A = prob\ A\ (c\ des\ A + d) \tag{6}$$

Under further plausible conditions, (5) and (6) are given either exactly (as in Ramsey's theory) or approximately by

[4] Jeffrey, *op. cit.*, chaps. 6, 8.

$$DES\ A = a\ des\ A + b, \tag{7}$$

$$PROB\ A = prob\ A \tag{8}$$

I take the principal advantage of the present theory over Ramsey's to be that here we work with the subject's actual beliefs, whereas Ramsey needs to know what the subject's preference ranking of relevant propositions would be if his views of what the world is were to be changed by virtue of his having come to believe that various arbitrary and sometimes bizarre causal relationships had been established via gambles.[5]

To see more directly how preferences may reflect beliefs in the present system, observe that by (2) we must have $prob\ A > prob\ B$ if the relevant portion of the preference ranking is

$$\begin{array}{c} A,\ B \\ T \\ \bar{B} \\ \bar{A} \end{array}$$

In particular, suppose that A and B are the propositions that the subject will get job 1 and that he will get job 2, respectively. Pay, working conditions, etc., are the same, so that he ranks A and B together. Now if he thinks himself more likely to get job 1 than job 2, he will prefer a guarantee of ($\bar{B}$) not getting job 2 to a guarantee of ($\bar{A}$) not getting job 1; for he thinks that an assurance of not getting job 2 leaves him more likely to get one or the other of the equally liked jobs than would an assurance of not getting job 1.

2. PROBABILISTIC ACTS AND OBSERVATIONS

We might call a proposition *observational* for a certain person at a certain time if at that time he can make an observation of which the *direct* effect will be that his degree of belief in the proposition will change to 0 or to 1. Similarly, we might call a proposition *actual* for a certain person at a certain time if at that time he can perform an act of which the *direct* effect will be that his degree of belief in the proposition will change to 0 or to 1. Under ordinary circumstances, the proposition that the sun is shining is observational and the proposition that the agent blows his nose is actual. Performance of an act may give the agent what Anscombe calls[6] 'knowledge without observation' of the truth of an appropriate actual proposition. Apparently, a proposition can be actual or observational without the agent's knowing that it is; and the agent can be mistaken in thinking a proposition actual or observational.

[5] Jeffrey, *op. cit.*, pp. 145–150.

[6] G. E. M. Anscombe, *Intention*, § 8, Oxford, 1957; 2nd ed., Ithaca, N.Y., 1963.

The point and meaning of the requirement that the effect be 'direct', in the definitions of 'actual' and 'observational', can be illustrated by considering the case of a sleeper who awakens and sees that the sun is shining. Then one might take the observation to have shown him, directly, that the sun is shining, and to have shown him indirectly that it is daytime. In general, an observation will cause numerous changes in the observer's belief function, but many of these can be construed as consequences of others. If there is a proposition E such that the *direct* effect of the observation is to change the observer's degree of belief in E to 1, then for any proposition A in the observer's preference ranking, his degree of belief in A after the observation will be the conditional probability

$$prob_E\ A = prob\ (A/E) = prob\ AE/prob\ E, \tag{9}$$

where *prob* is the observer's belief function before the observation. And conversely, if the observer's belief function after the observation is $prob_E$ and $prob_E$ is not identical with *prob*, then the *direct* effect of the observation will be to change the observer's degree of belief in E to 1. This completes a definition of *direct*.

But from a certain strict point of view, it is rarely or never that there is a proposition for which the direct effect of an observation is to change the observer's degree of belief in that proposition to 1; and from that point of view, the classes of propositions that count as observational or actual in the senses defined above are either empty or as good as empty for practical purposes. For if we care seriously to distinguish between 0.999 999 and 1.000 000 as degrees of belief, we may find that, after looking out the window, the observer's degree of belief in the proposition that the sun is shining is not quite 1, perhaps because he thinks there is one chance in a million that he is deluded or deceived in some way; and similarly for acts where we can generally take ourselves to be at best *trying* (perhaps with very high probability of success) to make a certain proposition true.

One way in which philosophers have tried to resolve this difficulty is to postulate a phenomenalistic language in which an appropriate proposition E can always be expressed, as a report on the immediate content of experience; but for excellent reasons, this move is now in low repute.[7] The crucial point is not that 0.999 999 is so close to 1.000 000 as to make no odds, practically speaking, for situations abound in which the gap is more like one half than one millionth. Thus, in examining a piece of cloth by candlelight one might come to attribute probabilities 0.6 and 0.4 to the propositions G that the cloth is green and B that it is blue, without there being any proposition E for which the direct effect of the observation is anything near changing the observer's degree of belief in E to 1. One might think of some such proposition as that (E) *the cloth looks green or possibly blue*, but this is

[7] See, e.g., J. L. Austin, *Sense and Sensibilia*, Oxford, 1962.

far too vague to yield *prob* $(G/E) = 0.6$ and *prob* $(B/E) = 0.4$. Certainly, there is *something* about what the observer sees that leads him to have the indicated degrees of belief in G and in B, but there is no reason to think the observer can express this something by a statement in his language. And physicalistically, there is some perfectly definite pattern of stimulation of the rods and cones of the observer's retina which prompts his belief, but there is no reason to expect him to be able to describe that pattern or to recognize a true description of it, should it be suggested.

As Austin[8] points out, the crucial mistake is to speak seriously of the *evidence* of the senses. Indeed the relevant experiences have perfectly definite characteristics by virtue of which the observer comes to believe as he does, and by virtue of which in our example he comes to have degree of belief 0.6 in G. But it does not follow that there is a proposition E of which the observer is certain after the observation and for which we have *prob* $(G/E) = 0.6$, *prob* $(B/E) = 0.4$, etc.

In part, the quest for such phenomenological certainty seems to have been prompted by an inability to see how uncertain evidence can be used. Thus C. I. Lewis:

> If anything is to be probable, then something must be certain. The data which themselves support a genuine probability, must themselves be certainties. We do have such absolute certainties, in the sense data initiating belief and in those passages of experience which later may confirm it. But neither such initial data nor such later verifying passages of experience can be phrased in the language of objective statement—because what can be so phrased is never more than probable. Our sense certainties can only be formulated by the expressive use of language, in which what is signified is a content of experience and what is asserted is the givenness of this content.[9]

But this motive for the quest is easily disposed of.[10] Thus, in the example of observation by candlelight, we may take the direct result of the observation (in a modified sense of 'direct') to be, that the observer's degrees of belief in G and B change to 0.6 and 0.4. Then his degree of belief in any proposition A in his preference ranking will change from *prob* A to

$$PROB\ A = 0.6\ prob\ (A/G) + 0.4\ prob\ (A/B)$$

In general, suppose that there are propositions $E_1, E_2, \ldots, E_n$, in which the observer's degrees of belief after the observation are $p_1, p_2, \ldots, p_n$; where the E's are pairwise incompatible and collectively exhaustive; where for each i, *prob* E_i is neither 0 nor 1; and where for each proposition A in the pref-

[8] Austin, *op. cit.*, chap. 10.

[9] C. I. Lewis, *An Analysis of Knowledge and Valuation*, La Salle, Illinois, 1946, p. 186.

[10] Jeffrey, *op. cit.*, chap. 11.

erence ranking and for each i the conditional probability of A on E_i is unaffected by the observation:

$$PROB\ (A/E_i) = prob\ (A/E_i) \tag{10}$$

Then the belief function after the observation may be taken to be *PROB*, where

$$PROB\ A = p_1\ prob\ (A/E_1) + p_2\ prob\ (A/E_2) + \ldots + p_n\ prob\ (A/E_n), \tag{11}$$

if the observer's preference rankings before and after the observation are both coherent. Where these conditions are met, the propositions $E_1, E_2, \ldots, E_n$, may be said to form a *basis* for the observation; and the notion of a basis will play the role vacated by the notion of *directness*.

The situation is similar in the case of acts. A marksman may have a fairly definite idea of his chances of hitting a distant target, e.g. he may have degree of belief 0.3 in the proposition H that he will hit it. The basis for this belief may be his impressions of wind conditions, quality of the rifle, etc.; but there need be no reason to suppose that the marksman can express the relevant data; nor need there be any proposition E in his preference ranking in which the marksman's degree of belief changes to 1 upon deciding to fire at the target, and for which we have $prob\ (H/E) = 0.3$. But the pair $H, \overline{H}$ may constitute a *basis* for the act, in the sense that for any proposition A in the marksman's preference ranking, his degree of belief after his decision is

$$PROB\ A = 0.3\ prob\ (A/H) + 0.7\ prob\ (A/\overline{H})$$

It is correct to describe the marksman as *trying* to hit the target; but the proposition that he is trying to hit the target can not play the role of E above. Similarly, it was correct to describe the cloth as *looking* green or possibly blue; but the proposition that the cloth looks green or possibly blue does not satisfy the conditions for directness.

The notion of directness is useful as well for the resolution of unphilosophical posers about probabilities, in which the puzzling element sometimes consists in failure to think of an appropriate proposition E such that the direct effect of an observation is to change degree of belief in E to 1, e.g. in the following problem reported by Mosteller.[11]

> Three prisoners, *a*, *b*, *and c*, with apparently equally good records have applied for parole. The parole board has decided to release two of the three, and the prisoners know this but not which two. A warder friend of prisoner *a* knows who are to be released. Prisoner *a* realizes that it would be unethical to ask the warder if he, *a*, is to be released, but thinks

[11] Problem 13 of Frederick Mosteller, *Fifty Challenging Problems in Probability*, Reading, Mass., Palo Alto, and London, 1965.

> of asking for the name of *one* prisoner *other than himself* who is to be released. He thinks that before he asks, his chances of release are $\frac{2}{3}$. He thinks that if the warder says '*b* will be released,' his own chances have now gone down to $\frac{1}{2}$, because either *a* and *b* or *b* and *c* are to be released. And so *a* decides not to reduce his chances by asking. However, *a* is mistaken in his calculations. Explain.

Here indeed the possible cases (in a self-explanatory notation) are

AB, AC, BC,

and these are viewed by *a* as equiprobable. Then *prob* A is $\frac{2}{3}$ but *prob* (A/B) $=$ *prob* $(A/C) = \frac{1}{2}$, and, since the warder must answer either '*b*' or '*c*' to *a*'s question, it looks as if the direct result of the 'observation' will be that *a* comes to attribute probability 1 either to the proposition B that *b* will be released, or to the proposition C that *c* will be released. But this is incorrect. The relevant evidence-proposition would be more like the proposition *that the warder says*, '*b*', or *that the warder says*, '*c*', even though neither of these will quite do. For it is only in cases AB and AC that the warder's reply is dictated by the facts: in case BC, where *b* and *c* are both to be released, the warder must somehow choose *one* of the two true answers. If *a* expects the warder to make the choice by some such random device as tossing a coin, then we have *prob* (A/the warder says, '*b*') $=$ *prob* (A/the warder says, '*c*') $=$ *prob* $A = \frac{2}{3}$; while if *a* is sure that the warder will say '*b*' if he can, we have *prob* (A/the warder says '*b*') $= \frac{1}{2}$ but *prob* (A/the warder says '*c*') $= 1$.

3. BELIEF: REASONS VS. CAUSES

Indeed it is desirable, where possible, to incorporate the results of observation into the structure of one's beliefs via a basis of form $E, \overline{E}$ where the probability of E after the observation is nearly 1. For practical purposes, E then satisfies the conditions of directness, and the 'direct' effect of the observation can be described as informing the observer of the truth of E. Where this is possible, the relevant passage of sense experience *causes* the observer to believe E; and if *prob* (A/E) is high, his belief in E may be a *reason* for his believing A, and E may be spoken of as (inconclusive) *evidence* for A. But the sense experience is evidence neither for E nor for A. Nor does the situation change when we speak physicalistically in terms of patterns of irritation of our sensory surfaces, instead of in terms of sense experience: such patterns of irritation *cause* us to believe various propositions to various degrees; and sometimes the situation can be helpfully analyzed into one in which we are caused to believe $E_1, E_2, \ldots, E_n$, to degrees $p_1, p_2, \ldots, p_n$, whereupon those beliefs provide *reasons* for believing other propositions to other degrees. But patterns of irritation of our sensory surfaces are not reasons or evidence for any of our beliefs, any more than irritation of the mucous membrane of the nose is a *reason* for sneezing.

When I stand blinking in bright sunlight, I can no more believe that the hour is midnight than I can fly. My degree of belief in the proposition that the sun is shining has two distinct characteristics. (a) It is 1, as close as makes no odds. (b) It is compulsory. Here I want to emphasize the second characteristic, which is most often found in conjunction with the first, but not always. Thus, if I examine a normal coin at great length, and experiment with it at length, my degree of belief in the proposition that the next toss will yield a head will have two characteristics. (a) It is $\frac{1}{2}$. (b) It is compulsory. In the case of the coin as in the case of the sun, I cannot decide to have a different degree of belief in the proposition, any more than I can decide to walk on air.

In my scientific and practical undertakings I must make use of such compulsory beliefs. In attempting to understand or to affect the world, I cannot escape the fact that I am part of it: I must rather make use of that fact as best I can. Now where epistemologists have spoken of observation as a source of *knowledge*, I want to speak of observation as a source of compulsory *belief* to one or another degree. I do not propose to identify a very high degree of belief with knowledge, any more than I propose to identify the property of being near 1 with the property of being compulsory.

Nor do I postulate any *general* positive or negative connection between the characteristic of being compulsory and the characteristic of being sound or appropriate in the light of the believer's experience. Nor, finally, do I take a compulsory belief to be necessarily a permanent one: new experience or new reflection (perhaps, prompted by the arguments of others) may loosen the bonds of compulsion, and may then establish new bonds; and the effect may be that the new state of belief is sounder than the old, or less sound.

Then why should we trust our beliefs? According to K. R. Popper,

> ...the decision to accept a basic statement, and to be satisfied with it, is causally connected with our experiences—especially with our *perceptual experiences.* But we do not attempt to *justify* basic statements by these experiences. Experiences can *motivate a decision*, and hence an acceptance or a rejection of a statement, but a basic statement cannot be *justified* by them—no more than by thumping the table.[12]

I take this objection to be defective, principally in attempting to deal with basic statements (observation reports) in terms of *decisions* to *accept* or to *reject* them. Here acceptance parallels belief, rejection parallels disbelief (belief in the denial), and tentativeness or reversibility of the decision parallels degree of belief. Because logical relations hold between statements, but not between events and statements, the relationship between a perceptual experience (an event of a certain sort) and a basic statement cannot

[12] K. R. Popper, *The Logic of Scientific Discovery*, London, 1959, p. 105.

be a logical one, and therefore, Popper believes, cannot be of a sort that would justify the statement:

> Basic statements are accepted as the result of a decision or agreement; and to that extent they are conventions.[13]

But in the absence of a positive account of the nature of acceptance and rejection, parallel to the account of partial belief given in section 1, it is impossible to evaluate this view. Acceptance and rejection are apparently acts undertaken as results of decisions; but somehow the decisions are conventional—perhaps only in the sense that they may be *motivated* by experience, but not *adequately* motivated, if adequacy entails justification.

To return to the question, 'Why should we trust out beliefs?' one must ask what would be involved in *not* trusting one's beliefs, if belief is analyzed as in section 1 in terms of one's preference structure. One way of mistrusting a belief is declining to act on it, but this appears to consist merely in lowering the degree of that belief: to mistrust a partial belief is then to alter its degree to a new, more suitable value.

A more hopeful analysis of such mistrust might introduce the notion of sensitivity to further evidence or experience. Thus, agents 1 and 2 might have the same degree of belief—$\frac{1}{2}$—in the proposition H_1 that the first toss of a certain coin will yield a head, but agent 1 might have this degree of belief because he is convinced that the coin is normal, while agent 2 is convinced that it is either two-headed or two-tailed, he knows not which.[14] There is no question here of agent 2's expressing his mistrust of the figure $\frac{1}{2}$ by lowering or raising it, but he can express that mistrust quite handily by aspects of his belief function. Thus, if H_i is the proposition that the coin lands head up the ith time it is tossed, agent 2's beliefs about the coin are accurately expressed by the function $prob_2$ where

$$prob_2\, H_i = \tfrac{1}{2}, prob_2\, (H_i/H_j) = 1,$$

while agent 1's beliefs are equally accurately expressed by the function $prob_1$ where

$$prob_1\, (H_{i_1}, H_{i_2}, \ldots, H_{i_n}) = 2^{-n},$$

if $i_1 < i_2 < \ldots < i_n$. In an obvious sense, agent 1's beliefs are *firm* in the sense that he will not change them in the light of further evidence, since we have

$$prob_1\, (H_{n+1}/H_1, H_2, \ \ldots, H_n) = prob_1\, H_{n+1} = \tfrac{1}{2},$$

while agent 2's beliefs are quite tentative and in that sense, mistrusted by their holder. Still, $prob_1\, H_i = prob_2\, H_i = \frac{1}{2}$.

[13] Popper, *op. cit.*, p. 106.

[14] This is a simplified version of 'the paradox of ideal evidence', Popper, *op. cit.*, pp. 407–409.

After these defensive remarks, let me say how and why I take compulsive belief to be sound, under appropriate circumstances. Bemused with syntax, the early logical positivists were chary of the notion of truth; and then, bemused with Tarski's account of truth, analytic philosophers neglected to inquire how we come to believe or disbelieve simple propositions. Quite simply put, the point is: coming to have suitable degrees of belief in response to experience is a matter of training—a *skill* which we begin acquiring in early childhood, and are never quite done polishing. The skill consists not only in coming to have appropriate degrees of belief in appropriate propositions under paradigmatically good conditions of observation, but also in coming to have appropriate degrees of belief between zero and one when conditions are less than ideal.

Thus, in learning to use English color words correctly, a child not only learns to acquire degree of belief 1 in the proposition that the cloth is blue, when in bright sunlight he observes a piece of cloth of uniform hue, the hue being squarely in the middle of the blue interval of the color spectrum: he also learns to acquire appropriate degrees of belief between 0 and 1 in response to observation under bad lighting conditions, and when the hue is near one or the other end of the blue region. Furthermore, his understanding of the English color words will not be complete until he understands, in effect, that blue is between green and violet in the color spectrum: his understanding of this point or his lack of it will be evinced in the sorts of mistakes he does and does not make, e.g. in mistaking green for violet he may be evincing confusion between the meanings of 'blue' and of 'violet', in the sense that his mistake is linguistic, not perceptual.

Clearly, the borderline between factual and linguistic error becomes cloudy, here: but cloudy in a perfectly realistic way, corresponding to the intimate connection between the ways in which we experience the world and the ways in which we speak. It is for this sort of reason that having the right language can be as important as (and can be in part identical with) having the right theory.

Then learning to use a language properly is in large part like learning such skills as riding bicycles and flying aeroplanes. One must train oneself to have the right sorts of responses to various sorts of experiences, where the responses are degrees of belief in propositions. This may, but need not, show itself in willingness to utter or assent to corresponding sentences. Need not, because e.g. my cat is quite capable of showing that it thinks it is about to be fed, just as it is capable of showing what its preference ranking is, for hamburger, tuna fish, and oat meal, without saying or understanding a word. With people as with cats, evidence for belief and preference is behavioral; and speech is far from exhausting behavior.[15]

Our degrees of beliefs in various propositions are determined jointly by

[15] Jeffrey, *op. cit.*, pp. 57–59.

our training and our experience, in complicated ways that I cannot hope to describe. And similarly for conditional subjective probabilities, which are certain ratios of degrees of belief: to some extent, these are what they are because of our training—because we speak the languages we speak. And to this extent, conditional subjective probabilities reflect *meanings*. And in this sense, there can be a theory of degree of confirmation which is based on analysis of meanings of sentences. Confirmation theory is therefore semantical and, if you like, logical.[16]

[16] Support of U.S. Air Force Office of Scientific Research is acknowledged, under Grant AF-AFOSR-529-65.

5

THE AIM OF INDUCTIVE LOGIC

Rudolf Carnap

By inductive logic I understand a theory of logical probability providing rules for inductive thinking. I shall try to explain the nature and purpose of inductive logic by showing how it can be used in determining rational decisions.

I shall begin with the customary schema of decision theory involving the concepts of utility and probability. I shall try to show that we must understand "probability" in this context not in the objective sense, but in the subjective sense, i.e., as the degree of belief. This is a psychological concept in empirical decision theory, referring to actual beliefs of actual human beings. Later I shall go over to rational or normative decision theory by introducing some requirements of rationality. Up to that point I shall be in agreement with the representatives of the subjective conception of probability. Then I shall take a further step, namely, the transition from a quasi-psychological to a logical concept. This transition will lead to the theory which I call "inductive logic."

We begin with the customary model of decision making. A person X at a certain time T has to make a choice between possible acts A_1, A_2, $\cdots$. X knows that the possible states of nature are S_1, S_2, $\cdots$; but he does not know which of them is the actual state. For simplicity, we shall here assume

---

Reprinted from *Logic, Methodology and Philosophy of Science*, edited by Ernest Nagel, Patrick Suppes and Alfred Tarski, with the permission of the publishers, Stanford University Press. The author [of this article] is indebted to the National Science Foundation for the support of research in inductive probability.

that the number of possible acts and the number of possible states of nature are finite. X knows the following: if he were to carry out the act A_m and if the state S_n were the actual state of nature, then the outcome would be $O_{m,n}$. This outcome $O_{m,n}$ is uniquely determined by A_m and S_n; and X knows how it is determined. We assume that there is a utility function U_X for the person X and that X knows his utility function so that he can use it in order to calculate subjective values.

Now we define the *subjective value* of a possible act A_m for X at time T:

(1) Definition.

$$V_{X,T}(A_m) = \sum_n U_X(O_{m,n}) \times P(S_n),$$

where $P(S_n)$ is the probability of the state S_n, and the sum covers all possible states S_n.

In other words, we take as the subjective value of the act A_m for X the *expected utility* of the outcome of this act. (1) holds for the time T before any act is carried out. It refers to the contemplated act A_m; therefore it uses the utilities for the possible outcomes $O_{m,n}$ of act A_m in the various possible states S_n. [If the situation is such that the probability of S_n could possibly be influenced by the assumption that act A_m were carried out, we should take the conditional probability $P(S_n|A_m)$ instead of $P(S_n)$. Analogous remarks hold for our later forms of the definition of V].

We can now formulate the customary *decision principle* as follows:

(2) Choose an act so as to maximize the subjective value V.

This principle can be understood either as referring to *actual* decision making, or to *rational* decisions. In the first interpretation it would be a psychological law belonging to *empirical* decision theory as a branch of psychology; in the second interpretation, it would be a normative principle in the theory of *rational* decisions. I shall soon come back to this distinction. First we have to remove an ambiguity in the definition (1) of value, concerning the interpretation of the probability P. There are several conceptions of probability; thus the question arises which of them is adequate in the context of decision making.

The main conceptions of probability are often divided into two kinds, objectivistic and subjectivistic conceptions. In my view, these are not two incompatible doctrines concerning the same concept, but rather two theories concerning two different probability concepts, both of them legitimate and useful. The concept of *objective* (or statistical) *probability* is closely connected with relative frequencies in mass phenomena. It plays an important role in mathematical statistics, and it occurs in laws of various branches of empirical science, especially physics.

The second concept is *subjective* (or personal) *probability*. It is the probability assigned to a proposition or event H by a subject X, say a person or a group of persons, in other words, the degree of belief of X in H. Now it seems to me that we should clearly distinguish two versions of subjective probability, one representing the *actual* degree of belief and the other the *rational* degree of belief.

Which of these two concepts of probability, the objective or the subjective, ought to be used in the definition of subjective value and thereby in the decision principle? At the present time, the great majority of those who work in mathematical statistics still regard the statistical concept of probability as the only legitimate one. However, this concept refers to an objective feature of nature; a feature that holds whether or not the observer X knows about it. And in fact, the numerical values of statistical probability are in general not known to X. Therefore this concept is unsuitable for a decision principle. It seems that for this reason a number of those who work in the theory of decisions, be it actual decisions or rational decisions, incline toward the view that some version of the subjective concept of probability must be used here. I agree emphatically with this view.

The statistical concept of probability remains, of course, a legitimate and important concept both for mathematical statistics and for many branches of empirical science. And in the special case that X knows the statistical probabilities for the relevant states S_n but has no more specific knowledge about these states, the decision principle would use these values. There is general agreement on this point. And this is not in conflict with the view that the decision principle should refer to subjective probability, because in this special situation the subjective probability for X would be equal to the objective probability.

Once we recognize that decision theory needs the subjective concept of probability, it is clear that the theory of *actual* decisions involves the first version of this concept, i.e., the *actual* degree of belief, and the theory of *rational* decisions involves the second version, the *rational* degree of belief.

Let us first discuss the theory of *actual* decisions. The concept of probability in the sense of the *actual* degree of belief is a psychological concept; its laws are empirical laws of psychology, to be established by the investigation of the behavior of persons in situations of uncertainty, e.g., behavior with respect to bets or game of chance. I shall use for this psychological concept the technical term "*degree of credence*" or shortly "*credence*." In symbols, I write '$Cr_{X,T}(H)$' for "the (degree of) credence of the proposition H for the person X at the time T." Different persons X and Y may have different credence functions $Cr_{X,T}$ and $Cr_{Y,T}$. And the same person X may have different credence functions Cr_{X,T_1} and Cr_{X,T_2} at different times T_1 and T_2; e.g., if X observes between T_1 and T_2 that H holds, then $Cr_{X,T_1}(H) \neq Cr_{X,T_2}(H)$. (Let the ultimate possible cases be represented by the points of a logical space, usually called the probability space. Then a proposition or event is understood, not as a sentence, but as the range of a sentence, i.e., the set of points representing those possible cases in which the sentence holds. To the conjunction of two sentences corresponds the intersection of the propositions.)

On the basis of credence, we can define *conditional credence*, "the credence of H with respect to the proposition E" (or "· · · given E"):

(3) DEFINITION.

$$Cr'_{X,T}(H|E) = \frac{Cr_{X,T}(E \cap H)}{Cr_{X,T}(E)},$$

provided that $Cr_{X,T}(E) > 0$. $Cr'_{X,T}(H|E)$ *is the credence which* H *would have for* X *at* T *if* X *ascertained that* E *holds.*

Using the concept of credence, we now replace (1) by the following:

(4) DEFINITION.

$$V_{X,T}(A_m) = \sum_n U_X(O_{m,n}) \times Cr_{X,T}(S_n)$$

As was pointed out by Ramsey, we can determine X's credence function by his betting behavior. A bet is a contract of the following form. X pays into the pool the amount u, his partner Y pays the amount v; they agree that the total stake $u+v$ goes to X if the hypothesis H turns out to be true, and to Y if it turns out to be false. If X accepts this contract, we say that he bets on H with the total stake $u+v$ and with the betting quotient $q = u/(u+v)$ (or, at odds of u to v). If we apply the decision principle with the definition (4) to the situation in which X may either accept or reject an offered bet on H with the betting quotient q, we find that X will accept the bet if q is not larger than his credence for H. Thus we may interpret $Cr_{X,T}(H)$ as the highest betting quotient at which X is willing to bet on H. (As is well known, this holds only under certain conditions and only approximately.)

Utility and credence are psychological concepts. The utility function of X represents the system of valuations and preferences of X; his credence function represents his system of beliefs (not only the content of each belief, but also its strength). Both concepts are theoretical concepts which characterize the state of mind of a person; more exactly, the non-observable micro-state of his central nervous system, not his consciousness, let alone his overt behavior. But since his behavior is influenced by his state, we can indirectly determine characteristics of his state from his behavior. Thus experimental methods have been developed for the determination of some values and some general characteristics of the utility function and the credence function ("subjective probability") of a person on the basis of his behavior with respect to bets and similar situations. Interesting investigations of this kind have been made by F. Mosteller and P. Nogee [13], and more recently by D. Davidson and P. Suppes [4], and others.

Now we take the step from empirical to *rational decision theory*. The latter is of greater interest to us, not so much for its own sake (its methodological status is in fact somewhat problematic), but because it is the connecting link between empirical decision theory and inductive logic. Rational decision theory is concerned not with actual credence, but with *rational* credence. (We should also distinguish here between actual utility and rational utility; but we will omit this.) The statements of a theory of this kind are not found by experiments but are established on the basis of requirements of rationality; the formal procedure usually consists in deducing theorems from axioms which are justified by general considerations of rationality, as we shall see. It seems fairly clear that the probability concepts used by the following authors are meant in the sense of rational credence (or rational credibility, which I shall explain presently): John Maynard Keynes (1921), Frank P. Ramsey (1928), Harold Jeffreys (1931), B. O. Koopman (1940), Georg Henrik von Wright (1941), I. G. Good (1950), and Leonard J. Savage (1954). I am inclined to include here also those authors who do not declare explicitly that their concept refers to rational rather than actual beliefs, but who accept general axioms and do not base their theories on psychological results. Bruno De Finetti (1931) satisfies these conditions; however, he says explicitly that his concept of "subjective probability" refers not to rational, but to actual beliefs. I find this puzzling.

The term "subjective probability" seems quite satisfactory for the actual degree of credence. It is frequently applied also to a probability concept interpreted as something like rational credence. But here the use of the word "subjective" might be misleading (comp. Keynes [9, p. 4] and Carnap [1, § 12A]). Savage has suggested the term "personal probability."

Rational credence is to be understood as the credence function of a completely rational person X; this is, of course, not any real person, but an imaginary, idealized person. We carry out the idealization step for step, by introducing *requirements of rationality* for the credence function. I shall now explain some of these requirements.

Suppose that X makes n simultaneous bets; let the ith bet ($i = 1, \cdots, n$) be on the proposition H_i with the betting quotient q_i and the total stake s_i. Before we observe which of the propositions H_i are true and which are false, we can consider the *possible* cases. For any possible case, i.e., a logically possible distribution of trust-values among the H_i, we can calculate the gain or loss for each bet and hence the total balance of gains and losses from the n bets. If in *every* possible case X suffers a net loss, i.e., his total balance is negative, it is obviously unreasonable for X to make these n bets. Let X's credence function at a given time be Cr. By a (finite) betting system in accordance with Cr we mean a finite system of n bets on n arbitrary propositions H_i ($i = 1, \cdots, n$) with n arbitrary (positive) stakes s_i, but with the betting quotients $q_i = Cr(H_i)$.

(5) Definition. *A function Cr is coherent if and only if there is no betting system in accordance with Cr such that there is a net loss in every possible case.*

For X to make bets of a system of this kind would obviously be unreasonable. Therefore we lay down the *first requirement* as follows:

R1. *In order to be rational, Cr must be coherent.*

Now the following important result holds:

(6) A function *Cr* from propositions to real numbers is coherent if and only if *Cr* is a normalized probability measure.

(A real-valued function of propositions is said to be a probability measure if it is a non-negative, finitely additive set function; it is normalized if its value for the necessary proposition is 1. In other words, a normalized probability measure is a function which satisfies the basic axioms of the calculus of probability, e.g., the axioms I through V in Kolmogoroff's system [10, § 1].)

The first part of (6) ("...coherent if...") was stated first by Ramsey [15] and was later independently stated and proved by De Finetti [5]. The much more complicated proof for the second part ("...only if...") was found independently by John G. Kemeny [8, p. 269] and R. Sherman Lehman [12, p. 256].

Let Cr' be the conditional credence function defined on the basis of *Cr* by (3). As ordinary bets are based on *Cr*, conditional bets are based on Cr'. The concept of coherence can be generalized so as to be applicable also to conditional credence functions. (6) can then easily be extended by the result that a conditional credence function Cr' is coherent if and only if Cr' is a normalized conditional probability measure, in other words, if and only if Cr' satisfies the customary basic axioms of conditional probability, including the general multiplication axiom.

Following Shimony [17], we introduce now a concept of coherence in a stronger sense, for which I use the term "strict coherence":

(7) Definition. *A function Cr is strictly coherent if and only if Cr is coherent and there is no (finite) system of bets in accordance with Cr on molecular propositions such that the result is a net loss in at least one possible case, but not a net gain in any possible case.*

It is clear that it would be unreasonable to make a system of bets of the kind just specified. Therefore we lay down the *second requirement*:

R2. *In order to be rational, a credence function must be strictly coherent.*

We define *regular credence function* (essentially in the sense of Carnap [1, § 55A]):

(8) Definition. *A function Cr is regular if and only if Cr is a normalized probability measure and, for any molecular proposition H,* $Cr(H) = 0$ *only if H is impossible.*

By analogy with (6) we have now the following important theorem; its first part is due to Shimony, its second part again to Kemeny and Lehman:

(9) A function *Cr* is strictly coherent if and only if *Cr* is regular.

Most of the authors of systems for subjective or logical probability adopt only the basic axioms; thus they require nothing but coherence. A few go one step further by including an axiom for what I call regularity; thus they require in effect strict coherence, but nothing more. Axiom systems of both kinds are extremely weak; they yield no result of the form "$P(H|E) = r$," except in the trivial cases where *r* is 0 or 1. In my view, much more should be required.

The two previous requirements apply to any credence function that holds for *X* at any time *T* of his life. We now consider two of these functions, Cr_n for the time T_n *and* Cr_{n+1} for a time T_{n+1} shortly after T_n. Let the proposition *E* represent the observation data received by *X* between these two time points. The *third requirement* refers to the transition from Cr_n to Cr_{n+1}:

R3. (a) *The transformation of* Cr_n *into* Cr_{n+1} *depends only on the proposition E.*
(b) *More specifically,* Cr_{n+1} *is determined by* Cr_n *and E as follows*: for any *H*, $Cr_{n+1}(H) = Cr_n(E \cap H)/Cr_n(E)$ (*hence* $= Cr'_n(H|E)$ *by definition* (3)).

Part (a) is of course implied by (b). I have separated part (a) from (b) because *X*'s function *Cr* might satisfy (a) without satisfying (b). Part (a) requires merely that *X* be rational to the extent that changes in his credence function are influenced only by his observational results, but not by any other factors, e.g., feelings like his hopes or fears concerning a possible future event *H*, feelings which in fact influence the beliefs of all actual human beings. Part (b) specifies exactly the transformation of Cr_n into Cr_{n+1}; the latter is the conditional credence Cr'_n with respect to *E*. The rule (b) can be used only if $Cr_n(E) \neq 0$; this condition is fulfilled for any possible observational result, provided that Cr_n satisfies the requirement of strict coherence.

Let the proposition E_{n+2} represent the data obtained between T_{n+1} and a later time point T_{n+2}. Let Cr_{n+2} be the credence function at T_{n+2} obtained by R3b from Cr_{n+1} with respect to E_{n+2}. It can easily be shown that the same function Cr_{n+2} results if R3b is applied to Cr_n with respect to the combined data $E_{n+1} \cap E_{n+2}$. In the same way we can determine any later credence function Cr_{n+m} from the given function Cr_n either in m steps, applying the rule R3b in each step with one datum of the sequence E_{n+1}, $E_{n+2}, \cdots, E_{n+m}$, or in one step with the intersection $\bigcap_{p=1}^{m} E_{n+p}$. If m is large so that the intersection contains thousands of single data, the objection might be raised that it is unrealistic to think of a procedure of this kind, because a man's memory is unable to retain and reproduce at will so many items. However, since our goal is not the psychology of actual human behavior in the field of inductive reasoning, but rather inductive logic as a system of rules, we do not aim at realism. We make the further idealization that X is not only perfectly rational but has also an infallible memory. Our assumptions deviate from reality very much if the observer and agent is a natural human being, but not so much if we think of X as a robot with organs of perception, data processing, decision making, and acting. Thinking about the design of a robot will help us in finding rules of rationality. Once found, these rules can be applied not only in the construction of a robot but also in advising human beings in their effort to make their decisions as rational as their limited abilities permit.

Consider now the whole sequence of data obtained by X up to the present time T_n: $E_1, E_2, \cdots, E_n$. Let K_{X,T_n} or, for short, K_n be the proposition representing the combination of all these data:

(10) Definition.

$$K_n = \bigcap_{i=1}^{n} E_i$$

Thus K_n represents, under the assumption of infallible memory, the total observational knowledge of X at the time T_n. Now consider the sequence of X's credence functions. In the case of a human being we would hesitate to ascribe to him a credence function at a very early time point, before his abilities of reason and deliberate action are sufficiently developed. But again we disregard this difficulty by thinking either of an idealized human baby or of a robot. We ascribe to him a credence function Cr_1 for the time point T_1; Cr_1 represents X's personal probabilities based upon the datum E_1 as his only experience. Going even one step further, let us ascribe to him an *initial credence function* Cr_0 for the time point T_0 before he obtains his first datum E_1. Any later function Cr_n for a time point T_n is uniquely determined by Cr_0 and K_n:

(11) For any H, $Cr_n(H) = Cr'_0(H|K_n)$, where Cr'_0 is the conditional function based on Cr_0.

$Cr_n(H)$ is thus seen to be the *conditional initial credence of* H given K_n.

How can we understand the function Cr_0? In terms of the robot, Cr_0 is the credence function that we originally build in and that he transforms step for step, with regard to the incoming data, into the later credence functions.

In the case of a human being X, suppose that we find at the time T_n his credence function Cr_n. Then we can, under suitable conditions, reconstruct a sequence $E_1, \cdots, E_n$, the proposition K_n, and a function Cr_0 such that (a) $E_1, \cdots, E_n$ are possible observation data, (b) K_n is defined by (10), (c) Cr_0 satisfies all requirements of rationality for initial credence functions, and (d) the application of (11) to the assumed function Cr_0 and K_n would lead to the ascertained function Cr_n. We do not assert that X actually experienced the data $E_1, \cdots, E_n$, and that he actually had the initial credence function Cr_0, but merely that, under idealized conditions, his function Cr_n could have evolved from Cr_0 by the effect of the data $E_1, \cdots, E_n$.

For the conditional initial credence (Cr'_0) we shall also use the term "*credibility*" and the symbol '*Cred*'. As an alternative to defining '*Cred*' on the basis of 'Cr_0', we could introduce it as a primitive term. In this case we may take the following universal statement as the main postulate for the theoretical primitive term '*Cred*':

(12) Let *Cred* be any function from pairs of propositions to real numbers, satisfying all requirements which we have laid down or shall lay down for credibility functions. Let H and A be any propositions (A not empty). Let X be any observer and T any time point. If X's credibility function is *Cred* and his total observational knowledge at T is A, then his credence for H and T is $Cred(H|A)$.

Note that (12) is much more general than (11). There the function *Cred* (or Cr'_0) was applied only to those pairs H, A, in which A is a proposition of the sequence $K_1, K_2, \cdots$, and thus represents the actual knowledge of X at some time point. In (12), however, A may be any non-empty proposition. Let A_1 be a certain proposition which does not occur in the sequence $K_1, K_2, \cdots$, and H_1 some proposition. Then the statement

$$Cr_T(H_1) = Cred(H_1|A_1)$$

is to be understood as a counterfactual conditional as follows:

(13) If the total knowledge of X at T had been A_1, then his credence for H_1 at T would have been equal to $Cred(H_1|A_1)$.

This is a true counterfactual based on the postulate (12), analogous to ordinary counterfactuals based on physical laws.

Applying (12) to X's actual total observational knowledge $K_{X,T}$ at time T, we have:

(14) For any H, $Cr_{X,T}(H) = Cred_X(H|K_{X,T})$

Now we can use credibility instead of credence in the definition of the subjective value of an act A_m, and thereby in the decision rule. Thus we have instead of (4):

(15) DEFINITION.

$$V_{X,T}(A_m) = \sum_n U_X(O_{m,n}) \times Cred_X(S_n|K_{X,T})$$

(If the situation is such that the assumption of A_m could possibly change the credence of S_n, we have to replace '$K_{X,T}$' by '$K_{X,T} \cap A_m$'; see the remark on (1).)

If *Cred* is taken as primitive, Cr_0 can be defined as follows:

> (16) DEFINITION. *For any H,* $Cr_0(H) = Cred\ (H/Z)$, *where Z is the necessary proposition* (*the tautology*).

This is the special case of (12) for the initial time T_0, when X's knowledge K_0 is the tautology.

While $Cr_{X,T}$ characterizes the momentary state of X at time T with respect to his beliefs, his function $Cred_X$ is a trait of his underlying permanent intellectual character, namely his permanent disposition for forming beliefs on the basis of his observations.

Since each of the two functions Cr_0 and *Cred* is definable on the basis of the other one, there are two alternative equivalent procedures for specifying a basic belief-forming disposition, namely either by Cr_0 or by *Cred*.

Most of those who have constructed systems of subjective or personal probability (in the narrower sense, in contrast to logical probability), e.g., Ramsey, De Finetti, and Savage, have concentrated their attention on what we might call "adult" credence functions, i.e., those of persons sufficiently developed to communicate by language, to play games, make bets, etc., hence persons with an enormous amount of experience. In empirical decision theory it has great practical advantages to take adult persons as subjects of investigation, since it is relatively easy to determine their credence functions on the basis of their behavior with games, bets, and the like. When I propose to take as a basic concept, not adult credence but either initial credence or credibility, I must admit that these concepts are less realistic and remoter from overt behavior and may therefore appear as elusive and dubious. On the other hand, when we are interested in *rational* decision theory, these concepts have great methodological advantages. Only for these concepts, not for credence, can we find a sufficient number of requirements of rationality as a basis for the construction of a system of inductive logic.

If we look at the development of theories and concepts in various branches of science, we find frequently that it was possible to arrive at powerful laws of great generality only when the development of concepts, beginning with directly observable properties, had progressed step by step to more abstract concepts, connected only indirectly with observables. Thus physics proceeds from concepts describing visible motion of bodies to the concept of a momentary electric force, and then to the still more abstract concept of a permanent electric field. In the sphere of human action we have first concepts describing overt behavior, say of a boy who is offered the choice of an apple or an ice cream cone and takes the latter; then we introduce the concept of an underlying momentary inclination, in this case the momentary preference of ice cream over apple; and finally we form the abstract concept of an underlying permanent disposition, in our example the general utility function of the boy.

What I propose to do is simply to take the same step from momentary inclination to the permanent disposition for forming momentary inclinations also with the second concept occurring in the decision principle, namely, personal probability or degree of belief. This is the step from credence to credibility.

When we wish to judge the morality of a person, we do not simply look at some of his acts, we study rather his character, the system of his moral values, which is part of his utility function. Single acts without knowledge of motives give little basis for a judgment. Similarly, if we wish to judge the rationality of a person's beliefs, we should not simply look at his present beliefs. Beliefs without knowledge of the evidence out of which they arose tell us little. We must rather study the way in which the person forms his beliefs on the basis of evidence. In other words, we should study his credibility function, not simply his present credence function. For example, let X have the evidence E that from an urn containing white and black balls ten balls have been drawn, two of them white and eight black. Let Y have the evidence E' which is similar to E, but with seven balls white and three black. Let H be the prediction that the next ball drawn will be white. Suppose that for both X and Y the credence of H is $\frac{2}{3}$. Then we would judge this same credence $\frac{2}{3}$ to be unreasonable for X, but reasonable for Y. We would condemn a credibility function *Cred* as non-rational if $Cred\ (H|E) = \frac{2}{3}$; while the result $Cred(H|E') = \frac{2}{3}$ would be no ground for condemnation.

Suppose X has the credibility function *Cred*, which leads him, on the basis of his knowledge K_n at time T_n to the credence function Cr_n, and thereby, with his utility function U, to the act A_m. If this act seems to us unreasonable in view of his evidence K_n and his utilities, we shall judge that *Cred* is non-rational. But for such a judgment on *Cred* it is not necessary that X is actually led to an unreasonable act. Suppose that for E and H as in the above example, K_n contains E and otherwise only evidence irrelevant for H. Then we have $Cr_n(H) = Cred(H|K_n) = Cred(H|E) = \frac{2}{3}$; and this result seems unreasonable on the given evidence. If X bets on H with betting quotient $\frac{2}{3}$, this bet is unreasonable, even if he wins it. But his credence $\frac{2}{3}$ is anyway unreasonable, no matter whether he acts on it or not. It is unreasonable because there are possible situations, no matter whether real or not, in which the result $Cred(H|E) = \frac{2}{3}$ would lead him to an unreasonable act. Furthermore, it is not necessary for our condemnation of the function *Cred* that it actually leads to unreasonable *Cr*-values. Suppose that another man X' has the same function *Cred*, but is not led to the unreasonable *Cr*-value in the example, because he has an entirely different life history, and at no time is his total knowledge either E or a combination of E with data irrelevant for H. Then we would still condemn the function *Cred* and the man X' characterized by this function. Our argument would be as follows: if the total knowledge of X' had at some time been E, or E together with irrelevant data, then his credence for H would have had the unreasonable value $\frac{2}{3}$. The same considerations hold, of course, for the initial credence function Cr_0 corresponding to the function *Cred*; for, on the basis of any possible knowledge proposition K, Cr_0 and *Cred* would lead to the same credence function.

The following is an example of a requirement of rationality for Cr_0 (and hence for *Cred*) which has no analogue for credence functions. As we shall see later, this requirement leads to one of the most important axioms of inductive logic. (The term "individual" means "element of the universe of discourse," or "element of the population" in the terminology of statistics.)

> R4. *Requirement of symmetry. Let a_i and a_j be two distinct individuals. Let H and H' be two propositions such that H' results from H by taking a_j for a_i and vice versa. Then Cr_0 must be such that $Cr_0(H)$ $= Cr_0(H')$.* (In other words, Cr_0 must be invariant with respect to any finite permutation of individuals.)

This requirement seems indispensable. H and H' have exactly the same logical form; they differ merely by their reference to two distinct individuals. These individuals may happen to be quite different. But since their differences are not known to X at time T_0, they cannot have any influence on the Cr_0-values of H and H'. But suppose that at a later time T_n, X's knowledge K_n contains information E relevant to H and H', say information making H more probable than H' (as an extreme case, E may imply that H is true and H' is false). Then X's credence function Cr_n at T_n will have different values for H and for H'. Thus it is clear that R4 applies only to Cr_0, but is not generally valid for other credence functions $Cr_n(n > 0)$.

Suppose that X is a robot constructed by us. Because H and H' are alike in all their logical properties, it would be entirely arbitrary and therefore inadmissible for us to assign to them different Cr_0-values.

A function Cr_0 is suitable for being built into a robot only if it fulfills the requirements of rationality; and most of these requirements (e.g., R4 and all those not yet mentioned) apply only to Cr_0 (and *Cred*) but not generally to other credence functions.

Now we are ready to take the step to *inductive logic*. This step consists in the transition from the concepts of the Cr_0-function and the *Cred*-function of an imaginary subject X to corresponding purely logical concepts. The former concepts are quasi-psychological; they are assigned to an imaginary subject X supposed to be equipped with perfect rationality and an unfailing memory; the logical concepts, in contrast, have nothing to do with observers and agents, whether natural or constructed, real or imaginary. For a logical function corresponding to Cr_0, I shall use the symbol '$\mathcal{M}$' and I call such functions (inductive) measure functions or $\mathcal{M}$-functions; for a logical function corresponding to *Cred*, I shall use the symbol '$\mathcal{C}$', and I call these functions (inductive) confirmation functions or $\mathcal{C}$-functions. I read '$\mathcal{C}$ $(H|E)$' as "the degree of confirmation (or briefly "the confirmation") of H with respect to E" (or: "... given E"). An $\mathcal{M}$-function is a function from propositions to real numbers. A $\mathcal{C}$-function is a function from pairs of propositions to real numbers. Any $\mathcal{M}$-function $\mathcal{M}$ is supposed to be defined in a purely logical way, i.e., on the basis of concepts of logic (in the wide sense, including set-theory and hence the whole of pure mathematics). Therefore the value $\mathcal{M}$ (A) for any proposition A depends merely on the logical (set-theoretic) properties of A (which is a set in a probability space) but not on any contingent facts of nature (e.g., the truth of A or of other contingent propositions). Likewise any $\mathcal{C}$-function is supposed to be defined in purely logical terms.

Inductive logic studies those $\mathcal{M}$-functions which correspond to rational Cr_0-functions, and those $\mathcal{C}$-functions which correspond to rational *Cred*-functions. Suppose $\mathcal{M}$ is a logically defined $\mathcal{M}$-function. Let us imagine a subject X whose function Cr_0 corresponds to $\mathcal{M}$, i.e., for every proposition H, $Cr_0(H) = \mathcal{M}(H)$. If we find that Cr_0 violates one of the rationality requirements, say R4, then we would reject this function Cr_0, say for a robot we plan to build. Then we wish also to exclude the corresponding function $\mathcal{M}$ from those treated as admissible in the system of inductive logic we plan to construct. Therefore, we set up axioms of inductive logic about $\mathcal{M}$-functions so that these axioms correspond to the requirements of rationality which we find in the theory of rational decision making about Cr_0-functions.

For example, we shall lay down as the basic axioms of inductive logic those which say that $\mathcal{M}$ is a non-negative, finitely additive, and normalized measure function. These axioms correspond to the requirement R1 of coherence, by virtue of theorem (6). Further we shall have an axiom saying that $\mathcal{M}$ is regular. This axiom corresponds to the requirement R2 of strict coherence by theorem (9).

Then we shall have in inductive logic, in analogy to the requirement R4 of symmetry, the following:

(17) AXIOM OF SYMMETRY. *$\mathcal{M}$ is invariant with respect to any finite permutation of individuals.*

All axioms of inductive logic state relations among values of $\mathcal{M}$ or $\mathcal{C}$ as dependent only upon the logical properties and relations of the propositions involved (with respect to language-systems with specified logical and semantical rules). Inductive logic is the theory based upon these axioms.

It may be regarded as a part of logic in view of the fact that the concepts occurring are logical concepts. It is an interesting result that this part of the theory of decision making, namely, the logical theory of the $\mathcal{M}$-functions and the $\mathcal{C}$-functions, can thus be separated from the rest. However, we should note that this logical theory deals only with the abstract, formal aspects of probability, and that the full meaning of (subjective) probability can be understood only in the wider context of decision theory through the connections between probability and the concepts of utility and rational action.

It is important to notice clearly the following distinction. While the *axioms* of inductive logic themselves are formulated in purely logical terms and do not refer to any contingent matters of fact, the *reasons* for our choice of the axioms are not purely logical. For example, when you ask me why I accept the axiom of symmetry (17), then I point out that if X had a Cr_0-function corresponding to an $\mathcal{M}$-function violating (17), then this function Cr_0 would violate R4, and I show that therefore X, in a certain possible knowledge situation, would be led to an unreasonable decision. Thus, in order to give my reasons for the axiom, I move from pure logic to the context of decision theory and speak about beliefs, actions, possible losses, and the like. However, this is not in the field of empirical, but of rational decision theory. Therefore, in giving my reasons, I do not refer to particular empirical results concerning particular agents or particular states of nature and the like. Rather, I refer to a *conceivable* series of observations by X, to conceivable sets of possible acts, of possible states of nature, of possible outcomes of the acts, and the like. These features are characteristic for an analysis of *reasonableness* of a given function Cr_0, in contrast to an investigation of the *successfulness* of the (initial or later) credence function of a given person in the real world. Success depends upon the particular contingent circumstances, rationality does not.

There is a class of axioms of inductive logic which I call *axioms of invariance*. The axiom of symmetry is one of them. Another one says that $\mathcal{M}$ is invariant with respect to any finite permutation of attributes belonging to a family of attributes, e.g., colors, provided these attributes are alike in their logical (including semantical) properties. Still another one says that if E is a proposition about a finite sample from a population, then $\mathcal{M}(E)$ is independent of the size of the population. These and other invariance axioms may be regarded as representing the valid core of the old *principle of indifference* (or principle of insufficient reason). The principle, in its original form, as used by Laplace and other authors in the classical period of the theory of probability, was certainly too strong. It was later correctly criticized by showing that it led to absurd results. However, I believe that the basic idea of the principle is sound. Our task is to restate it by specific restricted axioms.

It seems that most authors on subjective probability do not accept any axioms of invariance. In the case of those authors who take credence as their basic concept, e.g., Ramsey, De Finetti, and Savage, this is inevitable, since the invariance axioms do not hold for general credence functions. In order to obtain a stronger system, it is necessary to take as the basic concept either initial credence or credibility (or other concepts in terms of which these are definable).

When we construct an axiom system for $\mathcal{M}$, then the addition of each new axiom has the effect of excluding certain $\mathcal{M}$-functions. We accept an axiom if we recognize that the $\mathcal{M}$-functions excluded by it correspond to non-rational Cr_0-functions. Even on the basis of all axioms which I would accept at the present time for a simple qualitative language (with one-place predicates only, without physical magnitudes), the number of admissible $\mathcal{M}$-functions, i.e., those which satisfy all accepted axioms, is still infinite; but their class is immensely smaller than that of all coherent $\mathcal{M}$-functions. There will presumably be further axioms, justified in the same way by considerations of rationality. We do not know today whether in this future development the number of admissible $\mathcal{M}$-functions will always remain infinite or will become finite and possibly even be reduced to one. Therefore, at the present time I do not assert that there is only one rational Cr_0-function.

I think that the theory of the $\mathcal{M}$- and $\mathcal{C}$-functions deserves the often misused name of "*inductive logic*." Earlier I gave my reasons for regarding this theory as a part of logic. The epithet "inductive" seems appropriate because this theory provides the foundation for inductive reasoning (in a wide sense). I agree in this view with John Maynard Keynes and Harold Jeffreys. However, it is important that we recognize clearly the essential form of inductive reasoning. It seems to me that the view of almost all writers on induction in the past and including the great majority of contemporary writers, contains one basic mistake. They regard inductive reasoning as an *inference* leading from some known propositions, called the premisses or evidence, to a new proposition, called the conclusion, usually a law or a singular prediction. From this point of view the result of any particular inductive reasoning is the *acceptance* of a new proposition (or its rejection, or its suspension until further evidence is found, as the case may be). This seems to me wrong. On the basis of this view it would be impossible to refute Hume's dictum that there are no rational reasons for induction. Suppose that I find in earlier weather reports that a weather situation like the one we have today has occurred one hundred times and that it was followed each time by rain the next morning. According to the customary view, on the basis of this evidence the "inductive method" entitles me to accept the prediction that it will rain tomorrow morning. (If you demur because the number one hundred is too small, change it to one hundred thousand or any number you like.) I would think instead that inductive reasoning about a proposition should lead, not to acceptance or rejection, but to the assignment of a number to the proposition, viz., its $\mathcal{C}$-value. This difference may perhaps appear slight; in fact, however, it is essential. If, in accordance with the customary view, we accept the prediction, then Hume is certainly right in protesting that we have no rational reason for doing so, since, as everybody will agree, it is still possible that it will not rain tomorrow.

If, on the other hand, we adopt the new view of the nature of inductive reasoning, then the situation is quite different. In this case X does not assert the hypothesis H in question, e.g., the prediction "it will rain tomorrow"; he asserts merely the following statements:

(18) (a) At the present moment T_n, the totality of X's observation results is Kn.
(b) $\mathcal{C}\,(H|K_n) = 0.8$.
(c) $Cred_X(H|K_n) = 0.8$.
(d) $Cr_{X,T_n}(H) = 0.8$.

(a) is the statement of the evidence at hand, the same as in the first case. But now, instead of accepting H, X asserts the statement (c) of the *Cred*-value for H on his evidence. (c) is the result of X's inductive reasoning. Against this result Hume's objection does not hold, because X can give rational reasons for it. (c) is derived from (b) because X has chosen the function $\mathcal{C}$ as his credibility function. (b) is an analytic statement based on the definition of $\mathcal{C}$. X's choice of $\mathcal{C}$ was guided by the axioms of inductive logic. And for each of the axioms we can give reasons, namely, rationality requirements for credibility functions. Thus $\mathcal{C}$ represents a reasonable credibility function. Finally, X's credence value (d) is derived from (c) by (14).

Now some philosophers, including some of my empiricist friends, would raise the following objection. If the result of inductive reasoning is merely an analytic statement (like (b) or (c)), then induction cannot fulfill its purpose of guiding our practical decisions. As a basis for a decision we need a statement with factual content. If the prediction H itself is not available, then we must use a statement of the *objective* probability of H. In answer to this objection I would first point out that X has a factual basis in his evidence, as stated in (a). And for the determination of a rational decision neither the acceptance of H nor knowledge of the objective probability of H is needed. The rational subjective probability, i.e., the credence as stated in (d), is sufficient for determining first the rational subjective value of each possible act by (15), and then a rational decision. Thus in our example, in view of (b) X would decide to make a bet on rain tomorrow if it were offered to him at odds of four to one or less, but not more.

The old puzzle of induction consists in the following dilemma. On the one hand we see that inductive reasoning is used by the scientist and the man in the street every day without apparent scruples; and we have the feeling that it is valid and indispensable. On the other hand, once Hume awakens our intellectual conscience, we find no answer to his objection.

Who is right, the man of common sense or the critical philosopher? We see that, as so often, both are partially right. Hume's criticism of the customary forms of induction was correct. But still the basic idea of common sense thinking is vindicated: induction, if properly reformulated, can be shown to be valid by rational criteria.

REFERENCES

[1] R. Carnap, *Logical foundations of probability*. Chicago, 1950.

[2] R. Carnap, *The continuum of inductive methods*. Chicago, 1952.

[3] R. Carnap, "Inductive logic and rational decisions." (This is an expanded version of the present paper.) To appear as the introductory article in: *Studies in probability and inductive logic*, vol. I, R. Carnap, ed. Forthcoming.

[4] D. Davidson and P. Suppes, *Decision making: An experimental approach*. Stanford, 1957.

[5] B. De Finetti, "La prévision: ses lois logiques, ses sources subjectives." *Annales de l'Institut Henri Poincaré*, vol. 7 (1937), pp. 1–68.

[6] I. J. Good, *Probability and the weighing of evidence*. London and New York, 1950.

[7] H. Jeffreys, *Theory of probability*. Oxford (1939), 2nd ed. 1948.

[8] J. Kemeny, "Fair bets and inductive probabilities." *Journal of Symbolic Logic*, vol. 20 (1955), pp. 263–273.

[9] J. M. Keynes, *A treatise on probability*. London and New York, 1921.

[10] A. N. Kolmogoroff, *Foundations of the theory of probability*. New York (1950), 2nd ed. 1956.

[11] B. O. Koopman, "The bases of probability." *Bull. Amer. Math. Soc.*, vol. 46 (1940), pp. 763–774.

[12] R. S. Lehman, "On confirmation and rational betting." *Journal of Symbolic Logic*, vol. 20 (1955), pp. 251–262.

[13] F. C. Mosteller and P. Nogee, "An experimental measurement of utility." *Journal of Political Economy*, vol. 59 (1951), pp. 371–404.

[14] J. von Neumann and O. Morgenstern, *Theory of games and economic behavior*. Princeton (1944), 2nd ed. 1947.

[15] F. P. Ramsey, *The foundations of mathematics and other logical essays*. London and New York, 1931.

[16] L. J. Savage, *The foundations of statistics*. New York and London, 1954.

[17] A. Shimony, "Coherence and the axioms of confirmation." *Journal of Symbolic Logic,* vol. 20 (1955), pp. 1–28.

[18] G. H. von Wright, *The logical problem of induction*. Oxford and New York (1941), 2nd ed., 1957.

6

PROBABILITIES AND THE PROBLEM OF INDIVIDUATION

Bas C. van Fraassen

In philosophy of science, the theory of probability is usually discussed in connection with the problems of induction and confirmation, which belong to the methodology of science. Here we wish to approach the subject of probability in a different context: we wish to discuss probabilistic or statistical theories, that is, theories about the statistical behavior of certain kinds of physical entities or systems. Such theories are to be found both in the physical sciences and in the social sciences. We shall not discuss specific examples to any large extent, but when we do, the reference will be to physics. In the course of this discussion, we shall connect a problem in the logical theory of probability with a metaphysical problem that goes back at least to Aristotle, and which was discussed by, among others, Aquinas, Leibniz, and (in a contemporary form) by Margenau and Reichenbach.

1. THE LOGICAL THEORY OF PROBABILITY

To begin, we shall give a short summary of the logical theory of probability as applied to a very simple case.[1] We suppose that we wish to

This paper was read at the APA (East) Annual Meeting, December 29, 1969. The author wishes to acknowledge his debt to Prof. A. Grünbaum, University of Pittsburgh; Prof. K. Lambert, University of California (Irvine); Prof. W. Salmon, Indiana University; and Prof. C. Glymour, Princeton University, for their helpful comments on earlier drafts (1963 and 1968). The research for this paper was partially supported by NSF grant GS-1566.

[1] Cf. R. Carnap, *The Logical Foundations of Probability* (Chicago: University of Chicago Press, 1950), sections 14–18 and 54–55; also B. Skyrms, *Choice and*

Original publication by kind permission of the author.

describe a collection of n objects to be named $a_1, \ldots, a_n$. We describe them by the use of m predicates, $F_1, F_2, \ldots, F_m$; each of these is a one-place (that is, non-relational) predicate. Thus, by asserting the sentence $F_j a_i$, we assert that the object named a_i has the property expressed by F_j—for example, the property of being red.

We allow ourselves to make up complex assertions of these: if A and B are sentences, $\neg A$ will be the sentence which is false exactly when A is true, $(A \mathbin{\&} B)$ will be the sentence which is true exactly when both A and B are true, and $(A \vee B)$ is the sentence which is true exactly when at least one of A and B is true. That is all there is to our language. (We will allow ourselves some innocent liberties, such as omitting parentheses when convenient, and referring to the object named a_i as the object a_i, or even as a_i, and to the property expressed by F_j as F_j, and so on. That is, we shall ignore the mention/use distinction when this cannot lead to confusion.)

As an example, suppose we discuss only two objects, a_1 *and* a_2, each of which is a coin. And suppose we use only one predicate, F_1, which expresses the property of landing heads up. The simple sentences $F_1 a_1$, $F_1 a_2$, are called atomic sentences; and clearly each possible situation can be described (as completely as that is possible in our limited language) by asserting or denying each atomic sentence. That is, there are exactly four possibilities:

S_1: $F_1 a_1 \mathbin{\&} F_1 a_2$
S_2: $F_1 a_1 \mathbin{\&} \neg F_1 a_2$
S_3: $\neg F_1 a_1 \mathbin{\&} F_1 a_2$
S_4: $\neg F_1 a_1 \mathbin{\&} \neg F_1 a_2$

Each of these sentences S_1, S_2, S_3, S_4 is called a state-description. More generally, given that the language has n atomic sentences $A_1, \ldots, A_n$, we say that S is a *state-description* if and only if S has the form

$$B_1 \mathbin{\&} \cdots \mathbin{\&} B_n$$

where B_i is either A_i or $\neg A_i$, for $i = 1, 2, \ldots, n$. (Note: $(F_1 a_2 \mathbin{\&} F_1 a_1)$ is of course logically equivalent to S_1, but we shall reserve the name "state-description" for the case in which the subscripts are placed in the right order.)

Now each of these possibilities must have a certain probability, and the sum of all these probabilities must equal 1. For example, if the coins have not been tampered with, we would say that the probability that S_i is true, in the above example, is $\frac{1}{4}$. But if the coins are trick coins, the different state-descriptions may not be equally likely. An assignment of probabilities to the state-descriptions we shall call a probability metric (the term is Reichenbach's; see note [9]):

Chance (Belmont, Calif.: Dickenson Publishing Co., Inc., 1966), section V.3. and further references therein. We shall limit our exposition to the simple case of finitely many names and finitely many predicates, each of degree one. [Ed. note: Footnotes have been renumbered for use in this book.]

> m is a *probability metric* if m is a function which assigns to each state-description S a non-negative real number, and the sum of these numbers for all the state-descriptions equals 1.

Given a probability metric m, each sentence A has a definite probability $pm(A)$, which we can calculate. The method of calculation does not matter for our purposes, so we shall not discuss it. It is clear that conversely, if every sentence A has a probability $p(A)$, we can define a probability metric mp by: $mp(S) = p(S)$, for each state-description S. (For the method of calculation to be a good one, we must have $pmp(A) = p(A)$ of course.) So the basic problem of the logical theory of probability is: are some probability metrics better than others, and if so, is there a *single* probability metric which is the right one?

2. THE PROBLEM OF ALTERNATIVE PROBABILITY METRICS

Carnap attempted to answer this problem by arguing that a probability metric (he used the term "measure-function") must satisfy certain criteria, which correspond to some extent to the classical 'principle of indifference' (*op. cit.*, pp. 488–489). He even proposed a single metric, m^*, as candidate for the role of the best (or single correct) probability metric—and we shall discuss this metric below.

But we wish first to point out that this problem, in the form in which Carnap conceived it, does not concern us. For Carnap was basically concerned with the problem: to what extent does given evidence confirm (or make probable) a given hypothesis? This is a problem within the methodology of science. Our concern is with scientific theories concerning statistical behavior: theories which describe systems or aggregates of systems in statistical terms. Such a theory can be thought of as providing a language suited to the description of such a system or aggregate of systems, and specifying a probability metric for it. And we hold that the question of which is the correct probability metric—once the kind of system and the descriptive language are specified—is a purely empirical question.[2]

To make this clear, we shall briefly sketch two examples.

EXAMPLE 1. System s is a finite probabilistic automaton. This means that it is a system which has finitely many states, but it change state in an unpredictable manner. (For example, a human being, and especially the female of that species, seems to be like that.) Suppose that s has $m = 2^n$ different

[2] Even within the original context, some scepticism with respect to the question of the correct metric seems not to be misplaced: see the discussion of Carnap, Burks, and Lenz by H. Kyburg, *Probability and the Logic of Rational Belief*, Wesleyan University Press (Middletown, 1961), p. 51.

states, $s_1, \ldots, s_m$. We construct a language with only one name, a_1, used to designate system s, and with n independent predicates. This language has 2^n state-descriptions, and we may therefore choose the predicates in such a way that state-description S_i is true when and only when system s is in state s_i. Then the correct probability metric m is such that $m(S_i)$ is the probability that, at any given time, system s is found in state s_i. More precisely, over a sufficiently long period of time, system s must spend a fraction $m(S_i)$ of the total time interval in state s_i. (This is still not very precise, but could be made entirely precise by talking about limits.)

EXAMPLE 2. According to the kinetic theory developed in the 19th century, a gas is an assembly of molecules. If the gas is contained in a vessel, and left entirely undisturbed, we can say the following: there is here a finite number n of molecules, the position of each molecule is within a definite volume V, and the momentum of each molecule must be below a certain maximum (because the energy of the total gas is finite and will remain constant). The relevant language now has n names $a_1, \ldots, a_n$, each of which is used to refer to a distinct molecule of the gas. The predicates can express properties like that of having a position in region V_i and having a momentum within interval P_i. Let us suppose that F_i is the predicate expressing that property, and that we chose the regions V_i of equal magnitude and also the intervals P_i equal. In addition, let us so choose these properties that each molecule must have one and only one. The first principle of our theory must now be that not all state-descriptions describe possible states.[3] For example, $(F_1a_1 \ \& \ F_2a_1)$ is ruled out. The remaining state-descriptions can be simplified to have the form

$$S_1: \ F_{i_1}a_1 \ \& \ F_{i_2}a_2 \ \& \ \cdots \ \& \ F_{i_n}a_n$$

It is easily calculated that if the number of these predicates is m, the number of such state-descriptions is m^n.

In the terminology used here, each predicate is said to correspond to a 'cell', and F_ia_j is read as "molecule a_j is in cell C_i." Moreover, the above state-descriptions are said to describe the "micro-states" of the gas. Boltzmann postulated that each micro-state is equi-probable, so the correct metric here is $m^\dagger : m^\dagger \ (S_i) = 1/m^n$; i.e. $m^\dagger$ assigns equal probabilities to all admissible state-descriptions.

It is important to point out here that in example 1, the theory is not significant unless it postulates that the correct metric m is thus-and-so. For we cannot wait for the 'long run' to let experience tell us what it is; we formulate theories about what it will be, and then look to experience to

[3] This use of postulates to narrow down possibilities is described by Carnap in Appendix B to his *Meaning and Necessity*, 2nd ed. (Chicago: University of Chicago Press, 1956).

confirm or disconfirm them. In Boltzmann's case, it was later proved that his postulate was correct, on the assumption that molecules were what he (and classical physics) conceived them to be. Unfortunately it turned out that molecules, atoms, electrons and the like were not what classical physics had conceived them to be, and his postulate turned out not to carry over to the new case.[4]

Carnap considered the metric $m^\dagger$ defined by

$$m^\dagger(S) = (1/n)$$

where S is a state-description, and n the total number of state-descriptions (Carnap, *op. cit.*, pp. 562–567). He said that this metric has been favored in the history of the subject, referring to Peirce, Keynes, and Wittgenstein (he might have added Boltzmann). But for reasons concerning confirmation theory, he prefers a different metric. To define it, we need two preliminary notions. Let the state-descriptions be $S_1, \ldots, S_n$. We say that S_i and S_j are *isomorphic* if the one can be gotten from the other by a permutation of names. In our example of two coins,

$$F_1a_1 \,\&\, F_1a_2$$

is not isomorphic to any other state-description (check the list!), but

$$F_1a_1 \,\&\, \neg F_1a_2 \text{ and } \neg F_1a_1 \,\&\, F_1a_2$$

are isomorphic, because by permuting the names in $(F_1a_1 \,\&\, \neg F_1a_2)$ we get $(F_1a_2 \,\&\, \neg F_1a_1)$ which is logically equivalent to $(\neg F_1a_1 \,\&\, F_1a_2)$.

Now a *structure-description* is a disjunction $(S_i \vee S_j \vee \cdots \vee S_k)$ where each disjunct is a state-description isomorphic to each of the other disjuncts (and the subscripts are placed in the right order $i < j < \ldots < k$), and no state-description isomorphic to these has been left out. So in the case of the two coins we have three structure-descriptions:

$$T_1\text{: } (F_1a_1 \,\&\, F_1a_2) = S_1$$
$$T_2\text{: } (F_1a_1 \,\&\, \neg F_1a_2) \vee (\neg F_1a_1 \,\&\, F_1a_2) = (S_2 \vee S_3)$$
$$T_3\text{: } (\neg F_1a_1 \,\&\, \neg F_1a_2) = S_4$$

It is easily shown that if there are n objects and m 'cells' (see Example 2) there are $\binom{n+m-1}{n}$ structure-descriptions. The second metric m^* discussed by Carnap is such that each structure-description is assigned equal probability. Thus we have:

$$pm^*(T_1) = pm^*(T_2) = pm^*(T_3) = \tfrac{1}{3}$$

[4] For a relatively informal exposition, see A. d'Abro, *The Rise of the New Physics* (New York: Dover Publications, 1951), vol. II, Chap. XL; there are many textbooks dealing with this subject, e.g., D. Ter Haar, *Elements of Thermostatistics*, 2nd ed. (New York: Holt, Rinehart and Winston, 1966), Chaps. 1 and 4.

For the state-descriptions, we note that $T_2 = (S_2 \vee S_3)$, and that the method of calculation requires that $pm^* (T_2) = pm^* (S_2) + pm^* (S_3)$. In addition, Carnap had earlier laid down the requirement that any metric must assign the same value to isomorphic state-descriptions, so that we have $pm^* (S_2) = pm^* (S_3)$. Therefore we obtain:

$$m^* (S_1) = \tfrac{1}{3}$$
$$m^* (S_2) = \tfrac{1}{6}$$
$$m^* (S_3) = \tfrac{1}{6}$$
$$m^* (S_4) = \tfrac{1}{3}$$

From his experience with coins, the reader will be fairly certain that this is not correct. From this it follows that for the language which we constructed, m^* is not the correct metric. But the qualification concerning the language is essential here, as we shall see.

3. RELATIONS AMONG PROBABILITY METRICS

We have described two metrics, which in general yield different results. It might therefore seem to be a straightforward question which is correct in a given case. But this is not so, because there are significant cases in which they coincide.

Specifically, we *can* hold that m^* is the correct metric with respect to coin tossing. For when we describe two coins, we know very well that they have properties other than 'heads' and 'tails', and to describe them in terms of just one predicate may not be the right thing to do. (To put it in the terms which we used before, whether or not m^* is the correct metric depends to some extent on the structure of the language.) Suppose coin a_1 is scratched and coin a_2 is not scratched. Let F_2 express the property of being scratched. Then we have 16 different state-descriptions, the first one being

$$F_1a_1 \;\&\; F_1a_2 \;\&\; F_2a_1 \;\&\; F_2a_2$$

But some of these describe situations which we *know* cannot obtain (including the first one listed above), because a_1 is scratched and a_2 is not scratched. In fact, after we lay down the postulate

$$F_2a_1 \;\&\; \neg\, F_2a_2$$

we have only four state-descriptions left:

$$S'_1:\; F_1a_1 \;\&\; F_1a_2 \;\&\; F_2a_1 \;\&\; \neg\, F_2a_2$$
$$S'_2:\; F_1a_1 \;\&\; \neg\, F_1a_2 \;\&\; F_2a_1 \;\&\; \neg\, F_2a_2$$
$$S'_3:\; \neg\, F_1a_1 \;\&\; F_1a_2 \;\&\; F_2a_1 \;\&\; \neg\, F_2a_2$$
$$S'_4:\; \neg\, F_1a_1 \;\&\; \neg\, F_1a_2 \;\&\; F_2a_1 \;\&\; \neg\, F_2a_2$$

We note that no two of these are isomorphic (simply because each contains

F_2a_1 but not F_2a_2); therefore each of these is by itself a structure-description. In other words, $m^\dagger$ and m^* give the same result for this case ($m^\dagger$ $(S'_i) = m^*$ $(S'_i) = \frac{1}{4}$), which agrees exactly with our experience concerning coins.

Even if there are no scratches on either coin, we know very well that we can find some predicate here which can play the role of F_2, and still express a physical property. (At the very least we can use a movie camera to register the exact trajectory of each coin.) We shall call a predicate which plays this role an *individuating predicate*; if we discuss more than two objects, we shall need more than one predicate for this job, and we shall speak of an *individuating category* of predicates. To make this precise: we call the family of predicates $G_1, \ldots, G_m$ an individuating category for $a_1, \ldots, a_m$ if, in all possible situations, G_ia_j is true if and only if $i = j$. (We allow the predicates G_i to be complex; for example, G_2a_i may be $\neg G_1a_i$.)

Now one may ask whether such an individuating category may not always be found. For example, why don't we let G_i just express the property of being identical with the object a_i? Then we automatically have an individuating category, and $m^\dagger$ and m^* will always yield the same result. And indeed, G_i must necessarily have as extension (in our domain) the set of objects identical with a_i, so that the suggested individuating category seems to be essentially the only one available, and always to be available.

This trivializes the problem, unless we are able to lay down some restrictions on the properties to be expressed by the primitive predicates of the language. (The identity symbol is usually counted among the logical signs rather than among the predicates, and we must try to find a justification for some such procedure.) I think that this can be done, if not in general, at least in specific cases in which we have an explicit scientific theory dealing with the subject. For the probabilities which we are concerned with in science relate to frequencies which are empirically observable: for example, the frequency with which a given measurable physical magnitude has a certain value in a certain kind of experiment. So here the objects are described in terms of (measurable) physical properties (the 'observables' of physical theory) and we may restrict the predicates to those which express such properties.

Now the property of being identical with a_i may have the same extension as some such physical property; for example, that of having scratches, or that of having been tossed with the experimenter's right hand. But there is no *a priori* reason why this should be the case always.

We have seen then that if there is an individuating category of properties, which are expressed by the predicates of the language, then the metrics $m^\dagger$ and m^* yield the same probabilities. Now these properties may be expressed by complex predicates; for example, (F_1x & F_2x) may express the property individuating object a_1. The extreme case is that in which all the predicates of the language together generate an individuating category for the domain. In that case, the postulate limiting the state-descriptions would simply be:

> If for every primitive predicate F, Fx is true if and only if Fy is true, then x and y are one and the same individual.

Recalling the 'cell' terminology of Example 2, this may also be phrased as:

> No two individuals can occupy the same cell at once.

We shall call this postulate the *exclusion principle* (not capitalized: the expression "Exclusion Principle" is to be reserved for the (similar) well-known principle of atomic physics).

When the exclusion principle is satisfied,[5] the state-descriptions ruled out as impossible are those in which two individuals are ascribed all the same properties. It now follows that each structure-description is a disjunction of equally many state-descriptions. (For in each case, any permutation whatsoever leads to a logically distinct isomorphic state-description.) It is easily shown that if there are n individuals and m cells, the number of structure-descriptions is $\binom{m}{n}$.

Because the number of state-descriptions corresponding to a single structure-description is always the same here, an assignment of equal probabilities to the structure-descriptions will amount to an assignment of equal probabilities to state-descriptions, and vice versa. Consider, for example, our original case of two individuals and one predicate; the limitation leaves here two state-descriptions, but only one structure-description:

$$S''_1\colon\ F_1a_1 \,\&\, \neg F_1a_2 \qquad T\colon\ (S''_1 \vee S''_2)$$
$$S''_2\colon\ \neg F_1a_1 \,\&\, F_1a_2$$

However, the metrics yield the same results:

$$m^\dagger\ (S''_1) = m^\dagger\ (S''_2) = \tfrac{1}{2}; \qquad pm^*\ (T) = 1;$$
hence $\qquad\qquad\qquad\qquad$ hence
$$pm^\dagger\ (T) = 1 \qquad m^*\ (S''_1) = m^*\ (S''_2) = \tfrac{1}{2}$$

So m^* will end up by assigning equal values to each micro-state after all,

[5] In fact it is not the case for elementary particles in general. By the Exclusion Principle, it is the case that no two electrons are in the same state of motion at the same time, but photons, for example, do not satisfy this principle. And with electrons and photons alike, it is impossible to trace their individual trajectory from emission to absorption. This subject is always discussed in connection with quantum statistics; see, for example, Ter Haar, *loc. cit.*; J. Jauch, *Foundations of Quantum Mechanics* (Reading, Mass.: Addison-Wesley, 1968), section 15–3; H. Margenau, *The Nature of Physical Reality* (New York: McGraw-Hill, 1950), Chap. 20, and "The Exclusion Principle and its Philosophical Importance," *Philosophy of Science* 11 (1944), pp. 187–208.

and m^* and $m^\dagger$ coincide again when the exclusion principle holds.[6] This is not surprising since we found this to be the case whenever the predicates yield some individuating category.

We have now considered individuating categories of physical properties, and seen that m^* yields a drastic departure from $m^\dagger$ only when no such category is generated by the predicates of the language. We might still hold in that case that the language describes all the (measurable) physical properties of the objects in its domain; then either there is no individuating category of predicates, or its members do not express physical properties. Such a possibility is certain to raise philosophical hackles, and I think it is time to discuss the philosophical problems in this area in some detail.

4. INDIVIDUATION AND GENIDENTITY

We naturally group physical objects into certain kinds: some are men, some are trees, some are rocks. We call two individuals men because they have something in common: certain properties in virtue of which they are men.[7] This something in common Aristotle called a *form*. Now, if two individuals belong to the same kind (have the same form), what makes them different? This is called the problem of individuation, for we may rephrase the question as follows: what individuates different individuals which belong to the same kind?

Aristotle had an answer to this question: an individual has both matter and form, and the matter individuates: "when we have a whole, such and such a form in this flesh and these bones, this is Callias or Socrates; and they are different in virtue of their matter...." (*Metaphysics* VII, 1034a, 5–8). Later philosophers discussed this problem and Aristotle's solution at length. Aquinas pointed out that Callias and Socrates *both* have flesh and bones: the mere fact that they have matter does not individuate them. Rather, what individuates is "designated matter," matter which is limited in

[6] The reader familiar with statistical mechanics will by now have seen the close connection between m^* and quantum statistics, with the Bose statistics in effect when any number of individuals can occupy the same cell, and Fermi statistics when the Exclusion Principle holds. The intermediate cases ("para-statistics") appear so far to have no physical significance.

[7] Nominalistically inclined philosophers argue that the fact that two individuals are of the same kind does not imply that there *exists* something which they have in common. This is quite true, but we may allow such terms as "the property of being a featherless biped" to function as singular terms in our language without thereby committing ourselves to the claim that they refer to something real or existent. And we may then say that X and Y have some property in common to signify simply that for some such term T, the sentence "X has T and Y has T" is true. There are still further dodges that would help to clear us of unnecessary ontological commitments here. So we shall talk quite naively without thereby admitting to the charge of naive realism.

its dimensions in some specific way. Thus the emphasis must be on the word "this": it is *this* flesh and *these* bones that individuate Socrates (*De Ente et Essentia*, Chap. 2, paragraph 23). Aquinas also drew the consequence that immaterial intelligences (angels, to use the vernacular) which have no matter, are not individuated within the same kind. From this it follows that "there are among them as many species as there are individuals" (*op. cit.*, Chap. 5, paragraph 75).

Now this may not seem very satisfactory. Why can't we just say something like: Socrates has a pug nose and Callias does not, and that is what individuates them. Well, the question is whether such accidental properties must always individuate; after all, Callias might have had a pug nose too. In other words, no matter how many properties we use to identify the species, we must either say that there is a further principle which individuates, or that the species cannot have more than one member (as in the case of angels). Aquinas put this point very strongly: "That by which Socrates is man can be communicated to many; but that by which he is this man can be communicated only to one. If, therefore, Socrates were a man by that by which he is this man just as there cannot be many Socrates, so there could not be many men" (*Summa Theologiae*, I, q. 11, a. 3c).

The question seems academic in the case of physical objects, because after all, we know of an individuating accident: no two things can occupy exactly the same place at the same time. But does this sort of answer settle the matter? After all, we can imagine interpenetration of objects, as in the familiar story of the "walker through walls" who used his ability to visit his mistress's boudoir without her husband's knowledge. In short, unless we can explain, or explain how one might be able to explain, why these accidental properties individuate, the problem remains.

But even if we do agree that there may be a genuine problem here, Aquinas' solution may not satisfy us. For what individuates *this* piece of matter, this flesh, from *that*? The reference here is to 'designated matter', that is, apparently, to matter of determinate spatial dimensions. Is it then the place occupied that individuates the matter? And does this not bring us back to individuation by accidental properties? Aquinas' answer would probably have been that this misconceives the problem: only for individuals, which have also a form, does the problem of individuation arise, not for matter. But in that case, what does the distinction between matter and designated matter (between flesh in general and *this* flesh) come to?

The Gordian knot was cut by Leibniz. In an early essay, *De Principio Individui* (1663), he argues, entirely within the scholastic frame, for the position that neither matter nor form alone can individuate. But his real contribution came at a mature stage, when he argued for the Principle of Identity of Indiscernibles: "It is not true that two substances may be exactly alike and differ only numerically, *solo numero*, and ... what St. Thomas says on this point regarding angels and intelligence (*quod ibi omne indi-*

viduum sit species infima) is true of all substances" (*Discourse on Metaphysics* (1686), Chap. IX).

But what could Leibniz say to the argument that accidental properties, such as a pug nose, could always be shared by two members of a given kind, so that they could be alike in all respects? Well, the argument is simply invalid. It is true, for any property, that Socrates and Callias might share that property and yet be distinct individuals. But it just does not follow from this that Socrates and Callias might share all properties and yet be distinct, as any text treating modal logic will show.

We must however be very careful not to infer from this that two entities cannot be exactly alike at a given time with respect to properties that relate to just that time. This Aquinas saw too: within his theory it does not follow that Socrates' soul ceases to be distinct from any other soul once it leaves the body. For this soul is individuated by its *history* of having been received in this flesh and these bones, rather than those of Callias.

In other words, one state or situation or event may be distinguished from another because it has different historical connections (to the past or to the future). This became important when philosophers began to talk about nature more in terms of states and events than in terms of objects. An object has, at any given time, a specific state; and its states are related to each other by all being states of one and the same object. Russell, Whitehead, and Reichenbach used the term "event" not just for a happening (like a collision) but also for a state (like a drought, or just a case of being dry). And when two events are related by belonging to the history of one and the same object, they are said to be *genidentical.*

Once we introduce the genidentity relation we clearly do not need to refer to objects at all any more; whatever we wish to say about an object can now be said about an exhaustive class of mutually genidentical events. But this raised the question: can we not do without even this relation? Or better, since we do often talk about objects, can we not define the genidentity relation in terms of some other relations? Russell certainly seems to have thought that the answer is affirmative, and the relations to be used for the definition are those of resemblance and contiguity, plus those described by certain causal laws.[8]

In the case of microphysical objects, such a definition does not seem feasible, and Reichenbach argued that the question of the dispensability of the genidentity relation is here closely related to statistical behavior.[9] Our dis-

[8] *Our Knowledge of the External World* (London: Allen and Unwin, 1922), Chap. IV; *Human Knowledge* (New York: Simon and Schuster, 1948), Part Four, Chaps. VIII and IX.

[9] *The Direction of Time* (Berkeley: University of California Press, 1956), section 26. Reichenbach is discussing Boltzmann, Bose, and Fermi statistics, and does not use the framework of the logical theory of probability. What follows

cussion of probability metrics above will make it possible for us to appraise his argument.

5. REICHENBACH ON GENIDENTITY

Reichenbach discusses a case which appears to be adequately characterizable in such terms as we have now at our disposal. He outlines two possible positions concerning this case, and argues that one is ruled out by a basic methodological principle. Our strategy will be to argue first that in one sense his argument does not succeed, and in another sense that supposedly basic principle is unacceptable. Then we shall furthermore argue that the position for which he argued by elimination is itself vulnerable to logical and methodological objections.

Let us suppose that we have an assembly of n individuals, and the language describes m 'cells', that is, m possible states for any given individual. Suppose that any number of these individuals may occupy any given cell at once. And finally, suppose that m^* is the correct probability metric, the metric which accords with experience. We have here clearly a case in which m^* and $m^\dagger$ do not coincide, a case in which the predicates of the language do not generate an individuating category. To make the case interesting rather than odd, let us suppose that the language completely describes the measurable physical properties of these individuals. Reichenbach argues that in this case, the individuals are "not genidentical." By this he means that an individual here corresponds to an arbitrarily or conventionally selected series of events, and that there really is no such relation as genidentity.

As we said, Reichenbach considers two interpretations, the one ascribing genidentity to the objects and the other denying them genidentity. He then rules out the former on the basis of the frankly methodological principle that a description is preferable if it supplies a normal system, that is, a system free from causal anomalies.[10]

The situation is complicated because Reichenbach does not draw all the distinctions we have made. But we can at least take up his argument that the ascription of genidentity to the particles in this case leads to causal anomalies. He shows essentially that under the measure m^*, when there is no individuating category, events involving different individuals are not statistically independent. Consider our example of the use of m^* for two coins, a_1 and a_2, using only the predicate F_1, at the end of section 2. We see there that F_1a_1 is more probable given that F_1a_2 than it is by itself: $m^*(F_1a_1) = \frac{1}{2}$ but the conditional probability of F_1a_1 given $F_1a_2 = \frac{2}{3}$. From this Reichenbach immediately infers causal relationships: "Assume that we could

---

is therefore a reconstruction of his central argument within our present framework.

[10] *The Direction of Time*, p. 235.

assign material genidentity to each particle of a Bose ensemble; then we would find that the particles are mutually dependent in their motions. . . . These causal relationships would represent action at a distance . . ." (*op. cit.*, p. 234). But Reichenbach's move is entirely unacceptable to my mind. First, if by "causal relationship" he means only "statistical dependence," then the principle to accept wherever possible only a normal system is the principle that measure $m^\dagger$ is to be accepted wherever possible. There seems to be no reason to adopt this principle, and it may be compared with Poincaré's principle that Euclidean geometry will be held whatever the cost, wherever possible. Second, if Reichenbach has in mind a more full-blooded sense of "causal relationship," then he is saying that any departure from measure $m^\dagger$ must be explained in terms of perturbational forces (i.e., that real forces cause particles to follow Bose or Fermi, rather than Boltzmann statistics). This can only be compared with the Aristotelian's question of what keeps a body free of impressed forces in its state of motion: there is behind it a presupposition concerning what is normal and what needs to be explained. And this presupposition we deny.

But if the argument from causal anomalies is spurious, can we not demonstrate that the objects are not genidentical by our preceding results concerning probability metrics? The argument would of course be that genidentity provides us with an individuating category. For instead of "a_1 is scratched and a_2 is not scratched," we would have, say, "e_1 is genidentical with e and e_2 is not genidentical with e" where e is an event. But if this individuating category were available, then m^* would yield results agreeing with those of $m^\dagger$, which is not the case here.

Now this argument is not valid; all that can be established this way is that the predicates which the language actually has do not generate an individuating category. Since we have assumed that these predicates describe all the physical properties of the given individuals, it follows furthermore that genidentity (to a particular event sometime during the assembly's history) can not be an observable.[11] But the argument cannot rule out that the individuals are individuated by non-physical properties.

So the arguments against the ascription of genidentity to these objects are not valid. But to show an argument invalid is rarely very illuminating. Is it philosophically tenable to hold that these individuals "are genidentical," although genidentity is not an observable? Is it philosophically tenable at all to admit properties of physical individuals which are not measurable physical properties? The only way to answer these questions is to develop the alternative to the account accepted by Reichenbach, and to compare their relative advantages and disadvantages.

[11] And this we already knew for the cases which Reichenbach is discussing (elementary particles); no argument from statistical behavior is necessary to establish it (see note 6).

6. GENIDENTITY AND STATISTICAL BEHAVIOR

In connection with Reichenbach's argument, we discussed a case in which the individuals are clearly not individuated by physical properties. So these individuals might be called "identical" in the sense in which physicists tend to say that photons are "identical particles" because they are not distinguishable by means of measurement. But they are not *numerically identical* in the sense of being one and the same individual. (For example, measurement can show that at a given time, there is more than one photon in a certain region: *number is an observable.*) This can be stated more clearly in the event terminology: at a given time there are several events which have all physical properties in common, but which are numerically distinct. We added the hypothesis that these several events are differentiated by being genidentical with different, distinguishable events at other times, although this is not a physical property.

Now Reichenbach might well have said against this that non-physical properties are nonsensical in this context: if two events have all physical properties in common, then they have all properties in common. But note that then he would be denying the principle of identity of indiscernibles! Margenau saw this consequence and accepted it (*The Nature of Physical Reality*, pp. 440–441). I think that Reichenbach might have been less inclined to do so; but then his only path of retreat seems to be to deny that the predicates describe all the physical properties of the individuals after all. This retreat would have involved Reichenbach in an empirical claim, and experiment might not confirm that claim.

There are therefore three alternatives: Reichenbach's, Margenau's, and ours. Of these, Reichenbach's seems now the most precarious. But our alternative might be said to be non-scientific, since we are relying on the presence of an admitted non-observable, namely genidentity. Against this I will now argue that the postulate that the individuals are genidentical (or more generally, that they are individuated by non-physical properties) does have empirical consequences within our present position.

For suppose that we have *n* individuals with *m* distinct states or 'cells' described by the language, and suppose that the language describes all the physical properties these individuals can have. Then if these individuals are not different with respect to any non-physical properties, that is, properties not described in the language, it follows by the identity of indiscernibles that the exclusion principle holds. That is, no two of these individuals can occupy the same cell at once. (More precisely, in the terminology of events, the objects correspond to arbitrarily or conventionally selected series of events. Each event is simply the "being-occupied" of a given cell (at a given time). To say that there are *n* objects means only to say that at any given time, *n* such events are occurring. And so, at any given time, *n* cells are 'occupied'.) It will be recalled from section 3 that in this case m^* and $m^\dagger$ yield the same results, so the statistical behavior is objectively different from the previous case.[12] (This does not make genidentity an observable: the observable here is *number*.)

It may now be asked whether, if we encountered this *second* case in nature, we could then conclude that these individuals are "not genidentical." For a while I thought so, but I was dissuaded of this when I came across an ingenious hypothesis concerning time travel.[13] The possibility raised is that all *n* events occurring at a given time are genidentical with each other. This clearly does not differentiate them, so that the exclusion principle still holds. But equally clearly, these different events occurring at the same time can be states of one and the same individual only if that individual is traveling backward and forward in time; otherwise, a single individual has only one state at any given time. Because of the many philosophical difficulties concerning the notion of time travel, I must admit that I am not yet certain whether to take this possibility seriously.

To come back to our main topic however, I think that we have shown that Reichenbach's arguments against the ascription of genidentity in the case considered are not cogent; that the interpretation he prefers is objectionable in that it conflicts with the identity of indiscernibles; and that the interpretation he rejects is not empirically vacuous on our reconstruction.

POSTSCRIPT

Professor Wesley Salmon, Indiana University, commented on this paper at the Annual Meeting of the American Philosophical Association (Eastern Division) in December 1969 and in correspondence following that discussion. I should like to report here informally on his stimulating comments and react to some of the issues he raised.

According to Professor Salmon, the path from Carnap to Reichenbach via Aristotle, Aquinas, and Leibniz was probably not a geodesic, but it was surely the scenic route, giving us, as it did, a glimpse of the angels, so appropriate to the season. He suggested that the exposition could be given in a form at once rigorous and more general by using Carnap's notion of *Q*-

[12] In other words, Fermi statistics rather than Bose statistics is in effect here.

[13] This is Feynman's hypothesis that a certain elementary particle may travel both "forward" and "backward" in time; the states during its "forward" journey are electron-states. See R. P. Feynman, *Quantum Electrodynamics* (New York: Benjamin, 1962), pp. 68–70. It is not clear whether Feynman thought of this hypothesis as possibly explaining the Exclusion Principle in the way that it explains the exclusion principle in the abstract case which we discuss.

predicate.[14] These are the strongest predicates in the language; each individual must satisfy exactly one *Q-predicate*, and for each predicate F we have $Qx \supset (Fx \equiv (y)\ (Qy \supset Fy))$ if Q is a Q-predicate. So if F_1 and F_2 are the only predicates in the language, there are exactly four forms that Qx may have: F_1x & F_2x, F_1x & $\neg F_2x$, $\neg F_1x$ & F_2x, $\neg F_1x$ & $\neg F_2x$. The Q-predicates are the linguistic counterpart of cells, of course, and the exclusion principle can be rephrased as: no Q-predicate applies truly to more than one individual. A state-description is then a conjunction of the form Q_1b_1 & $\cdots$ & Q_nb_n, and state-descriptions are isomorphic exactly if the one results from the other by a permutation of individual constants. Given the exclusion principle, a structure description for n individuals is determined by a choice of n distinct Q-predicates (the "instantiated" ones), and each structure description encompasses $n!$ distinct isomorphic state-descriptions. Hence it follows at once that, given the exclusion principle, $m^* = \mathrm{m}^\dagger$.

Professor Salmon then pointed out that Carnap's requirement of descriptive completeness may make the choice between m^* and $m^\dagger$ illusory. This requirement is the stipulation that "a system of [predicates] be taken which is sufficiently comprehensive for expressing all the qualitative attributes in the given universe."[15] If these attributes individuate the individuals—as seems always to be the case for the objects of everyday experience—the two measure functions coincide. (Carnap has recognized, at various points, the possibility that besides qualitative attributes also "positional" attributes may be needed for individuation.[16] Whether this possibility is ever realized, and if so, to what extent, are other questions.)

It is with respect to Reichenbach that Professor Salmon and I seem to be in disagreement. Reichenbach holds to the methodological principle that statistical dependence requires causal explanation. He holds in addition to the principle that causal explanation should be via action by contact propagated at a finite velocity. The anomalies that Reichenbach sees in quantum theory (for example, in connection with the double-slit experiment and the Einstein-Podolsky-Rosen paradox) arise from the fact that explanations of the form demanded by these methodological principles cannot be carried through (for the "inter-phenomena").

In the paper I maintained that Reichenbach's demand for causal explanations of any case of statistical dependence betrays a prejudice in favor of $m^\dagger$. Salmon's first argument is that the evidence does not entail the conclusion of a prejudice. For it is possible for individual behavior to be statistically inde-

14 Carnap, *The Logical Foundations of Probability*, pp. 124–125.

15 *Ibid.*, pp. 74–75.

16 Carnap considered this problem already in his early work; see his *The Logical Structure of the World* and *Pseudo-Problems in Philosophy*, tr. R. A. George (Berkeley: University of California Press, 1967), Secs. 13 and 15.

pendent for measures other than $m^{\dagger}$. For example, let there be two individual constants a and b and one predicate F, and define m by:

$$m(Fa \,\&\, Fb) = \frac{4}{9}$$

$$m(\neg Fa \,\&\, Fb) = m(Fa \,\&\, \neg Fb) = \frac{2}{9}$$

$$m(\neg Fa \,\&\, \neg Fb) = \frac{1}{9}$$

Then the corresponding absolute and conditional probabilities are $pm(Fa) = \frac{2}{3}$, $pm(\neg Fa) = \frac{1}{3}$, $pm(Fa/Fb) = \frac{2}{3}$, $pm(Fa/\neg Fb) = \frac{2}{3}$, and similarly for Fb. In this Salmon is clearly right, and I certainly have no alternative grounds (historical, psychoanalytic, or astrological) for imputing this prejudice to Reichenbach.

Salmon argues secondly that Reichenbach's demand is a reasonable one. If, for example, significant statistical dependencies appear in extrasensory perception experiments, we would not accept as a satisfactory explanation the thesis that this is normal because the effects follow exactly some suitably chosen, but rather unusual, probability measure function. Salmon notes that Reichenbach did not rule out that nature may make methodological principles untenable. Given a situation in which we cannot hold consistently to all of the principles that continuants are genidentical, statistical dependencies result from causal action by contact, and indiscernibles are identical, none of these principles are sacred.

I wish to take issue with this at two points; not with the interpretation of Reichenbach, which is likely more fair than my own, but with the position explained. First, I have great difficulty in accepting that the identity of indiscernibles is merely a methodological principle. I can conceive of action at a distance; I think that we may tenably be said to be able to conceive of continuants that are not genidentical and time travel. But I cannot conceive of distinct but indiscernible entities. The supposed examples of such purely numerical distinctness seem to me all to be of entities discernible only by counterfactual properties.[17]

Second, I differ with Reichenbach on theories and explanation. To have a theory is to be in a position to answer why-questions (requests for explanation: specifically, requests for causal explanation). But the answer may either be a *direct* answer (an assignment of causes) or a *corrective* answer (a refusal to assign causes on the ground that the phenomenon is normal).[18]

[17] Cf. my *An Introduction to the Philosophy of Time and Space* (New York: Random House, 1969), pp. 63–65.

[18] For similar views, see A. Grünbaum, *Philosophical Problems of Space and Time* (New York: Alfred A. Knopf, 1963), pp. 406–407; B. Ellis, "The Origin

This may explain my reference to Aristotelian physics in this connection: if the Newtonian is asked to explain accelerated motions, he assigns causes (perturbational forces), but if he is asked to explain the continuance of uniform rectilinear motion, he gives the corrective answer that this is normal.

Turning now to ESP and quantum phenomena, I would characterize the situation as follows. We have a theory about the transmission and collection of information by persons, and statistical dependencies in ESP experiments contravene what this theory declares to be normal. But this theory we have much reason to hold; indeed, it ties in with a large class of well-developed scientific theories. Hence what we look for is an assignment of causes that will explain these phenomena within this theoretical context. The hypothesis that we have hold of the wrong probability measure for these phenomena would deny much firmly held background theory and is therefore not reasonable at this point.

But the case of the statistical dependencies that lead Reichenbach to speak of anomalies in quantum theory is different. The interference of probabilities in the double-slit experiment and the statistical correlations between separated systems in the Einstein-Podolsky-Rosen paradox are normal according to quantum theory: they are entailed to occur by the theory without the presence of perturbational forces. The same is true for the statistical dependencies noted by Reichenbach in Bose statistics. Hence the theory provides a corrective answer to the request for an assignment of causes to explain these various statistical dependencies. The terms "paradox" and "anomaly" are apt only in that they indicate profound departures from the structure of classical physics.

My opinion is therefore that the demand for causal explanations in the case of statistical dependence is not a reasonable principle. For it seems to me that it is one function of a physical theory (about a given domain of phenomena) to declare exactly where causal explanations are to be had and where not. A given theory might be superseded because it declares that causal explanations may be had where in fact none are forthcoming. Or a theory may be superseded by a theory providing causal explanations for phenomena declared to be normal by the former. But theories are not superseded for solely that reason: if a new theory has nothing to recommend it except that it postulates causal mechanisms for what a previous theory declared to be normal, then that new theory is not taken seriously.

Professor Salmon's comments during the several phases of this paper that he witnessed were always encouraging and more than kind; and I want to thank him for initiating much stimulating discussion.

---

and Nature of Newton's Laws of Motion," pp. 29–68 in R. Colodny (ed.), *Beyond the Edge of Certainty* (Englewood Cliffs, N.J.: Prentice-Hall, 1965), especially pp. 43–45. The question/direct answer/corrective answer/presupposition terminology is adapted from N. D. Belnap Jr., "Questions: Their Presuppositions, and How They Can Fail to Arise," pp. 23–37 in K. Lambert (ed.), *The Logical Way of Doing Things* (New Haven: Yale University Press, 1969).

7

INQUIRIES INTO THE FOUNDATIONS OF SCIENCE

Wesley C. Salmon

In 1950, L. Ron Hubbard published his book *Dianetics*,[1] which purported to provide a comprehensive explanation of human behavior, and which recommended a therapy for the treatment of all psychological ills. According to Hubbard's theory, psychological difficulties stem from "engrams," or brain traces, that are the results of experiences the individual has undergone while unconscious due to sleep, anesthesia, a blow to the head, or any other cause. Of particular importance are those that occur before birth. Hubbard gives strikingly vivid accounts of life in the womb, and it is far from idyllic. There is jostling, sloshing, noise, and a variety of rude shocks. Any unpleasant behavior of the father can have serious lasting effects upon the child. On a Saturday night, for example, the father comes home drunk and in an ugly mood; he beats the mother and with each blow he shouts, "Take that, take that!" The child grows up and becomes a kleptomaniac.

It is perhaps worth remarking that the author of this work had no training whatsoever in psychology or psychiatry. The basic ideas were first published in an article in *Astounding Science Fiction*. In spite of its origins, this book was widely read, the theory was taken seriously by many people, and the therapy it recommended was practiced extensively. A psychologist friend of mine remarked at the time, "I can't condemn this theory before it is carefully tested, but afterwards I will."

In the same year—it seems to have been a vintage year for things of this

[1] L. Ron Hubbard, *Dianetics: The Modern Science of Mental Healing* (Hermitage House, 1950).

---

sort—Immanuel Velikovski published *Worlds in Collision*,[2] a book that attempted to account for a number of the miracles alleged in the *Old Testament*, such as the flood and the sun's standing still. This latter miracle, it was explained, resulted from a sudden stop in the earth's rotation about its axis which was brought about, along with the various other cataclysms, by the very close approach to the earth of a giant comet which later became the planet Venus. One of the chief difficulties encountered by Velikovski's explanation is that, on currently accepted scientific theory, the rotation of the earth simply would not stop as a result of the postulated close approach of another large body. In order to make good his explanation, Velikovski must introduce a whole body of physical theory which is quite incompatible with that which is generally accepted today, and for which he can summon no independent evidence. The probability that Velikovski's explanation is correct is, therefore, no greater than the probability that virtually every currently accepted physical theory is false.

Before the publication of the book, parts of Velikovski's theory were published serially in *Harper's* Magazine. When the astounding new theory did not elicit serious consideration from the scientific community, the editors of *Harper's* expressed outrage at the lack of scientific objectivity exhibited by the scientists.[3] They complained, in effect, of a scientific establishment with its scientific orthodoxy, which manifests such overwhelming prejudice against heterodox opinions that anyone like Velikovski, with radically new scientific ideas, cannot even get a serious hearing. They were not complaining that the scientific community rejected Velikovski's views, but rather that they dismissed them without any serious attempts at testing.

The foregoing are but two examples of scientific prejudgment of a theory; many other fascinating cases can be found in Martin Gardner's *Fads and Fallacies in the Name of Science*.[4] Yet, there is a disquieting aspect of this situation. We have been told on countless occasions that the methods of science depend upon the objective observational and experimental testing of hypotheses; science does not, to be sure, prove or disprove its results absolutely conclusively, but it does demand objective evidence to confirm or disconfirm them. This is the scientific ideal. Yet scientists in practice do certainly make judgments of plausibility or implausibility about newly suggested theories, and in cases like those of Hubbard and Velikovski, they judge the new hypotheses too implausible to deserve further serious consideration. Can it be that the editors of *Harper's* had a point, and that there is a large discrepancy between the ideal of scientific objectivity and the actual

[2] Immanuel Velikovski, *Worlds in Collision* (Doubleday and Co., 1950).

[3] *Harper's Magazine*, 202 (June 1951), 9–11.

[4] Martin Gardner, *Fads and Fallacies in the Name of Science* (Dover Publications, Inc., 1957). Gardner's excellent discussions of Hubbard and Velikovski provide many additional details.

practice of prejudgment on the basis of plausibility considerations alone? One could maintain, of course, that this is merely an example of the necessary compromise we make between the abstract ideal and the practical exigencies. Given unlimited time, talent, money, and material, perhaps we should test every hypothesis that comes along; in fact, we have none of these commodities in unlimited supply, so we have to make practical decisions concerning the use of our scientific resources. We have to decide which hypotheses are promising, and which are not. We have to decide which experiments to run, and what equipment to buy. These are all practical decisions that have to be made, and in making them, the scientist (or administrator) is deciding which hypotheses will be subjected to serious testing and which will be ignored. If *every* hypothesis that comes along had to be tested, I shudder to think how Air Force scientists would be occupied with anti-gravity devices and refutations of Einstein.

Granted that we do, and perhaps must, make use of these plausibility considerations, the natural question concerns their status. Three general sorts of answers suggest themselves at the outset. In the first place, they might be no more than expressions of the attitudes and prejudices of individual scientists or groups of scientists. The editors of *Harper's* might be right in claiming that they are mere expressions of prejudice against ideas that are too novel—the tool used by the scientific establishment to enforce its own orthodoxy. If that suggestion is too conspiratorial in tone, perhaps they arise simply from the personal attitudes of individual scientists. In the second place, they might be thought to have a purely practical function. Perhaps they constitute a necessary but undesirable compromise with the ideal of scientific objectivity for the sake of getting on with the practical work of science. Or maybe these plausibility considerations have a heuristic value in helping scientists discover new and promising lines of research, but their function is solely in relation to the discovery of hypotheses, not to their justification. In the third place, it might be held that somehow plausibility arguments constitute a proper and indispensable part of the very logic of the justification of scientific hypotheses. This is the view I shall attempt to elaborate and defend. I shall argue that plausibility arguments are objective in character, and that they must be taken into account in the evaluation of scientific hypotheses on the basis of evidence.[5]

The issue being raised is a logical one. We are asking what ingredients enter into the evaluation of scientific hypotheses in the light of evidence. In order to answer such questions, it is necessary to look at the logical schema that represents the logical relation between evidence and hypotheses in scientific inference. Many scientific textbooks, especially the introductory ones,

[5] In my book, *The Foundations of Scientific Inference* (University of Pittsburgh Press, 1967), I have discussed these issues in greater detail and have argued this case at greater length.

attempt to give a brief characterization of the process of confirming and disconfirming hypotheses. The usual account is what is generally known as the *hypothetico-deductive method*. As it is frequently described, the method consists in deducing consequences from the hypothesis in question, and checking by observation to determine whether these consequences actually occur. If they do, that counts as confirming evidence for the hypothesis; if they do not, the hypothesis is disconfirmed.

One immediate difficulty with the foregoing characterization of the hypothetico-deductive method is that from a general hypothesis it is impossible to deduce any observational consequences. Consider, for example, Kepler's first two laws of planetary motion: the first states that the orbits of the planets are elliptical, and the second describes the way the speed of the planet varies as it moves through the ellipse. With this general knowledge of the motion of Mars, for instance, it is impossible to deduce its location at midnight, and so to check by observation to see whether it fulfills the prediction or not. But with the addition of some further observational knowledge, it is possible to make such deductions—for instance, if we know its position and velocity at midnight last night. This additional observational evidence is often referred to as the "initial conditions;" from the hypothesis together with statements about initial conditions it is possible to deduce a concrete prediction that can be checked by observation. With this addition, the hypothetico-deductive method can be represented by the following simple schema:

H-D schema: H (hypothesis)
I (initial conditions)
———————————
O (observational prediction)

Although it is always possible for errors of observation or measurement to occur, and consequently for us to be mistaken about the initial conditions, I shall assume for purposes of the present discussion that we have correctly ascertained the initial conditions, so that the hypothesis is the only premise whose truth is in question. This is one useful simplifying assumption.

Another very important simplifying assumption is being made. In many cases the observational prediction does not follow from the hypothesis and initial conditions alone, but so-called "auxiliary hypotheses" are also required. For instance, if an astronomical observation is involved, optical theories concerning the behavior of telescopes may be implicitly invoked. In principle, a false prediction can be the occasion to call these auxiliary hypotheses into question, so that the most that can be concluded is that *either* an auxiliary hypothesis *or* the hypothesis up for testing is false, but we cannot say which. For purposes of this discussion, however, I shall assume that the truth of the auxiliary hypotheses is not in question, so that the hypothesis we are trying to test is still the only premise of the argument whose truth is open to question. Such simplifying assumptions are admittedly unrealistic, but things are difficult enough with them, and relinquishing them does not help with the problems we are discussing.

Under the foregoing simplifying assumptions a false prediction provides a decisive result: if the prediction is false the hypothesis is falsified, for a valid deduction with a false conclusion *must* have at least one false premise, and the hypothesis being tested is the only premise about which we are admitting any question. However, if the prediction turns out to be true, we certainly cannot conclude that the hypothesis is true, for to infer the truth of the premises from the truth of the conclusion is an elementary logical fallacy. And this fallacy is not mitigated in the least by rejecting the simplifying assumptions and admitting that other premises might be false. The fallacy, called *affirming the consequent,* is illustrated by the following example: If the patient has chickenpox, he will run a fever; the patient is running a fever; therefore, he has chickenpox. The difficulty is very fundamental and very general. Even though a hypothesis gives rise to a true prediction, there are always other different hypotheses that would provide the same prediction. This is the *problem of the alternative hypotheses.* It is especially apparent in any case in which one wishes to explain data that can be represented by points on a graph in terms of a mathematical function that can be represented by a curve. There are always many different curves that fit the data equally well; in fact, for any finite number of data, there are infinitely many such curves. Additional observations will serve to disqualify some of these (in fact, infinitely many), but infinitely many alternatives will still remain.

Since we obviously cannot claim that the observation of a true consequence establishes the truth of our hypothesis, the usual claim is that such observations tend to support or confirm the hypothesis, or to lend it probability. Thus, it is often said, the inference from the hypothesis and initial conditions to the prediction is deductive, but the inference in the opposite direction, from the truth of the prediction to the hypothesis is inductive. Inductive inferences do not pretend to establish their results with certainty; instead, they confirm them or make them probable. The whole trouble with looking at the matter this way is that it appears to constitute an automatic transformation of deductive fallacies into correct inductive arguments. When we discover, to our dismay, that our favorite deductive argument is invalid, we simply rescue it by saying that we never intended it to be deductive in the first place, but that it is a valid induction. With reference to this situation, the famous American logician Morris R. Cohen is said to have quipped, "A logic book is divided into two parts; in the first (on deduction) the fallacies are explained, and in the second (on induction) they are committed." Surely inductive logic, if it plays a central role in scientific method, must have better credentials than this.

When questions about deductive validity arise, they can usually be resolved in a formal manner by reference to an appropriate logical system. It has not always been so. Modern mathematical logic dates from the early nineteenth century, and it has undergone extraordinary development, largely in response to problems that arose in the foundations of mathematics. One such problem concerned the foundations of geometry, and it assumed critical importance with the discovery of non-Euclidean geometries. Another problem concerned the status of the infinitesimal in the calculus, a concept that was the center of utter confusion for two centuries after the discovery of the "infinitesimal" calculus. Thanks to extensive and fruitful investigations of the foundations of mathematics, we now have far clearer and more profound understanding of many fundamental mathematical concepts, as well as an extremely well-developed and intrinsically interesting discipline of formal deductive logic. The early investigators in this field could never have conceived in their wildest imaginings the kinds of results that have emerged.[6]

It is an unfortunate fact that far less attention has been paid to the foundational questions that arise in connection with the empirical sciences and their logic. When questions of inductive validity arise, there is no well-established formal discipline to which they can be referred for definitive solution. A number of systems of inductive logic have been proposed, some in greater and some in lesser detail, but none is more than rudimentary, and none is widely accepted as basically correct. Questions of inductive correctness are more often referred to scientific or philosophical intuitions, and these are notoriously unreliable guides.

We do have one resource which, although not overlooked entirely, is not exploited as fully as it could be. I refer to the mathematical calculus of probability. The probability calculus will not, by itself, solve all of our foundational problems concerning scientific inference, but it will provide us with a logical schema for scientific inference which is far more adequate than the *H-D* schema. And insofar as the probability calculus fails to provide the answers to foundational questions, it will at least help us to pose those problems in intelligible and, hopefully, more manageable form.

In order to show how the probability calculus can illuminate the kinds of questions I have been raising, I should like to introduce a very simple illustrative game. This game is played with two decks of cards composed as follows: deck 1 contains eight red cards and four black cards; deck 2 contains four red cards and eight black cards. A player begins by tossing a standard die; if the side one appears he draws a card from the first deck, and if any other side comes up he draws a card from the second deck. The draw of a red card constitutes a win. There is a simple formula for calculating the probability of a win resulting on a play of this game. Letting

[6] For a very readable account of recent developments, and a comparison with the earlier situation in the foundations of geometry see Paul J. Cohen and Reuben Hersh, "Non-Cantorian Set Theory," *Scientific American*, December 1967, vol. 217, no. 6.

"$P(A,B)$" stand for the probability *from A to B* (i.e., the probability of B, given A), and letting "A" stand for tosses of the die, "B" for draws from deck 1 (which occur when and only when an ace is tossed on the die), and "C" for draws of red cards, the following formula, which is a special case of the "theorem on total probability" yields the desired computation:

$$P(A,C) = P(A,B)\,P(A\,\&\,B,C) + P(A,\bar{B})\,P(A\,\&\,\bar{B},C) \tag{1}$$

The ampersand means "and" and the bar above a symbol negates it. Accordingly, the probabilities appearing in the formula are:

$P(A,C)$—probability of drawing a red card on a play of the game.
$P(A,B)$—probability of drawing from deck 1 on a play of the game ($=\frac{1}{6}$).
$P(A,\bar{B})$—probability of drawing from deck 2 on a play of the game ($=\frac{5}{6}$).
$P(A\,\&\,B,C)$—probability of drawing a red card if you play and draw from deck 1 ($=\frac{2}{3}$).
$P(A\,\&\,\bar{B},C)$—probability of drawing a red card if you play and draw from deck 2 ($=\frac{1}{3}$).

The theorem on total probability yields the result

$$P(A,C) = \frac{1}{6} \times \frac{2}{3} + \frac{5}{6} \times \frac{1}{3} = \frac{7}{18}$$

Suppose, now, that this game is being played, and you enter the room just in time to see that the player has drawn a red card, but you did not see from which deck it was drawn. Perhaps someone even offers you a wager on whether it came from deck 1 or deck 2. Again, the probability calculus provides a simple formula to compute the desired probability. This time it is a special form of "Bayes' theorem" and it can be written in either of two ways:

$$P(A\,\&\,C,B) = \frac{P(A,B)\,P(A\,\&\,B,C)}{P(A,C)} \tag{2}$$

$$= \frac{P(A,B)\,P(A\,\&\,B,C)}{P(A,B)\,P(A\,\&\,B,C) + (P(A,\bar{B})\,P(A\,\&\,\bar{B},C)} \tag{3}$$

The theorem on total probability (1) assures us that the denominators of the two fractions are equal; we must, of course, impose the restriction that $P(A,C) \neq 0$ in order to avoid an indeterminate fraction. The expression on the left evidently represents the probability that a draw which produced a red card was made from deck 1. Substituting known values in equation (2) yields

$$P(A \& C,B) = [\frac{1}{6} \times \frac{2}{3}] \Big/ [\frac{7}{18}] = \frac{2}{7}$$

There is nothing controversial about either of the foregoing theorems or their applications to simple games of chance of the type just described.

In order to get at our logical questions about the nature of scientific inference, let me redescribe the game and what we learned about it, and in so doing I shall admittedly be stretching some meanings. It is nevertheless illuminating. We can think of the drawing of a red card as an effect that can be produced in either of two ways, by tossing an ace and drawing from the first deck or by tossing a number other than one and drawing from the second deck. When we asked for the probability that a red card had been drawn from deck 1, we were asking for the probability that the first of the two possible causes rather than the second was operative in bringing about this effect. In fact, there are two causal hypotheses, and we were calculating the probability that was to be assigned to one of them, namely, the hypothesis that the draw came from the first deck. Notice that the probability that the draw came from the first deck is considerably less than one-half, making it much more likely that the draw came from the second deck, even though the probability that you will get a red card if you draw from the first deck is much greater than the probability that you will get a red card if you draw from the second deck. The reason, obviously, is that many more draws are made from the second deck, so even though many more black than red cards are drawn from the second deck, still the preponderance of red cards also comes from the second deck. This point has fundamental philosophical importance.

Continuing with the bizarre use of terms, let us look at the probabilities used to carry out the computation via Bayes' theorem. $P(A,B)$ and $P(A,\overline{B})$ are known as *prior probabilities*; they are the probabilities, respectively, that the particular cause is operative or not, regardless of the result of the draw. These probabilities are obviously linked in a simple manner,

$$P(A,\overline{B}) = 1 - P(A,B),$$

so that knowledge of one of them suffices. $P(A \& B,C)$ and $P(A \& \overline{B},C)$ are usually known as *likelihoods*. $P(A \& B,C)$ is the likelihood of the causal hypothesis that the draw came from deck 1 given that the draw was red, while $P(A \& \overline{B},C)$ is the likelihood that the hypothesis is false (i.e., the likelihood of an alternative) given the same result. Note, however, that *the likelihood of a hypothesis is not a probability of that hypothesis*; it is, instead, the probability of a result given that the hypothesis holds. Note, also, that the two likelihoods need not add up to one; they are logically independent of one another and both need to be known—knowledge of one only does not suffice. These are the probabilities that appear on the right hand side of the second form of Bayes' theorem (3). In the first form of Bayes' theorem (2) we do not need the second likelihood, $P(A \& \overline{B},C)$, but we require $P(A,C)$ instead. This probability has no common name, but it is the probability that the effect in question occurs regardless of which cause is operative. But whichever form of the theorem is used, we need three logically distinct probabilities in order to carry out the calculation. $P(A \& C,B)$, the probability we endeavor to establish, is known as the ***posterior probability*** of the hypothesis. When we entertain the two causal hypotheses about the draw of the card, we may take the fact that the draw produced a red card as observational evidence relevant to the causal hypotheses. (A rapid calculation will show that the probability that the draw came from deck 1 if it was a black card $= 1/11$.) Thus, we may think of our posterior probability, $P(A \& C,B)$, as the probability of a hypothesis in the light of observational evidence. This is precisely the kind of question which arose in connection with the hypothetico-deductive method, and in connection with our attempt to understand how evidence confirms or disconfirms scientific hypotheses. Bayes' theorem therefore constitutes a logical schema, found in the mathematical calculus of probability, that shows some promise of incorporating the main logical features of the kind of inference the hypothetico-deductive schema is intended to describe.

The striking difference between Bayes' theorem and the *H-D* schema is the relative complexity of the former compared with the latter. In fact, in some special cases the *H-D* schema provides just one of the probabilities required in Bayes' theorem, but never does it yield either of the other two required. Thus, the *H-D* schema is inadequate as an account of scientific inference because it is a gross oversimplification which omits reference to essential logical features of the inference. Bayes' theorem fills these gaps. The *H-D* schema describes a situation in which an observable result is deducible from a hypothesis (in conjunction with initial conditions, and possibly auxiliary hypotheses, all of which we are assuming to be true); thus, if the hypothesis is correct, the result *must* occur and cannot fail to occur. In this special case, $P(A \& B,C) = 1$, but without two other probabilities, say $P(A,B)$ and $P(A \& \overline{B},C)$, no conclusion at all can be drawn regarding the posterior probability. Inspection of Bayes' theorem makes it evident that $P(A \& B,C) = 1$ is completely compatible with $P(A \& C,B) = 0$. At best, the *H-D* schema yields the likelihood of the hypothesis for that given evidence, but we need a prior probability and the likelihood of an alternative hypothesis on the same evidence.

That these other probabilities are indispensable, and the manner in which they function in scientific reasoning, can be indicated by examples. Consider *Dianetics* once more. As remarked above, this book contained not only a theory to explain behavior, but also it contained recommendations for a therapy to be practiced for the treatment of psychological disturbances. The therapeutic procedure bears strong resemblances to psychoanalysis; it consists of the elimination of those "engrams" that are causing trouble by bringing to consciousness, through a process of free association, the unconscious experiences that produced the engrams in the first place. The theory, presumably, enables us to deduce that practice of the recommended therapy will produce cures of psychological illness. At the time the theory was in vogue, this therapy was practiced extensively, and there is every reason to believe that "cures" did occur. There were unquestionably cases in which people with various neurotic symptoms were treated, and they experienced a remission of their symptoms. Such instances would seem to count, according to the hypothetico-deductive method, as confirming instances. That they cannot actually be so regarded is due to the fact that there is a far better explanation of these "cures." We know that there is a phenomenon of "faith-healing" that consists in the efficacy of any treatment the patient sincerely believes to be effective. Many neurotic symptoms are amenable to such treatment, so anyone with such symptoms who believed in the soundness of the dianetic approach could be "cured" regardless of the truth or falsity of the theory upon which it is based. The reason, in terms of Bayes' theorem, is that the second likelihood—the probability $P(A \& \bar{B},C)$ that the same phenomenon would occur even if the hypothesis were false—is very high. Since this term occurs in the denominator, the value of the whole fraction tends to be small when the term is large.

A somewhat similar problem arises in connection with psychotherapy based upon more serious theoretical foundations. The effectiveness of any therapeutic procedure has to be compared with the so-called "spontaneous remission rate." Any therapy will produce a certain number of cases in which there is a remission of symptoms, but in a group of people with similar problems, but who undergo no therapy of any kind, there will also be a certain percentage who experience remission of symptoms. For a therapy to be judged effective, it has to improve upon the spontaneous remission rate; it is not sufficient that there be some remissions among those who undergo the treatment. In terms of Bayes' theorem, this means that we must look at both likelihoods, $P(A \& B,C)$ and $P(A \& \bar{B},C)$, not just the one we have been given in the standard *H-D* schema. This is just what experimental controls are all about. For instance, vitamin C has been highly touted as a cold remedy, and many cases have been cited of people recovering quickly from colds after taking massive doses. But in a *controlled* experiment in which two groups of people of comparable age, sex, state of general health, and severity of colds are compared, where one group is given vitamin C and the other is not, no difference in duration or severity of colds is detected.[7] This gives us a way of comparing the two likelihoods.

[7] The Editors of Consumer Reports, *The Medicine Show* (Simon and Schuster, 1961), chap. 2.

Let me mention, finally, an example of a strikingly successful confirmation, showing how the comparative likelihoods effect this sort of situation. At the beginning of the nineteenth century, two different theories of light were vying for supremacy: the wave theory and the corpuscular theory. Each had its strong advocates, and the evidence up to that point was not decisive. One of the supporters of the corpuscular theory was the mathematician Poisson, who deduced from the mathematical formulation of the wave theory that, if that theory were true, there should be a bright spot in the center of the shadow of a disk. Poisson declared that this absurd result showed that the wave theory is untenable, but when the experiment was actually performed the bright spot was there. Such a result was unthinkable on the corpuscular theory, so this turned into a triumph for the wave theory, because the probability on any other theory then available was negligible.[8] It was not until about a century later that the need for a combined wave-particle theory was realized. Arthimetically, the force of this dramatic confirmation is easily seen by noting that if $P(A \,\&\, \overline{B},C) = 0$ in (3), the posterior probability $P(A \,\&\, C,B)$ automatically becomes 1.

In addition to the two likelihoods, Bayes' theorem requires us to have a prior probability $P(A,B)$ *or* $P(A,\overline{B})$ in order to ascertain the posterior probability. These prior probabilities are probabilities of hypotheses without regard to the observational evidence provided by the particular test we are considering. In the card-drawing game described above, the prior probability was the probability of a draw from one particular deck regardless of whether the draw produced a red or black card. In the more serious cases of the attempt to evaluate scientific hypotheses, the probability of a hypothesis without regard to the test is precisely the sort of plausibility considered that was discussed at the outset. How plausible is a given hypothesis; what is its chance of being a successful one? This is the type of consideration that is demanded by Bayes' theorem in the form of a prior probability. The traditional stumbling-block to the use of Bayes' theorem as an account of the logic of scientific inference is the great difficulty of giving a description of what sort of things these prior probabilities could be.

It seems possible, nevertheless, to give many examples of plausibility arguments, and even to classify them into very general types. Such arguments may then be regarded as criteria which are used to evaluate prior probabilities—criteria that indicate whether a hypothesis is plausible or implausible, whether its prior probability is to be rated high or low. I shall mention three general types of criteria, and give some instances of each.

(1) Let us call criteria of the first general type *formal criteria*, for they involve formal logical relations between the hypothesis under consideration and other accepted parts of science. This kind of consideration was illustrated at the outset by Velikovski's theory, which contradicts virtually all

[8] See Max Born and Emil Wolf, *Principles of Optics* (Pergamon Press, 1964), p. 375.

of modern physics. Because of this formal relationship we can say that Velikovski's theory must have a very low prior probability, since it is incompatible with so much we accept as correct. Another example of the same type can be found in those versions of the theory of telepathy that postulate the *instantaneous* transference of thought from one person to another, regardless of the distance that separates them. For, the special theory of relativity stipulates that information cannot be transmitted at a speed greater than the speed of light, and so it would preclude instantaneous thought transmission. It would be even worse for precognition, the alleged process of direct perception of future occurrences, for this would involve messages being transmitted backward in time! Such parapsychological hypotheses must be given extremely low prior probabilities because of their logical incompatibility with well-established portions of physical science. A hypothesis could, of course, achieve a high prior probability on formal grounds by being the logical consequence of a well-established theory. Kepler's laws, for example, are extremely probable (as approximations) because of their relation to Newtonian gravitational theory.

(2) I shall call criteria of the second type *pragmatic criteria.* Such criteria have to do with the evaluation of hypotheses in terms of the circumstances of their origin—for example, the qualifications of the author. This sort of consideration has already been amply illustrated by the example of *Dianetics.* Whenever a hypothesis is dismissed as being a "crank" hypothesis, pragmatic criteria are being brought to bear. In his fascinating *Fads and Fallacies in the Name of Science,* Martin Gardner offers some general characteristics by which cranks can be identified.[9]

One might be tempted to object to the use of pragmatic criteria on the ground, as we have all been taught, that it is a serious fallacy to confuse the *origin* of a theory with its *justification*. Having been told the old story about how Newton was led to think of universal gravitation by seeing an apple fall, we are reminded that that incident has nothing to do with the truth or justification of Newton's gravitational theory. That issue must be decided on the evidence.[10] Quite so. But there are factors in the origin of a hypothesis, such as the qualifications of the author, which have an *objective* probability relationship to the hypothesis and its truth. Crank hypotheses seldom, if ever, turn out to be sound; they are based upon various misunderstandings, prejudices, or sheer ignorance. It is *not* fallacious to conclude that they have low prior probabilities.

(3) Criteria of the third type are by far the most interesting and important; let us call them *material criteria.* They make reference, in one way or another, to what the hypothesis actually says, rather than to its formal relation

[9] See p. 12–14, Reference 4.

[10] An elementary account of the distinction between discovery and justification is given in my *Logic* (Prentice-Hall, Inc., 1963), § 3.

to other theories, or to the circumstances surrounding its origins. These criteria do, however, depend upon comparisons of various theories or hypotheses; they make reference to analogies or similarities among different ones. Again, a few examples may be helpful.

Perhaps the most frequently cited criterion by which to judge the plausibility of hypotheses is the property of simplicity. Curve drawing illustrates this point very aptly. Given data which can be represented graphically, we generally take the smoothest curve—the one with the simplest mathematical expression—which comes sufficiently near the data points as representing the best explanatory hypothesis for those data. This factor was uppermost with Kepler, who kept searching for the simplest orbits to account for planetary motion, and finally settled upon the ellipse as filling the bill. Yet, we do not *always* insist upon the simplest explanation. We do not take seriously the "hypothesis" that television is solely responsible for the breakdown of contemporary morals, assuming that there is such a breakdown, for it is an obvious oversimplification. It may be that simplicity is more to be prized in the physical than in the social sciences, or in the advanced than in the younger sciences. But it does seem that we need to exercise reasonable judgment as to just what degree of simplicity is called for in any given situation.

Another consideration that may be used in plausibility arguments concerns causal mechanisms. There was a time when all scientific explanation was teleological in character; even the motion of inanimate objects was explained in terms of the endeavor to achieve their natural places. After the physics of Galileo and Newton had removed all reference to purpose from these realms, the remnants of teleological language remained: "Nature abhors a vacuum" and "Water seeks its own level." But though there have been a few attempts to read purpose into such laws as least action ("The Absolute is lazy"), it is for the most part fully conceded that physical explanation is nonpurposive.

The great success of Newtonian physics provided a strong plausibility argument for Darwin's account of the development of the biological species. The major difference between Darwin's evolutionary theory and its alternative contenders is the thoroughgoing rejection of teleological explanation by Darwin. Although teleological sounding language may sometimes creep in when we talk about natural selection, the concept is entirely nonpurposive. We ask, "Why is the polar bear white?" We answer, "Because that color provides a natural camouflage." It sometimes sounds a bit as if we are saying that the bear thinks the situation over and decides before he is born that white would be the best color, and so he chooses that color. But, of course, we mean no such thing. We are aware that, literally, no choice or planning is involved. There are chance mutations, some favorable to escaping from enemies and finding food. Those animals that have the favorable characteristics tend to survive and reproduce their kind, while those with unfavorable characteristics tend to die out without reproducing. The cause and effect relations in the evolutionary account are just as mechanical and without purpose as are those in Newtonian physics. This non-teleological theory is in sharp contrast to the theory of special creation according to which God created the various species because it somehow fit his plan.

The non-teleological character of Newton's theory surely must lend plausibility to a non-teleological biological theory such as Darwin's. If physics, which was far better developed and more advanced than any other science, got that way by abandoning teleological explanations for efficient causation, then it seems plausible for those sciences that are far less developed to try the same approach. When this approach paid off handsomely in the success of evolutionary theory, how much more plausible it becomes for other branches of science to follow the same line. Thus, for theories in psychology and sociology, for example, higher plausibility and higher probability would now attach to those hypotheses that are free from teleological components than to those that retain teleological explanation. When a biological hypothesis comes along that regresses to the pre-Darwinian teleology, such as Lecomte du Noüy's *Human Destiny*, it must be assigned a low prior probability.[11]

Let me give one final example of material criteria. Our investigations of the nature of physical space, extending over many centuries, have led to some rather sophisticated conceptions. To early thinkers, nothing could have been more implausible than to suppose that space is homogeneous and isotropic. Everyday experience seems clearly to demonstrate that there is a preferred direction—down. This view was expressed poetically by Lucretius in *The Nature of the Universe*, in which he describes the primordial state of affairs in which all the atoms are falling downward in space at a uniform speed.[12] On this view, not only was the downward direction preferred, but also, it was possible to distinguish absolute motion from absolute rest. By Newton's time it seemed clear that space had no preferred direction; rather, it was isotropic—possessed of the same structure in every direction. This consideration lent considerable plausibility to Newton's inverse square law, for if space is Euclidean and it has no preferred directions, then we should expect any force, such as gravitation, to spread out uniformly in all directions. In Euclidean geometry, the surface of a sphere varies with the square of the radius, so if the gravitational force spreads out uniformly in the surrounding space, it should diminish with the square of the distance.

Newton's theory, though it regards space as isotropic, still makes provision for absolute motion and rest. Einstein, reflecting on the homogeneity

[11] Pierre Lecomte du Noüy, *Human Destiny* (Longmans, Green and Co., 1947).

[12] Lucretius, *The Nature of the Universe*, trans. Ronald Latham (Penguin Books, 1951). Originally titled *De Rerum Natura*, and usually translated *On the Nature of Things*. The Latham translation is modern, and is far more intelligible than the older ones.

of space, enunciated a principle of relativity which precludes distinguishing physically between rest and uniform motion. In the beginning, if we believe Einstein's own autobiographical account, this principle recommended itself entirely on the grounds of its very great plausibility.[13] The matter does not rest there, of course, for it had to be incorporated into a physical theory that could be subjected to experimental test. His special theory of relativity has been tested and confirmed in a wide variety of ways, and it is now a well-established part of physics, but prior to the tests and its success in meeting them, it could be certified as highly plausible on the basis of very general characteristics of space.

Up to this point I have been attempting to establish two facts about prior probabilities: (1) Bayes' theorem shows that they are needed, and (2) scientific practice shows that they are used. But their status has been left very vague indeed. There is a fundamental reason. In spite of the fact that the probability calculus was established early in the seventeenth century, hardly any serious attention was given to the analysis of the meaning of the concept of probability until the latter part of the nineteenth century. There is nothing especially unusual about this situation. Questions about the meanings of fundamental concepts are foundational questions, and foundational investigations usually follow far behind the development of a discipline. Even today there is no real consensus on this question; there are, instead, three distinct interpretations of the probability concept, each with its strong adherents. A fortiori, there is no widely accepted answer to the question of the nature of the prior probabilities, for they seem to be especially problematic in character. Among the three leading probability theories, the *logical theory* regards probability as an *a priori measure* that can be assigned to propositions or states of affairs, the *personalistic theory* regards probability as a *subjective measure* of degrees of belief, and the *frequency theory* regards probability as a *physical characteristic* of types of events.

The logical theory is the direct descendent of the famous classical theory of Laplace. According to the classical theory, probability is the ratio of favorable to equally possible cases. The equi-possibility of cases, which is nothing other than the equal probability of these cases, is determined a priori on the basis of a *principle of indifference,* namely, two cases are equally likely if there is no reason to prefer one to the other. This principle gets into deep logical difficulty. Consider, for example, a car that makes a trip around a one mile track in a time somewhere between one and two minutes, but we know no more about it. It seems reasonable to say that the time could have been in the interval from one to one-and-one-half minutes, or it could have been in the interval of one-and-one-half to two minutes; we don't know which. Since these intervals are equal, we have no reason to prefer

[13] Albert Einstein, "Autobiographical Notes" in *Albert Einstein: Philosopher-Scientist,* ed., Paul Arthur Schilpp (The Library of Living Philosophers, 1949).

one to the other, and we assign a probability of one-half to each of them. Our information about this car can be put in other terms. We know that the car made its trip at an *average* speed somewhere in the range of 60 to 30 miles per hour. Again, it seems reasonable to say that the speed could have been in the range 60–45 miles per hour, or it could have been in the range 45–30 miles per hour; we don't know which. Since the two intervals are equal, we have no reason to prefer one to the other, and we assign a probability of one-half to each. But we have just contradicted our former result, for a time of one-and-one-half minutes corresponds with an average speed of forty, not forty-five, miles per hour.

This contradiction, known as the Bertrand paradox, brings out the fundamental difficulty with any method of assigning probabilities a priori. Such a priori decisions have an unavoidable arbitrary component to them, and in this case, the arbitrary component gives rise to two equally reasonable, but incompatible, ways of assigning the probabilities. Although the logical interpretation, in its current form, escapes this particular form of paradox, it is still subject to philosophical criticism because of the same general kind of aprioristic arbitrariness.

The personalistic interpretation is the twentieth century successor of an older and more naive subjective concept. According to the crude subjective view, a probability is no more nor less than a subjective degree of belief; it is a measure of our ignorance. If I assign the probability value one-half to an outcome of heads on a toss of the coin, this means that I expect heads just as often as I expect tails, and my uncertainty is equally divided between the two outcomes. If I expect twice as strongly as not that an American will be the first human to set foot on the moon, then that event has a probability of two-thirds.

The major difficulty with the old subjective interpretation arises because subjective states do not always come in sizes that will fit the mathematical calculus of probability. It is quite possible, for example, to find a person who believes to the degree one-sixth that a six will turn up on any toss of a given die, and who also believes that the tosses are independent of one another (the degree to which he believes in an outcome of six on a given toss is unaffected by the outcome of the previous toss). This same individual may also believe to the degree one-half, that he will get at least one six in three tosses of that die. There is, of course, something wrong here. If the probability of six on a given toss is one-sixth, and if the tosses are independent, this probability is considerably less than one-half (it is approximately 0.42). For four tosses, the probability of at least one six is well over one-half. This is a trivial kind of error that has been recognized as such for hundreds of years, but it is related to a significant error that led to the discovery of the mathematical calculus of probability. In the seventeenth century, the view was held that in 24 tosses of a pair of dice, there should be at least a fifty-fifty chance of tossing at least one double six. In fact, the probability is just under one-half in 24 tosses; in 25 it is just over one-half. The point of these examples is very simple. If probabilities are just subjective degrees of belief, the mathematical calculus of probability is mistaken, because it specifies certain relations among probabilities that do not obtain among degrees of belief.

Modern personalists do not interpret probabilities merely as subjective degrees of belief, but rather, as *coherent* degrees of belief. To say that degrees of belief are coherent means that they are related in such manner as to satisfy the conditions imposed by the mathematical calculus of probability. The personalists have seen that degrees of belief that violate the mathematical calculus involve some sort of error or blunder that is analogous to a logical inconsistency. Hence, when a combination of degrees of belief is incoherent, some adjustment or revision is called for in order to bring these degrees into conformity with the mathematical calculus. The chief objection to the personalist view is that it is not objective; we shall have to see whether and to what extent the lack of objectivity is actually noxious.

The frequency interpretation goes back to Aristotle who characterized the probable as that which happens often. More exactly, it regards a probability as a relative frequency of occurrence in a large sequence of events. For instance, a probability of one-half for heads on tosses of a coin would mean that in the long run the ratio of the number of heads to the number of tosses approaches and remains close to one-half. To say that the probability of getting a head on a particular toss is one-half means that this toss is a member of an appropriately selected large class of tosses within which the overall relative frequency of heads is one-half. It seems evident that there are many contexts in which we deal with large aggregates of phenomena, and in these contexts the frequency concept of probability seems well suited to the use of statistical techniques—e.g., in quantum mechanics, kinetic theory, sociology, and the games of chance, to mention just a few. But it is much more dubious that the frequency interpretation is at all applicable to such matters as the probability of a scientific hypothesis in the light of empirical evidence. In this case where are we to find the large classes and long sequences to which to refer our probabilities of hypotheses? This difficulty has seemed insuperable to most authors who have dealt with the problem. The general conclusion has been that the frequency interpretation is fine in certain contexts, but we need a radically different probability concept if we are to deal with the probability of hypotheses.

Returning to our main topic of concern, we easily see that each of the foregoing three probability theories provides an answer to the question of the nature of plausibility considerations and prior probabilities. According to the logical interpretation, hypotheses are plausible or not on the basis of certain a priori considerations; on this view, reason dictates which hypotheses are to be taken seriously and which not. According to the personalistic interpretation, prior probabilities represent the prior opinion or attitude of the investigator toward the hypothesis before he sets about testing it. Different investigators may, of course, have different views of the same hypothesis, so prior probabilities may vary from individual to individual. According to the frequency interpretation, prior probabilities arise from experience with scientific hypotheses, and they reflect this experience in an objective way. To say that a hypothesis is plausible, or has a high prior probability, means that it is of a type that has proved successful in the past. We have found by experience that hypotheses of this general type have often worked well in science.

From the outset, the personalistic interpretation enjoys a major advantage over the other two. It is very difficult to see how we are to find non-arbitrary a priori principles to use as a basis for establishing prior probabilities of the a priori type for the logical interpretation, and it is difficult to see how we are reasonably to define classes of hypotheses and count frequencies of success for the frequency interpretation. But personal probabilities are available quite unproblematically. Each individual has his degree of belief in the hypothesis, and that's all there is to it. Coherence demands that degrees of belief conform to the mathematical calculus, and Bayes' theorem is one of the important relations to be found in the calculus. Bayes' theorem tells us how, if we are to avoid incoherence, we must modify our degrees of belief in the light of new evidence. The personalists, who constitute an extremely influential school of contemporary statisticians, are indeed so closely wedded to Bayes' theorem that they have even taken its name and are generally known as "bayesians."

The chief objection to the personalist approach is that it injects a purely subjective element into the testing and evaluation of scientific hypotheses; we feel that science should have a more objective foundation. The bayesians have a very persuasive answer. Even though two people may begin with radically different attitudes toward a hypothesis, accumulating evidence will force a convergence of opinion. This is a basic mathematical fact about Bayes' theorem; it is easily seen by an example. Suppose a coin which we cannot examine is being flipped; but we are told the results of the tosses. We know that it is either a fair coin or a two-headed coin, we don't know which, and we have very different prior opinions on the matter. Suppose your prior probability for a two-headed coin is 1/100 while mine is one-half. Then as we learn that various numbers of heads have been tossed (without any tails, of course), our opinions come closer and closer together as follows:

Number of tosses resulting in head	Prior probability that coin has two heads	
	1/100	1/2
	Posterior probability on given evidence	
1	2/101	2/3
2	4/103	4/5
10	$1024/1123 \simeq .91$	$1024/1025 \simeq .99$

After only ten tosses, we both find it overwhelmingly probable that the coin that produced this sequence of results is a two-headed one. This phenomenon is sometimes called "swamping of the priors," for their influence on the posterior probabilities becomes smaller and smaller as evidence accumulates. The only qualification is that we must begin with somewhat open minds. If we begin with the certainty that the coin is two-headed or with the certainty that it is not, i.e., with prior probability of zero or one, evidence will not change that opinion. But if we begin with prior probabilities differing ever so little from those extremes convergence will sooner or later occur. As L. J. Savage remarked, it is not necessary to have an open mind, it is sufficient to have one that is slightly ajar.

The same consideration about the swamping of prior probabilities also enables the frequentist to overcome the chief objection to his approach. If it were necessary to have clearly defined classes of hypotheses, within which exact values of frequencies of success had to be ascertained, the situation would be pretty hopeless, but because of the swamping phenomenon, it is sufficient to have only the roughest approximation. All that is really needed is a reasonable guess as to whether the value is significantly different from zero. In the artificial coin tossing example, where there are only two hypotheses, it is possible to be perfectly open-minded and give each alternative a non-negligible prior probability, but in the serious cases of evaluation of scientific hypotheses, there are infinitely many alternative hypotheses, all in conflict with one another, and they cannot all have non-negligible prior probabilities. This is the problem of the alternative hypotheses again. For this reason, it is impossible to be completely open-minded, so we must find some basis for assigning negligible prior probabilities to some possible hypotheses. This is tantamount to judging some hypotheses to be too implausible to deserve further testing and consideration. It is my conviction that this is done on the basis of experience; it is not done by means of purely a priori considerations, nor is it a purely subjective affair. As I tried to suggest by means of the examples of plausibility arguments, scientific experience with the testing, acceptance, and rejection of hypotheses provides an objective basis for deciding which hypotheses deserve serious testing and which do not. I am not suggesting that we proceed on the basis of plausibility considerations to summary dismissal of almost every hypothesis that comes along; on the contrary, the recommendation would be for a high degree of open-mindedness. However, we need not and cannot be completely open-minded with regard to any and every hypothesis of whatever description that happens to be proposed by anyone. This approach shows how we can be reasonably open-minded in science without being stupid about it. It provides an answer to the kind of charge made by the editors of *Harper's*: Science *is* objective, but its objectivity embraces two aspects, objective testing and objective evaluation of prior probabilities. Plausibility arguments are used in science, and their use is justified by Bayes' theorem. In fact, Bayes' theorem shows that they are indispensable. The frequency interpretation of probability enables us to view them as empirical and objective.

It would be an unfair distortion of the situation for me to conclude without remarking that the view I have been advocating is very definitely a minority view among inductive logicians and probability theorists. There is no well agreed upon majority view. One of the most challenging aspects of this sort of investigation lies in the large number of open questions, and the amount that remains to be done. Whether my view is correct is not the main issue. Of far greater importance is the fact that there are many fundamental problems that deserve extensive consideration, and we cannot help but learn a great deal about the foundations of science by pursuing them.

II

Problems and Paradoxes of Inductive Logic

8

RECENT PROBLEMS OF INDUCTION

Carl G. Hempel

> It is true that from truths we can conclude only truths; but there are certain falsehoods which are useful for finding the truth.
> —Leibniz, Letter to Canon Foucher (1692)

THE CLASSICAL PROBLEM OF INDUCTION

In the philosophical discussion of induction, one problem has long occupied the center of the stage—so much so, indeed, that it is usually referred to as *the* problem of induction. That is the problem of justifying the way in which, in scientific inquiry and in our everyday pursuits, we base beliefs and assertions about empirical matters on logically inconclusive evidence.

This classical problem of justification, raised by Hume and made famous by his skeptical solution, is indeed of great philosophical importance. But more recent studies, most of which were carried out during the past two or three decades, have given rise to new problems of induction, no less perplexing and important than the classical one, which are logically prior to it in the sense that the classical problem cannot even be clearly stated—let alone solved—without some prior clarification of the new puzzles.

In this paper, I propose to discuss some of these recent problems of induction.

Induction may be regarded as effecting a transition from some body of empirical information to a hypothesis which is not logically implied by it, and for this reason it is often referred to as nondemonstrative *inference.* This characterization has to be taken with a grain of salt; but it is suggestive and convenient, and in accordance with it, I will therefore sometimes

From Robert G. Colodny (ed.), *Mind and Cosmos, Essays in Contemporary Science and Philosophy* (University of Pittsburgh Press, 1966), pp. 112–134. Reprinted by permission of the University of Pittsburgh Press.

refer to the sentences specifying the evidence as the *premises* and to the hypothesis based on it as the conclusion of an "*inductive inference.*"

Among the simplest types of inductive reasoning are those in which the evidence consists of a set of examined instances of a generalization, and the hypothesis is either the generalization itself or a statement about some unexamined instances of it. A standard example is the inference from the evidence statement that all ravens so far observed have been black to the generalization that all ravens are black or to the prediction that the birds now hatching in a given clutch of raven eggs will be black or to the retrodiction that a raven whose skeleton was found at an archeological site was black. As these examples show, induction does not always proceed from the particular to the general or from statements about the past or present to statements about the future.

The inductive procedures of science comprise many other, more complex and circumstantial, kinds of nondemonstrative reasoning, such as those used in making a medical diagnosis on the basis of observed symptoms, in basing statements about remote historical events on presently available evidence, or in establishing a theory on the basis of appropriate experimental data.

However, most of the problems to be considered here can be illustrated by inductions of the simple kind that proceed from instances of a generalization, and in general I will use these as examples.

THE NARROW INDUCTIVIST VIEW OF SCIENTIFIC INQUIRY

It should be stressed at the outset that what we have called inductive inference must not be thought of as an effective method of discovery, which by a mechanical procedure leads from observational data to appropriate hypotheses or theories. This misconception underlies what might be called the narrow inductivist view of scientific inquiry, a view that is well illustrated by the following pronouncement:

> If we try to imagine how a mind of superhuman power and reach, but normal so far as the logical processes of its thought are concerned . . . would use the scientific method, the process would be as follows: First, all facts would be observed and recorded, *without selection* or *a priori* guess as to their relative importance. Second, the observed and recorded facts would be analyzed, compared, and classified, *without hypothesis or postulates* other than those necessarily involved in the logic of thought. Third, from this analysis of the facts, generalization would be inductively drawn as to the relations, classificatory or causal, between them. Fourth, further research would be deductive as well as inductive, employing inferences from previously established generalizations.[1]

[1] A. B. Wolfe, "Functional Economics," *The Trend of Economics*, ed. R. G. Tugwell (New York: Knopf, 1924), p. 450 (author's italics).

It need hardly be argued in detail that this conception of scientific procedure, and of the role induction plays in it, is untenable; the reasons have been set forth by many writers. Let us just note that an inquiry conforming to this idea would never go beyond the first stage, for—presumably to safeguard scientific objectivity—no initial hypotheses about the mutual relevance and interconnections of facts are to be entertained in this stage, and as a result, there would be no criteria for the selection of the facts to be recorded. The initial stage would therefore degenerate into an indiscriminate and interminable gathering of data from an unlimited range of observable facts, and the inquiry would be totally without aim or direction.

Similar difficulties would beset the second stage—if it could ever be reached—for the classification or comparison of data again requires criteria. These are normally suggested by hypotheses about the empirical connections between various features of the "facts" under study. But the conception just cited would prohibit the use of such hypotheses, and the second stage of inquiry as here envisaged would again lack aim and direction.

It might seem that the quoted account of inductive scientific procedure could be rectified by simply adding the observation that any particular scientific investigation is aimed at solving a specified problem, and that the initial selection of data should therefore be limited to facts that are relevant to that problem. But this will not do, for the statement of a problem does not generally determine what kinds of data are relevant to its solution. The question as to the causes of lung cancer does not by itself determine what sorts of data would be relevant—whether, for example, differences in age, occupation, sex, or dietary habits should be recorded and studied. The notion of "relevant" facts acquires a clear meaning only when some specific answer to the problem has been suggested, however tentatively, in the form of a hypothesis: an observed fact will then be favorably or unfavorably relevant to the hypothesis according as its occurrence is by implication affirmed or denied by the hypothesis. Thus, the conjecture that smoking is a potent causative factor in lung cancer affirms by implication a higher incidence of the disease among smokers than among nonsmokers. Data showing for a suitable group of subjects that this is the case or that it is not would therefore constitute favorably relevant (confirming) or unfavorably relevant (disconfirming) evidence for the hypothesis. Generally, then, those data are relevant and need to be gathered which can support or disconfirm the contemplated hypothesis and which thus provide a basis for testing it.

Contrary to the conception quoted above, therefore, hypotheses are put forward in science as tentative answers to the problem under investigation. And contrary to what is suggested by the description of the third stage of inquiry above, such answers in the form of hypotheses or theories cannot be inferred from empirical evidence by means of some set of mechanically applicable rules of induction. There is no generally applicable mechanical routine of "inductive inference" which leads from a given set of data to a corresponding hypothesis or theory somewhat in the way in which the familiar routine of multiplication leads from any two given integers, by a finite number of mechanically performable steps, to the corresponding product.

To be sure, mechanical induction routines can be specified for certain special kinds of cases, such as the construction of a curve, and of an analytic expression for the corresponding function, which will fit a finite set of points. Given a finite set of measurements of associated values of temperature and volume for a given body of gas under constant pressure, this kind of procedure could serve mechanically to produce a tentative general law connecting temperature and volume of the gas. But for generating scientific theories, no such procedure can be devised.

Consider, for example, a theory, such as the theory of gravitation or the atomic theory of matter, which is introduced to account for certain previously established empirical facts, such as regularities of planetary motion and free fall, or certain chemical findings such as those expressed by the laws of constant and of multiple proportions. Such a theory is formulated in terms of certain concepts (those of gravitational force, of atom, of molecule, etc.) which are novel in the sense that they had played no role in the description of the empirical facts which the theory is designed to explain. And surely, no set of induction rules could be devised which would be generally applicable to just any set of empirical data (physical, chemical, biological, etc.) and which, in a sequence of mechanically performable steps, would generate appropriate novel concepts, functioning in an explanatory theory, on the basis of a description of the data.[2]

Scientific hypotheses and theories, then, are not mechanically inferred from observed "facts": *They are invented by an exercise of creative imagination*. Einstein, among others, often emphasized this point, and more than a century ago William Whewell presented the same basic view of induction. Whewell speaks of scientific discovery as a "process of invention, trial, and acceptance or rejection" of hypotheses and refers to great scientific advances as achieved by "Happy *Guesses*," by "felicitous and inexplicable strokes of inventive talent," and he adds: "No rules can ensure to us similar success in new cases; or can enable men who do not possess similar endowments, to make like advances in knowledge."[3] Similarly, Karl Popper has characterized scientific hypotheses and theories as conjectures, which must then

[2] This argument does not presuppose a fixed division of the vocabulary of empirical science into observational and theoretical terms; it is quite compatible with acknowledging that as a theory becomes increasingly well established and accepted, certain statements couched in terms of its characteristic concepts may come to be qualified as descriptions of "observed facts."

[3] William Whewell, *The Philosophy of the Inductive Sciences*, 2d ed. (London: John W. Parker, 1847), II, 41 (author's italics).

be subjected to test and possible falsification.[4] Such conjectures are often arrived at by anything but explicit and systematic reasoning. The chemist Kékulé, for example, reports that his ring formula for the benzene molecule occurred to him in a reverie into which he had fallen before his fireplace. Gazing into the flames, he seemed to see snakes dancing about; and suddenly one of them moved into the foreground and formed a ring by seizing hold of its own tail. Kékulé does not tell us whether the snake was forming a *hexagonal* ring, but that was the structure he promptly ascribed to the benzene molecule.

Although no restrictions are imposed upon the *invention* of theories, scientific objectivity is safeguarded by making their *acceptance* dependent upon the outcome of careful tests. These consist in deriving, from the theory, consequences that admit of observational or experimental investigation, and then checking them by suitable observations or experiments. If careful testing bears out the consequences, the hypothesis is accordingly supported. But normally a scientific hypothesis asserts more than (i.e., cannot be inferred from) some finite set of consequences that may have been put to test, so that even strong evidential support affords no conclusive proof. It is precisely this fact, of course, that makes inductive "inference" nondemonstrative and gives rise to the classical problem of induction.

Karl Popper, in his analysis of this problem, stresses that the inferences involved in testing a scientific theory always run deductively from the theory to implications about empirical facts, never in the opposite direction; and he argues that therefore "Induction, i.e., inference based on many observations, is a myth. It is neither a psychological fact, nor a fact of ordinary life, nor one of scientific procedure";[5] and it is essentially this observation which, he holds, "solves . . . Hume's problem of induction."[6] But this is surely too strong a claim, for although the procedure of empirical science is not inductive in the narrow sense we have discussed and rejected, it still may be said to be *inductive in a wider sense,* referred to at the beginning of this paper: While scientific hypotheses and theories are not *inferred* from empirical data by means of some effective inductive procedure, they are *accepted* on the basis of observational or experimental findings which afford no deductively conclusive evidence for their truth. Thus, the classical problem of induction retains its import: What justification is there for accepting hypotheses on the basis of incomplete evidence?

[4] See, for example, Popper's essay, "Science: Conjectures and Refutations," in his book, *Conjectures and Refutations* (New York and London: Basic Books, 1962).

[5] Karl Popper, "Philosophy of Science: A Personal Report," *British Philosophy in the Mid-Century*, ed. C. A. Mace (London: Allen and Unwin, 1957), pp. 155–91, quotation from p. 181.

[6] Popper, "Philosophy of Science," p. 183.

The search for an answer to this question will require a clearer specification of the procedure that is to be justified; for while the hypotheses and theories of empirical science are not deductively implied by the evidence, it evidently will not count as inductively sound reasoning to accept a hypothesis on the basis of just any inconclusive evidence. Thus, there arises the logically prior problem of giving a more explicit characterization and precise criteria of what counts as sound inductive reasoning in science.

It may be instructive briefly to consider the analogue to this problem for deductive reasoning.

DEDUCTION AND INDUCTION; DISCOVERY AND VALIDATION

Deductive soundness, of course, is tantamount to deductive validity. This notion can be suggestively although imprecisely characterized by saying that an argument is deductively valid if its premises and its conclusion are so related that if all the premises are true, then the conclusion cannot fail to be true as well.[7]

As for *criteria* of deductive validity, the theory of deductive logic specifies a variety of forms of inference which are deductively valid, such as, for example, *modus ponens:*

$$\frac{\begin{array}{l} p \supset q \\ p \end{array}}{q}$$

or the inference rules of quantificational logic. Each of these represents a sufficient but not necessary condition of deductive validity. These criteria have the important characteristic of being expressible by reference to the syntactical structure of the argument, and thus without any reference to the meanings of the extralogical terms occurring in premises and conclusion. As we will see later, criteria of inductive soundness cannot be stated in purely syntactical terms.

We have already noted that whatever the rules of induction may be, they cannot be expected to specify mechanical routines leading from empirical evidence to appropriate hypotheses. Are the rules of deductive inference superior in this respect? Consider their role in logic and mathematics.

A moment's reflection shows that no interesting theorem in these fields is discovered by a mechanical application of the rules of deductive inference. Unless a putative theorem has first been put forward, such application

[7] Precise general characterizations of deductive validity, for arguments in languages of certain specified forms, will be found, e.g., in W. V. O. Quine, *Methods of Logic*, rev. ed. (New York: Holt, Rinehart & Winston, 1959).

would lack direction. Discovery in logic and mathematics, no less than in empirical science, *calls for imagination and invention;* it does not follow any mechanical rules.

Next, even when a putative theorem has been proposed, the rules of deduction do not, in general, provide a mechanical routine for proving or disproving it. This is illustrated by the famous arithmetical conjectures of Goldbach and of Fermat, which were proposed centuries ago but have remained undecided to this day. Mechanical routines for proving or disproving any given conjecture can be specified only for systems that admit of a decision procedure; and even for first-order quantificational logic and for elementary arithmetic, it is known that there can be no such procedure. In general, then, the construction of a proof or a disproof for a given logical or mathematical conjecture requires ingenuity.

But when a putative theorem has been proposed and a step-by-step argument has been offered as a presumptive proof for it, then the rules of deductive logic afford a means of establishing the validity of the argument: If each step conforms to one of those rules—a matter which can be decided by mechanical check—then the argument is a valid proof of the proposed theorem.

In sum, the formal rules of deductive inference are not rules of discovery leading mechanically to correct theorems or even to proofs for conjectured theorems which are in fact provable; rather, they provide criteria of soundness or of validity for proposed deductive proofs.

Analogously, rules of inductive inference will have to be conceived, not as canons of discovery, but as criteria of validation for proposed inductive arguments; far from generating a hypothesis from given evidence, they will *presuppose* that, in addition to a body of evidence, a hypothesis has been put forward, and they will then serve to appraise the soundness of the hypothesis on the basis of the evidence.

Broadly speaking, inductive arguments might be thought of as taking one of these forms:

$$\begin{array}{l} e \\ = \\ h \end{array} \quad \text{(i.e., evidence } e \text{ supports hypothesis } h\text{)}$$

$$\begin{array}{l} e \\ = [r] \\ h \end{array} \quad \text{(i.e., evidence } e \text{ supports hypothesis } h \text{ to degree } r\text{)}$$

Here, the double line is to indicate that the relation of *e* to *h* is not that of full deductive implication but that of partial inductive support.

The second of these schemata incorporates the construal of inductive support as a quantitative concept. Rules of induction pertaining to it would provide criteria determining the degree of support conferred on certain kinds of hypotheses by certain kinds of evidence sentences; these criteria might even amount to a general definition assigning a definite value of r to any given e and h; this is one objective of Carnap's inductive logic.[8]

The first schema treats inductive support or confirmation as a qualitative concept; the corresponding inference rules would specify conditions under which a given evidence sentence supports, or confirms, a given hypothesis.[9]

The formulation of rules of these or similar kinds will be required to explicate the concept of inductive inference in terms of which the classical problem of justification is formulated. And it is in this context of explication that the newer problems of induction arise. We now turn to one of those problems; it concerns the qualitative concept of confirmation.

THE PARADOXES OF QUALITATIVE CONFIRMATION

The most familiar rules of induction concern generalizations of the simple form "All F are G." According to one widely asserted rule, a hypothesis of this kind receives support from its positive instances—i.e., from cases of F that have been found also to be G. For example, the hypothesis "All ravens are black," or

$$(x)\,(Rx \supset Bx) \qquad (h)$$

is supported, or confirmed, by any object i such that

$$Ri \cdot Bi \qquad (I)$$

or, as we will say, by any evidence sentence of the form "$Ri \cdot Bi$." Let us refer to such instances as *positive instances of type I for h*. Similarly, h is disconfirmed (invalidated) by any evidence sentence of the form $Ri \cdot -Bi$. This criterion was explicitly discussed and advocated by Jean Nicod;[10] I will therefore call it Nicod's criterion.

[8] See especially the following publications by Rudolf Carnap: *Logical Foundations of Probability*, 2d ed. (Chicago: University of Chicago Press, 1962); "The Aim of Inductive Logic," *Logic, Methodology and Philosophy of Science: Proceedings of the 1960 International Congress*, eds. E. Nagel, P. Suppes, and A. Tarski (Stanford: Stanford U. Press, 1962), pp. 303–18.

[9] It seems to me, therefore, that Popper begs the question when he declares: "But it is obvious that this rule or craft of 'valid inducton'... simply does not exist. No rule can ever guarantee that a generalization inferred from true observations, however often repeated, is true" ("Philosophy of Science," p. 181). That inductive reasoning is not *deductively* valid is granted at the outset; the problem is that of constructing a concept of *inductive* validity.

[10] Jean Nicod, *Foundations of Geometry and Induction* (New York: Harcourt, Brace & World, 1930), p. 219. Nicod here speaks of "truths or facts," namely, "the presence or absence of B in a case of A," as confirming or invalidating "the law *A entails B*" (author's italics). Such confirmatory and disconfirmatory facts can be thought of as described by corresponding evidence sentences. Nicod

Now, the hypothesis *h* is logically equivalent to, and thus makes exactly the same assertion as, the statement that all nonblack things are nonravens, or

$$(x)\ (-Bx \supset -Rx) \qquad (h')$$

According to Nicod's criterion, this generalization is confirmed by *its* instances—i.e., by any individual *j* such that

$$-Bj \cdot -Rj \qquad (II)$$

But since *h'* expresses exactly the same assertion as *h,* any such individual will also confirm *h*. Consequently, such things as a yellow rose, a green caterpillar, or a red herring confirm the generalization "All ravens are black," by virtue of being nonblack nonravens. I will call such objects *positive instances of type II for h.*

Next, the hypothesis *h* is logically equivalent also to the following statement:

$$(x)\ [(Rx \vee -Rx) \supset (-Rx \vee Bx)] \qquad (h'')$$

in words: Anything that is a raven or not a raven—i.e., anything at all—either is not a raven or is black. Confirmatory instances for this version which I will call *positive instances of type III for h,* consist of individuals *k* such that

$$-Rk \vee Bk \qquad (III)$$

This condition is met by any object *k* that is not a raven (no matter whether it is black) and by any object *k* that is black (no matter whether it is a raven). Any such object, then, affords a confirmatory instance in support of the hypothesis that all ravens are black.

On the other hand, the hypothesis *h* can be equivalently expressed by the sentence

$$(x)\ [(Rx \cdot -Bx) \supset (Rx \cdot -Rx)] \qquad (h''')$$

for which nothing can possibly be a confirmatory instance in the sense of Nicod's criterion, since nothing can be both a raven and not a raven.

These peculiarities, and some related ones, of the notion of confirmatory instance of a generalization have come to be referred to as the *paradoxes of*

remarks about his criterion: "We have not seen it stated in any explicit manner. However, we do not think that anything ever written on induction is incompatible with it" (p. 220). Whether Nicod regards the specified conditions as necessary and sufficient or merely as sufficient for confirmation or invalidation is not entirely clear, although he does say: "It is conceivable that we have here the only two direct modes in which a fact can influence the probability of a law" (p. 219). We will construe his criteria simply as *sufficient* conditions of confirmation and invalidation.

confirmation.[11] And indeed, at first glance they appear to be implausible and perhaps even logically unsound. But on further reflection one has to conclude, I think, that they are perfectly sound, that it is our intuition in the matter which leads us astray, so that the startling results are paradoxical only in a psychological, but not in a logical sense.

To see this, let us note first that the results in question follow deductively from two simple basic principles, namely: (*A*) A generalization of the form "All *F* are *G*" is confirmed by its positive instances—i.e., by cases of *F* that have been found also to be cases of *G*. (*B*) Whatever confirms a hypothesis also confirms any logically equivalent one.

Principle (*A*) is, in effect, part of Nicod's criterion, of which Nicod himself remarks that it "cannot claim the force of an axiom. But it offers itself so naturally and introduces such great simplicity, that reason welcomes it without feeling any imposition."[12] We will encounter some surprising exceptions to it in Sections 5 and 6, but it does indeed seem very reasonable in cases of the kind we have considered so far—i.e., in reference to generalizations of universal conditional form containing exclusively property terms (one-place predicates).

Principle (*B*) may be called the equivalence condition. It simply reflects the idea that whether given evidence confirms a hypothesis must depend only on the content of the hypothesis and not on the way in which it happens to be formulated.

And once we accept these principles, we must also accept their surprising logical consequences.

Let us look at these consequences now from a different point of view, which will support the claim that they are sound. Suppose we are told that in the next room there is an object i which is a raven. Our hypothesis h then tells us about i that it is black, and if we find that this is indeed the case, so that we have $Ri \cdot Bi$, then this must surely count as bearing out, or confirming, the hypothesis.

Next, suppose we are told that in the adjoining room there is an object j that is not black. Again, our hypothesis tells us something more about it, namely, that it is not a raven. And if we find that this is indeed so—i.e., that $-Bj \cdot -Rj$, then this bears out, and thus supports, the hypothesis.

Finally, even if we are told only that in the next room there is an object k, the hypothesis still tells us something about it, namely, that either it is no raven or it is black—i.e., that $-Rk \vee Bk$; and if this is found to be the case, it again bears out the hypothesis.

[11] These paradoxes were first noted in my essay "Le problème de la vérité," *Theoria* (Göteborg), 3 (1937), 206–46 (see especially p. 222) and were discussed in greater detail in my articles "Studies in the Logic of Confirmation," *Mind,* 54 (1945), 1–26, 97–121, and "A Purely Syntactical Definition of Confirmation," *The J. of Symbolic Logic,* 8 (1943), 122–43.

[12] Nicod, *Geometry and Induction,* pp. 219–20.

Thus, our three types of positive instance must indeed be counted as confirmatory or supporting evidence for the generalization that all ravens are black.

Finally, the fact that the formulation *h‴* of our generalization admits of no confirming instances in the sense of Nicod's criterion presents no serious problem if, as here has been done, that criterion is stated as a sufficient but not necessary condition of confirmation.

But why does it seem implausible or paradoxical in the first place that positive instances of types *II* and *III* should be confirmatory for the generalization *h?* One important reason seems to lie in the assumption that the hypothesis "All ravens are black" is a statement about ravens and not about nonravens, let alone about all things in general. But surely, such a construal is untenable; anyone who accepts *h* would be bound to accept also the sentences *h′* and *h″,* which by the same token would have to be viewed as statements about nonravens and about all things, respectively. The use made of some statements of the form "All *F* are *G*" illustrates the same point. The Wassermann test, for example, is based, roughly speaking, on the generalization that any person infected with syphilis has a positive Wassermann reaction; but in view of its diagnostic implications for cases yielding a negative test result, this generalization surely cannot be said to be about syphilitically infected persons only.

To say that positive instances of types *I, II,* and *III* all confirm the hypothesis *h* is not to say, however, that they confirm the generalization to the same extent. Indeed, several writers have argued that the different types differ greatly in this respect and that, in particular, a positive instance of type *I,* i.e., a black raven, lends much stronger support to our generalization than a positive instance of type *II,* i.e., a nonblack object that is not a raven; and they have suggested that this is the objective basis for the first impression that instances of type *I* alone can count as confirmatory for our hypothesis.

This view can be made plausible by the following suggestive but imprecise consideration: Let *k* be the hypothesis "All marbles in this bag are red," and suppose that there are twenty marbles in the bag. Then the generalization *k* has twenty instances of type *I,* each being provided by one of the marbles. If we had checked each of the twenty objects that are marbles in the bag, we have exhaustively tested the hypothesis. And roughly speaking we might say that if we have examined one of the marbles and found it red, we have shown one twentieth of the total content of the hypothesis to be true.

Now consider the contrapositive of our generalization—i.e., the statement, "Any object that is not red is not a marble in this bag." Its instances are provided by all nonred objects. There are a large number of these in the world—perhaps infinitely many of them. Examining one of them and averring that it is not a marble in the bag is therefore to check, and corroborate, only a tiny portion of all that the hypothesis affirms. Hence, a positive finding of type *II* would indeed support our generalization, but only to a very small extent.

Analogously in the case of the ravens. If we may assume that there are vastly more nonblack things than there are ravens, then the observation of one nonblack thing that is not a raven would seem to lend vastly less support to the generalization that all ravens are black than would the observation of one raven that *is* black.

This argument might serve to mitigate the paradoxes of confirmation.[13] But I have stated it here only in an intuitive fashion. A precise formulation would require an explicit quantitative theory of degrees of confirmation or of inductive probability, such as Carnap's. Even within the framework of such a theory, the argument presupposes further assumptions, and the extent to which it can be sustained is not fully clear as yet.

Let us now turn to another perplexing aspect of induction. I will call it Goodman's riddle, because it was Nelson Goodman who first called attention to this problem and proposed a solution for it.[14]

GOODMAN'S RIDDLE: A FAILURE OF CONFIRMATION BY "POSITIVE INSTANCES"

One of the two basic principles from which we deduced the paradoxes of confirmation stated that a generalization of the form "All *F* are *G*" is confirmed, or supported, by its positive instances of type *I*—i.e., by objects which are *F* and also *G*. Although this principle seems entirely obvious, Goodman has shown that there are generalizations that derive no support at all from their observed instances. Take for example the hypothesis

All ravens are blite (*h*)

where an object is said to be blite if it is either examined before midnight tonight and is black or is not examined before midnight and is white.

Suppose now that all the ravens examined so far have been found to be black; then, by definition, all ravens so far examined are also blite. Yet this latter information does not support the generalization *h*, for that generalization implies that all ravens examined after midnight will be white—and

[13] It was first offered by Janina Hosiasson-Lindenbaum in her article "On Confirmation," *The J. of Symbolic Logic,* 5 (1940), 133–48. Similar ideas were proposed by, among others, D. Pears, "Hypotheticals," *Analysis,* 10 (1950), 49–63; I. J. Good, "The Paradoxes of Confirmation," Pts. I and II, *The British J. for the Philosophy of Science,* 11 (1960), 145–48; 12 (1961) 63–64. A detailed and illuminating study of qualitative confirmation and its paradoxes is offered in sec. 3, Pt. I of Israel Scheffler, *The Anatomy of Inquiry* (New York: Knopf, 1963).

[14] Nelson Goodman, *Fact, Fiction, and Forecast* (Cambridge: Harvard U. Press, 1955); 2d, rev. ed. (Indianapolis: Bobbs-Merrill, 1965).

surely our evidence must be held to militate against this forecast rather than to support it.

Thus, some generalizations do derive support from their positive instances of type *I;* for example, "All ravens are black," "All gases expand when heated," "In all cases of free fall from rest, the distance covered is proportional to the square of the elapsed time," and so forth; but other generalizations, of which "All ravens are blite" is an example, are not supported by their instances. Goodman expresses this idea by saying that the former generalizations can, whereas the latter cannot, be *projected* from examined instances to as yet unexamined ones.

The question then arises how to distinguish between projectible and nonprojectible generalizations. Goodman notes that the two differ in the character of the terms employed in their formulation. The term "black," for example, lends itself to projection; the term "blite" does not. He traces the difference between these two kinds of term to what he calls their *entrenchment*—i.e., the extent to which they have been used in previously projected hypotheses. The word "blite," for example, has never before been used in a projection, and is thus much less entrenched than such words as "black," "raven," "gas," "temperature," "velocity," and so on, all of which have served in many previous inductive projections—successful as well as unsuccessful ones. What Goodman thus suggests is that our generalizations are chosen not only in consideration of how well they accord with the available evidence, but also in consideration of how well entrenched are their constituent extralogical terms.

By reference to the relative entrenchment of those terms, Goodman then formulates criteria for the comparison of generalizations in regard to their projectibility, and he thus constructs the beginnings of a theory of inductive projection.

I cannot enter into the details of Goodman's theory here, but I do wish to point out one of its implications which is, I think, of great importance for the conception of inductive inference.

As we noted earlier, the standard rules of deductive inference make reference only to the syntactical form of the sentences involved; the inference rules of quantification theory, for example, apply to all premises and conclusions of the requisite form, no matter whether the extralogical predicates they contain are familiar or strange, well entrenched or poorly entrenched. Thus,

All ravens are blite

and

r is a raven

deductively implies

r is blite

no less than

All ravens are black

and

r is a raven

deductively implies

r is black

But on Goodman's conception of projectibility, even elementary rules of induction cannot be similarly stated in purely syntactical terms. For example, the rule that a positive instance confirms a generalization holds only for generalizations with adequately entrenched predicates; and entrenchment is neither a syntactical nor even a semantic property of terms, but a pragmatic one; it pertains to the actual use that has been made of a term in generalizations projected in the past.

A FURTHER FAILURE OF CONFIRMATION BY "POSITIVE INSTANCES"

Goodman's riddle shows that Nicod's criterion does not offer a generally adequate sufficient condition of confirmation: Positive instances do not confirm nonprojectible hypotheses.

But the criterion fails also in cases of a quite different kind, which do not hinge on the use of predicates such as "blite." Consider the hypothesis, "If for any two persons *x,y* it is not the case that each likes the other, then the first likes the second, but not vice versa"; in symbolic notation:

$$(x)\,(y)[-(Lxy \cdot Lyx) \supset (Lxy \cdot -Lyx)] \qquad (h)$$

Let *e* be the information that *a,b* are two persons such that *a* likes *b* but not vice versa, i.e. that

$$Lab \cdot -Lba \qquad (e)$$

This information can equivalently be stated as follows:

$$-(Lab \cdot Lba) \text{ and } (Lab \cdot -Lba) \qquad (e')$$

for the first of these two sentences is a logical consequence of the second one. The sentence *e′* then represents a positive instance of type *I* for *h;* hence, on Nicod's criterion, *e′* should confirm *h.*[15]

[15] Nicod does not explicitly deal with hypotheses which, like *h,* contain relational terms rather than only property terms such as "raven" and "black"; but the application here suggested certainly seems to be in full accord with his basic conception.

But *e'* is equivalent to

$$-(Lba \cdot Lab) \text{ and } (-Lba \cdot Lab) \qquad (e'')$$

and this, on Nicod's criterion, disconfirms *h*. In intuitive terms, the preceding argument is to this effect: If *a* is counted as the first person and *b* as the second, then the information provided by *e* shows that, as *e'* makes explicit, *a* and *b* satisfy both the antecedent and the consequent of *h* and thus confirm the hypothesis; but if *b* is counted as the first person and *a* as the second one, then by virtue of the same information, *b* and *a* satisfy the antecedent but not the consequent of *h,* as is made explicit in *e''*. Thus, on Nicod's criterion, *e* constitutes both confirming and invalidating evidence for *h*.

Incidentally, *h* can be thrown into the form

$$(x)\,(y)\,(Lxy \cdot Lyx)\,, \qquad (h')$$

which makes it obvious that the evidence *e* logically contradicts the given hypothesis; hence, the same is true of *e'*, although Nicod's criterion qualifies *e'* as confirming *h*.[16]

Hypotheses of the form illustrated by *h* can be formulated in terms of well-entrenched predicate expressions, such as "*x* likes *y*" and "*x* is soluble in *y*"; the difficulty here illustrated does not, therefore, spring from the use of ill-behaved predicates of the Goodmanian variety.

The difficulty rather shows that the intuition which informs the Nicod criterion simply fails when the hypotheses under consideration include relational terms rather than only property terms. If one considers, in addition, that the Nicod criterion it limited to hypotheses of universal conditional form, then it becomes clear that it would be of great interest to develop a general characterization of qualitative confirmation which (1) affords a full definition rather than only partial criteria for the confirmation of a hypothesis *h* by an evidence sentence *e,* (2) is applicable to any hypothesis, of whatever logical form, that can be expressed within a specified language, and (3) avoids the difficulties of the Nicod criterion which have just been pointed out.

An explicit definition of this kind for the concept "*h* qualitatively confirms *e*" has in fact been constructed for the case where *h* and *e* are formulated in a formalized language that has the structure of a first-order functional calculus without identity; *h* may be any sentence whatsoever in such a language, and *e* may be any consistent sentence containing no quantifiers. The concept thus defined demonstrably avoids the difficulties encountered by the Nicod criterion in the case of hypotheses with relational predicates; and it implies the Nicod criterion in reference to those hypoth-

[16] This further paradox of qualitative confirmation was briefly noted in my article, "Studies in the Logic of Confirmation," p. 13.

eses of universal conditional form which contain only property terms. It has been argued, however, that the concept thus arrived at is not fully satisfactory as an explication of the vague idea of qualitative confirmation because it fails to capture certain characteristics which might plausibly be attributed to the relation of qualitative confirmation.[17]

THE AMBIGUITY OF INDUCTION

I now turn to a further basic problem, which I will call the problem of inductive ambiguity. This facet of induction, unlike those we have considered so far, is not a recent discovery; both the problem and a possible solution of it have been recognized, if not always very explicitly, by several writers on probability, past as well as contemporary. But certain aspects of the problem are of special interest in the context of our discussion, and I will therefore consider them briefly.

Suppose that we have the following information:

> Jones, a patient with a sound heart, has just had an appendectomy, and of all persons with sound hearts who underwent appendectomy in the past decade, 93% had an uneventful recovery. (e_1)

This information, taken by itself, would clearly lend strong support to the hypothesis

> Jones will have an uneventful recovery. (h_1)

But suppose that we also have the information:

> Jones is a nonagenarian with serious kidney failure; he just had an appendectomy after his appendix had ruptured; and in the past decade, of all cases of appendectomy after rupture of the appendix among nonagenarians with serious kidney failure only 8% had an uneventful recovery. (e_2)

This information by itself lends strong support to the contradictory of h_1:

> Jones will not have an uneventful recovery. $(-h_1)$

[17] The general definition is developed in "A Purely Syntactical Definition of Confirmation"; the gist of it is presented in sec. 9 of my article essay, "Studies in the Logic of Confirmation." The objections in question were raised especially by R. Carnap in *Logical Foundations of Probability,* secs. 86–88. Briefly, Carnap's principal objection is to the effect that under an adequate definition of qualitative confirmation, e should confirm h only if, in the sense of inductive probability theory, e raises the prior probability of h; and my definition of confirmation is not compatible with such a construal.

But e_1 and e_2 are logically compatible and may well both be part of the information available to us and accepted by us at the time when Jones' prognosis is being considered. In this case, our available evidence provides us with a basis for two rival arguments, both of them inductively sound, whose "conclusions" contradict each other. This is what I referred to above as the ambiguity of inductive reasoning: Inductively sound reasoning based on a consistent, and thus possibly true, set of "premises" may lead to contradictory "conclusions."

This possibility is without parallel in deductive reasoning: The consequences deducible from any premises selected from a consistent set of sentences form again a consistent set.

When two sound inductive arguments thus conflict, which conclusion, if any, is it reasonable to accept, and perhaps to act on? The answer, which has long been acknowledged, at least implicitly, is this: If the available evidence includes the premises of both arguments, it is irrational to base our expectations concerning the conclusions exclusively on the premises of one or the other of the arguments; the credence given to any contemplated hypothesis should always be determined by the support it receives from the *total* evidence available at the time. (Parts may be omitted if they are irrelevant in the sense that their omission leaves the inductive support of the contemplated hypothesis unchanged.) This is what Carnap has called the *requirement of total evidence*. According to it, an estimate of Jones' prospects of recovery should be based on all the relevant evidence at our disposal; and clearly, a physician trying to make a reasonable prognosis will try to meet this requirement as best he can.

What the requirement of total evidence demands, then, is that the credence given to a hypothesis h in a given knowledge situation should be determined by the inductive support, or confirmation, which h receives from the total evidence e available in that situation. Let us call this confirmation $c(h,e)$. Now for some brief comments on this maxim.

(1) In the form just stated, the requirement presupposes a quantitative concept of the degree, $c(h,e)$, to which the evidence e confirms or supports the hypothesis h. This raises the question how such a concept might be defined and whether it can be characterized so generally that $c(h,e)$ is determined for *any* hypothesis h that might be proposed, relative to *any* body of evidence e that might be available. This issue has been much discussed in recent decades. Carnap, in his theory of inductive logic, has developed an explicit and completely general definition of the concept for the case where e and h are any two sentences expressible in one or another of certain formalized languages of relatively simple logical structure.[18] Others have ar-

[18] See especially the following publications: "On Inductive Logic," *Philosophy of Science*, 12 (1945), 72–97; *Logical Foundations of Probability; The Continuum of Inductive Methods* (Chicago: U. of Chicago Press, 1952).

gued that the concept in question can be satisfactorily defined at best for certain special types of hypotheses and of evidential information. For example, if the total relevant evidence consists just of the sentences e_1 and e_2 listed above, certain analysts would hold that no probability or degree of confirmation can be significantly assigned to the hypothesis, "Jones will have an uneventful recovery," since the evidence provides no information about the percentage of uneventful recoveries among nonagenarians with sound hearts but seriously defective kidneys who undergo appendectomy after rupture of the appendix.

(2) Next, let us note that while the requirement of total evidence is a principle concerning induction, it is not a rule of inductive inference or, more precisely, of inductive support, for it does not concern the question whether, or how strongly, a given hypothesis is supported by given evidence. The requirement is concerned rather with the rational use, or application, of inductive reasoning in the formation of empirical beliefs. This observation suggests a distinction between two kinds of rules pertaining to inductive reasoning:

(*a*) *Rules of inductive support, or of valid inductive inference.* These would encompass, for example, all criteria concerning the qualitative confirmation or disconfirmation of generalizations by positive or negative instances; criteria determining degrees of confirmation; and also all general principles connecting degrees of confirmation with each other, such as the law, that the degrees of confirmation of a hypothesis and of its contradictory on the same evidence add up to unity.

(*b*) *Rules of application.* These concern the use of rules of the former kind in the rational formation of empirical beliefs. The requirement of total evidence is one such rule of application, but not the only one, as will soon be seen.

The distinction between rules of inference and rules of application can be made also in reference to deductive reasoning. The rules of inference, as we noted earlier, provide criteria of deductive validity; but they qualify as deductively valid many particular arguments whose conclusions are false, and they do not concern the conditions under which it is reasonable to believe, or to accept, the conclusion of a deductively valid argument. To do so would be the task of rules for the rational application of deductive inference.

One such rule would stipulate, for example, that if we have accepted a set of statements as presumably true, then any logical consequence of that set (or, perhaps rather, any statement that is known to be such a consequence) should equally be accepted as presumably true.

The two kinds of rules for deduction call for quite different kinds of justification. An inference rule such as *modus ponens* might be justified by showing that when applied to true premises it will invariably yield a true conclusion—which is what is meant by the claim that an argument conforming to the rule is deductively valid.

But in order to justify a rule of application, we will have to consider what ends the acceptance or rejection of deductive conclusions is to serve. For example, if we are interested in accepting a set of statements, or of corresponding beliefs, which will afford us an emotionally reassuring or esthetically satisfying account of the world, then it will not always be reasonable to accept, or to believe, the logical consequences of what we have previously accepted. If, on the other hand, truth is what we value in our accepted statements, and if we are accordingly concerned to give credence to all statements that are true as far as our information enables us to tell, then indeed we have to accept all the consequences of previously accepted statements; thus, justification of our rule of application requires reference to the objectives, or the values, that our acceptance procedure is meant to achieve.

INDUCTION AND VALUATION

Similarly, if we wish to devise rules for the rational application of valid inductive reasoning, or if we wish to appraise or justify such rules, we will have to take into account the objectives to be achieved by the inductive acceptance procedure, or the values or disvalues of the consequences that might result from correct or from incorrect acceptance decisions. In this sense, the construction and the justification of inductive acceptance rules for empirical statements presupposes judgments of value.

This is especially obvious when we wish to decide whether a given hypothesis is to be accepted in the strong sense of being relied on as a basis for practical action. Suppose, for example, that a new vaccine has been developed for immunization against a serious infectious disease that can afflict humans as well as chimpanzees. Let *h* be the hypothesis that the vaccine is both safe and effective in a sense specified by suitable operational criteria, and suppose that the hypothesis has been tested by examining a number of samples of the vaccine for safety and effectiveness. Let *e* be the evidence thus obtained.

Our rules of inductive support may then tell us how strongly the hypothesis is confirmed by the evidence; but in deciding whether to act on it we will have to consider, besides the strength of confirmation, also the kind of action that is contemplated, and what benefits might result from a correct decision, what harm from a mistaken one. For example, our standards of acceptance are likely to differ according as humans or chimpanzees are to be treated with the vaccine; and it may well happen that *on the same evidence* the given hypothesis is accepted as a basis of action in one case but rejected in the other.

Inductive decisions of this kind have been extensively studied in the mathematical theory of testing and decision-making. This theory deals in particular with the case where the values or disvalues attached to the possible consequences of the available decisions are expressible in numerical terms as so-called utilities. For such situations, the theory has developed a number of specific decision rules, which are rules of application in our sense. These rules—maximin, maximax, maximizing the expectable utility of the outcome, and others—make the acceptance or the rejection of the hypothesis contingent on the utilities assigned to the different possible consequences of acceptance or rejection; and when a measure for the evidential support of the hypothesis is available, that support is likewise taken into consideration.[19] In this fashion, the inductive decision rules combine empirical considerations with explicitly valuational ones.

That rules for the acceptance or rejection of empirical hypotheses thus presuppose valuational considerations has been emphasized by several writers. Some of these have made the stronger claim that the values in question are ethical values. Thus, Churchman asserts that "the simplest question of fact in science requires for even an approximation, a judgment of value," and that "the science of ethics . . . is *basic* to the meaning of any question the experimental scientist raises."[20] And in the context of a detailed study of the logic of testing statistical hypotheses, Braithwaite asserts, in a similar vein: "To say that it is 'practically certain' that the next 1000 births in Cambridge will include the birth of at least one boy includes a hedonic or ethical assessment."[21]

But while it is true that the justification of rules of acceptance for statements of fact requires reference to judgments of preference or of valuation, the claim that the values concerned are ethical values is, I think, open to question. Our argument about valuational presuppositions has so far been concerned only with the acceptance of hypotheses as a basis of specific *actions,* and in this case the underlying valuations may indeed be ethical in character. But what standards will govern the acceptance and rejection of hypotheses for which no practical application is contemplated? Braithwaite's statement about male births in Cambridge might well belong in that category, and surely so do the hypotheses examined in pure, or basic, scientific research; these might concern, for example, the rate of recession of distant galaxies or the spontaneous creation of hydrogen atoms in empty space. In such cases, it seems, we simply wish to decide, in consideration of the available evidence, whether to believe a proposed hypothesis; whether to record

[19] A lucid account of these rules and of their theoretical use will be found in R. D. Luce and H. Raiffa, *Games and Decisions* (New York: Wiley, 1957).

[20] C. W. Churchman, *Theory of Experimental Inference* (New York: Macmillan, 1948), pp. vii, viii (author's italics).

[21] R. B. Braithwaite, *Scientific Explanation* (Cambridge: Cambridge U. Press, 1953), p. 251.

it, so to speak, in our book of tentative scientific knowledge, without envisaging any technological application. Here, we cannot relevantly base our decisions on any utilities or disutilities attached to practical consequences of acceptance or rejection and, in particular, ethical considerations play no part.

What will have to be taken into account in constructing or justifying inductive acceptance rules for pure scientific research are the objectives of such research or the importance attached in pure science to achieving certain kinds of results. What objectives does pure scientific research seek to achieve? Truth of the accepted statements might be held to be one of them. But surely not truth at all costs. For then, the only rational decision policy would be never to accept any hypothesis on inductive grounds since, however well supported, it might be false.

Scientific research is not even aimed at achieving very high probability of truth, or very strong inductive support, at all costs. Science is willing to take considerable chances on this score. It is willing to accept a theory that vastly outreaches its evidential basis if that theory promises to exhibit an underlying order, a system of deep and simple systematic connections among what had previously been a mass of disparate and multifarious facts.

It is an intriguing but as yet open question whether the objectives, or the values, that inform pure scientific inquiry can all be adequately characterized in terms of such theoretical desiderata as confirmation, explanatory power, and simplicity and, if so, whether these features admit of a satisfactory combination into a concept of purely theoretical or scientific utility that could be involved in the construction of acceptance rules for hypotheses and theories in pure science. Indeed, it is by no means clear whether the conception of basic scientific research as leading to the provisional acceptance or rejection of hypotheses is tenable at all. One of the problems here at issue is whether the notion of accepting a hypothesis independently of any contemplated action can be satisfactorily explicated within the framework of a purely logical and methodological analysis of scientific inquiry[22] or whether, if any illuminating construal of the idea is possible at all, it will have to be given in the context of a psychological, sociological, and historical study of scientific research.[23]

[22] For a fuller discussion and bibliographic references concerning these issues, see, e.g., sec. 12 of C. G. Hempel, "Deductive-Nomological *vs.* Statistical Explanation" in *Scientific Explanation, Space, and Time,* eds. H. Feigl and G. Maxwell, Minnesota Studies in the Philosophy of Science, III (Minneapolis: U. of Minnesota Press, 1962), 98–169. Some of the basic issues are examined in R. B. Braithwaite's paper, "The Role of Values in Scientific Inference," and especially the discussion of that paper in *Induction: Some Current Issues,* eds. H. E. Kyburg, Jr., and E. Nagel (Middletown, Conn.: Wesleyan U. Press, 1963), pp. 180–204.

[23] Such an alternative conception is represented, e.g., by T. S. Kuhn's work, *The Structure of Scientific Revolutions* (Chicago: U. of Chicago Press, 1962).

To conclude with a summary that centers about the classical problem of induction: For a clear statement of the classical problem of justification, two things are required. First, the procedure to be justified must be clearly characterized—this calls for an explication of the rules governing the inductive appraisal of hypotheses and theories; second, the intended objectives of the procedure must be indicated, for a justification of any procedure will have to be relative to the ends it is intended to serve. Concerning the first of these tasks, we noted that while there are no systematic mechanical rules of inductive discovery, two other kinds of rule have to be envisaged and distinguished, namely, rules of support and rules of application. And in our discussion of the objectives of inductive procedures we noted certain connections between rational belief on one hand and valuation on the other.

Whatever insights further inquiry may yield, the recognition and partial exploration of these basic problems has placed the classical problem of induction into a new and clearer perspective and has thereby advanced its philosophical clarification.

9

DE PRINCIPIIS NON DISPUTANDUM . . . ?

[ON THE MEANING AND THE LIMITS OF JUSTIFICATION]

Herbert Feigl

Arguments purporting to justify beliefs or evaluations often proceed from specific to more general issues. Opposition and challenge tend to provoke critical reflection; through various dialectical moves higher levels of justification are reached and made explicit. Argument usually terminates with appeals to principles which are considered indisputable, at least by those who invoke them. But, notoriously, initial disagreements cannot always be removed by what is called "rational argument." Frequently enough, initial disagreement can be traced back to disagreement in basic presuppositions. It is a characteristic of those modern cultures which endorse freedom of thought that they countenance divergencies in religious, political, or economic positions. "It is all a matter of one's ultimate presuppositions"—this phrase and its variants indicate that enlightened common sense is aware of the limits of argument and justification. But on the other hand there is also the deep-rooted wish to be *right,* absolutely right, in one's basic outlook. When the disagreement concerns mere gastronomical matters, we are quite willing to reconcile ourselves with the saying, "*De gustibus non est disputandum.*" Art critics and aestheticians, however, do not unreservedly extend such tolerance to all issues of aesthetic evaluation. Most people, including the majority of philosophers, are still more reluctant to grant any relativity to the basic standards of moral evaluation. There is, at least in this age of sci-

---

From Max Black (ed.), *Philosophical Analysis, A Collection of Essays* (Prentice-Hall, Inc., Englewood Cliffs, N.J., 1963), pp. 113–116 and 123–131. Reprinted by permission of Max Black and the author.

ence, almost complete unanimity as regards the criteria by which we judge the claims of ordinary factual knowledge. And perhaps genuine indisputability is attributed to the principles of formal logic. At least the simplest canons of deductive reasoning, as they are exemplified, e.g., in some of the syllogisms or in elementary arithmetic, are quite generally accepted as indispensable presuppositions of any sort of argument.

While there is no intention here to cast doubt upon the particular gradation just sketched, it need not be taken for granted either. What we do wish to clarify is the status of the very principles which in each of these various fields constitute the standards of validity or the bases of justification. The question mark attached to the title of this essay is not only to indicate that I am going to raise more questions that I shall be able to answer, but also to stress the deeply troublesome and controversial character of the main issues of justification.

The present essay aims at the illumination and at least a partial resolution of the following puzzling questions:

(1) What are the meanings of the term "justification" and what are the logical structures of the corresponding procedures of reasoning?

(2) If justification consists in the stating of reasons, and if the fallacies of the petitio principii and of the infinite regress are to be avoided, what are the limits to which justification can legitimately be pursued? By what criteria do we know that we have reached the limits of justification?

(3) What is the nature of the "ultimate presuppositions" which serve as the uppermost principles of justification?

(4) Can disagreement with respect to these principles be settled only by such nonrational procedures as persuasion, indoctrination, propaganda, therapeutic influence, or coercion? Or else, *if* rational argument concerning first principles *is* possible, what are *its* standards of justification?

(5) How is the issue regarding the primacy of "theoretical reason" vs. the primacy of "practical reason" to be resolved?

In order to approach these intriguing questions with any hope for clarification we shall first have to make sure that we understand what we mean by "justification" and what major types of justification are employed in various contexts.

The search for justification, the capacity for critical reflection, are among the marks of the much vaunted rationality of man. He is sometimes able, and often willing, to state the reasons for accepting or repudiating knowledge claims and evaluations. The procedure of justification is here taken to be precisely this stating of reasons. More fully explicated, justification consists in the disclosure (exhibition, demonstration) of a conformity of that which is to be justified (the *justificandum*) with a certain principle or a set of principles which do the justifying (the *justificans*). We justify claims of factual knowledge by means of empirical confirmation. We cite evidence. But the facts that constitute what we call "evidence" have a bearing on our knowledge claim only by virtue of some principles of confirmation (or induction). We justify claims of mathematical truth by proof. But the validity of a proof depends upon conformity with the principles of deduction. We justify moral approvals or condemnations by reference to ethical principles, and so on. Justification as here understood thus invariably involves at least an implicit reference to some standards or norms which serve, in the given context, as principles of justification. When challenged to justify any one of these principles in turn, people are apt to get impatient or "probably blow up right in your face, because you have put your finger on one of [their] absolute presuppositions, and people are apt to be ticklish in their absolute presuppositions."[1] Indeed, if we ask a typical laboratory scientist what justifies him in his unquestioning acceptance of arithmetic or of the principle of empirical induction, he will, at the very best, tell us that he takes these principles for granted and that it is not part of his business to validate or justify them. We are apt to get an analogous answer from, say, a democratically minded statesman engaged in promoting some measure of social reform, if we ask him for a justification of the principle of justice for all. However, all this is psychology. The facts mentioned may be taken as symptoms of the ultimacy of the principles in the given context. A symptom of a distinctly logical character (but not decisive either, as regards logical ultimacy) is the circularities that are apt to arise on this level of argument. Requests for justification, if complied with at all, tend to elicit answers which are more or less disguised forms of question begging.

It is generally recognized that one of the major tasks of philosophical analysis consists in making explicit (i.e., formulating articulately) the more basic justifying principles. It is almost equally well recognized that *giving reasons* for our beliefs is something altogether different from pointing out their causes. He who has not grasped this difference has not even begun to understand what philosophy is all about. For philosophical analysis endeavors to reconstruct (explicate, clarify) the procedures of justification.

The uses of the word "reason" suffer unfortunately from a good deal of ambiguity. Besides naming a capacity of the human mind (part of which may be the ability to state reasons), it is used to designate causes and purposes, as well as grounds of validation. What a rich source of confusion lies in the little phrases "the reason why," "because," and "since"! Aristotle and Schopenhauer and many thinkers between and after have struggled to disentangle these and other meanings of "reason." Kant stated what concerns us here very clearly by distinguishing between the questions *quid facti* and *quid juris*. Husserl, at the beginning of this century, opposed most explicitly the psychologistic confusion of the two questions. But, even like Kant and Husserl, many other philosophers have not been free from serious relapses into the very confusion they had recognized and severely criticized.

[1] R. G. Collingwood, *An Essay on Metaphysics* (1940), p. 31, cf. also p. 44.

The word "justification" shares some of the ambiguities of the word "reason" (as used in phrases like "giving reasons"). As we proceed we shall find it not only indispensable but also highly clarifying to distinguish between justification in the sense of *validation* and justification in the more usual sense of an argument concerning ***means*** with respect to ***ends***. The type of justification which we wish to distinguish from validation may be called "pragmatic" or "instrumental" justification (*justificatio actionis* as contrasted with *justificatio cognitionis*). In what follows we shall take the terminological liberty of using the term "vindication" as a short expression for this second meaning.

Other terms related to "justification" are "criticism" and "appraisal." In their ordinary meanings these terms stand for reflective acts which are directed toward other acts or attitudes, rather than upon the cognitive content of propositions. If it is cognition that is under critical examination it is usually the process of reasoning, or the acceptance or rejection of knowledge claims, rather than the validity of the relevant propositions that is being "criticized." But here again the two meanings are usually so intimately intertwined that it takes a special analytic effort to separate them neatly from one another....

Justification involving appeal to principles of methodology. It is generally granted that consistency and conclusiveness of reasoning are necessary conditions for arguments purporting to substantiate knowledge claims. It is almost equally well agreed that while consistency and conclusiveness may be sufficient in the purely formal disciplines, they are not sufficient in the realm of factual knowledge. The sort of justification that the claims of factual knowledge require leads us then to a consideration of principles outside the domain of formal logic. These principles belong to the field of inductive logic.

One qualification may be in order here. Inductive logic is not required for a justification of factual-knowledge claims that involve no transcendence beyond the completely and directly given. Perhaps a better way to state this contention is in the subjunctive mood: If there were knowledge claims susceptible to complete and direct verification (or refutation), their justification would involve appeal only to immediate data, to designation rules, to definitions, and to the principles of formal logic. We shall not attempt here to clean up this particular corner of the Augean stables of philosophy. In any case the doctrine of immediate knowledge seems to me highly dubious. The very terms in which we formulate observation statements are used according to rules which involve reference beyond the occasion of direct experience to which they are applied. If they involve no such reference, then they are not terms of a language as we usually conceive languages. Terms referring exclusively to the data of the present moment of a stream of experience could not fulfill the function of symbols in observation statements that are connected with symbols in other statements (of laws and/or other observation statements). If we are not to be reduced to mere signals indicating the individual occurrences of direct experience, we must formulate our statements regarding these individual occurrences in such a manner that they are capable of revision on the basis of other observation statements and of laws that are confirmable by observation.

Appeal to the justifying principles of inductive logic is inevitably made if the dialectic question "How do you know?" is pursued to the limit. An engineer may rest satisfied with reference to specific physical laws when he justifies his claims as to the efficiency (or inefficiency) of a particular machine. A physicist in turn may justify those specific laws by deduction from very general and basic laws, such as those of thermodynamics or electromagnetics. But when pressed for the reasons of his acceptance of those more (or most) general laws, he will invariably begin to speak of verification, of generalization, or of hypothetico-deductive confirmation. A researcher in medicine will proceed similarly. In order to justify a particular hypothesis, e.g., one according to which a certain disease is caused by a virus, he will quote evidence and/or will reason by analogy and induction. Justification of knowledge claims in the historical disciplines (natural as well as social) conforms to the same pattern.

A given item of observation is evidence not in and by itself, but only if viewed in the light of principles of inductive inference. On the level of common sense and on the more "empirical" levels of scientific inquiry those principles are simply the accepted laws of the relevant field. Utilizing some items of evidence and some pertinent laws we justify our assertions concerning past events in history, concerning the causes of diseases in pathology, concerning the existence of as yet unobserved heavenly bodies in astronomy, concerning motivations or learning-processes in psychology, and so on. The search for causes quite generally presupposes the assumption that events do have causes. The investigator of a crime would give us a queer look (if nothing worse) if we asked him how he could be so sure that the death under investigation must have had *some* cause or causes. He takes this for granted and it is not *his* business to justify the principle of causality. Philosophers however have felt that it is their business to justify the belief in causality.

We need not review in too much detail the variety of attempts that have been made in its behalf. The assimilation of causal to logical necessity was definitely refuted by Hume. Kant's transcendental deduction of a causal order depends on the premise that the human understanding impresses this order upon the data of the senses. Part of this premise is the tacit assumption that reason will remain constant throughout time and that it therefore can be relied upon invariably to bestow (the same) causal order upon the data. In this psychologistic version of Kant's epistemology we thus find that the lawfulness of the world is demonstrable only at the price of an assumption concerning the lawfulness of Reason. (We shall not press any further questions concerning the credibility of this ingenious but phantastic account of the nature of cognition.) Turning to the presuppositional version of the *Critique* we gladly acknowledge that Kant, more incisively than any of his predecessors, disclosed the frame of justifying principles within which the questions of natural science are raised as well as answered. But the elevation of strict determinism and of Newtonian mechanics to the rank of synthetic a priori principles proved to be a mistake. The development of recent science provides, if not a fully cogent refutation, at least a most serious counter-argument against any such attempt at a rationalistic petrifaction of the laws of science of a given epoch. More crucial yet, Kant did not achieve what he proposed to do: overcome Hume's skepticism. The presuppositional analysis furnishes no more than this: "Knowledge" as we understand this term connotes explanation and prediction. Therefore, if knowledge of nature is to be possible, nature must be predictable. This is true enough. But do we establish in this fashion a synthetic a priori guarantee for the order of nature? Not in the least. All we have attained is an *analytic* proposition drawn from the definition of "knowledge."

Can *intuition* justify the belief in causality? Even if we granted that we have an intuitive acquaintance with causal necessity in some of its instantiations, how can we assure ourselves without *inductive* leap that the intuited samples are representative of the structure of the world at large?

It should scarcely be necessary to point out that the again fashionable attempts to rehabilitate the concept of Real Connections will not advance the problem of the justification of induction. We grant that problems such as those regarding the meaning of contrary-to-fact conditionals show that Hume's (and generally the radical empiricist) analysis of causal propositions is in need of emendation. Indeed, it seems inevitable to establish and clarify a meaning of "causal connection" that is stronger than Hume's constant conjunction and weaker than entailment or deducibility. Perhaps the distinction between laws (nomological statements) and initial conditions will help here. Our world, to the extent that it is lawful at all, is characterized by both: its laws and its initial condition. Counterfactual conditionals tamper with initial conditions. The frame of the laws is left intact in asking *this* kind of hypothetical question. The "real possibilities" which are correlative to the "real connections" are precisely the class of initial conditions compatible with the laws of our world (or of some other fancied world). These considerations show that the laws of a given world may be viewed as something stronger than can be formulated by means of general implication. They are the very principles of confirmation of any singular descriptive statement that is not susceptible to complete and direct verification. They are therefore constitutive principles of the conception of a given world, or rather of a class or family of worlds which are all characterized by the same laws.[2]

[2] Cp. W. Sellars, "Concepts as Involving Laws and Inconceivable without Them," *Phil. of Science,* XV (1948), p. 287.

But the reconstruction of laws in terms of modal implications does not alter one whit their status in the methodology of science. Clearly, besides counterfactual hypotheticals we can equally easily formulate counternomological ones. Here we tamper with the laws. And the question of which *family* of worlds our world is a member can be answered only on grounds of empirical evidence and according to the usual rules of inductive procedure. Upon return from this excursion we are then confronted, just as before, with the problem of induction.

Obviously the next step in the dialectic must be the lowering of our level of aspiration. We are told that it is the quest for certainty that makes the justification of induction an insoluble problem. But we are promised a solution if only we content ourselves with probabilities. Let us see. "Probability" in the sense of a degree of expectation will not do. This is the psychological concept to which Hume resorted in his account of belief, or that Santayana has in mind when he speaks of "animal faith." What we want is a justifiable degree of expectation. And how do we justify the assignment of probability ratios to predictions and hypotheses? That depends on how we explicate the *objective* concept of probability. But here we encounter the strife of two schools of thought. According to the frequency interpretation there is no other meaning of "probability" than that of the limit of relative frequency. According to the logical interpretation "probability" (in the sense of strength of evidence, weight, degree of confirmation) consists in a logical relation between the evidence which bestows and the proposition upon which there is bestowed a certain degree of credibility. The adherents of this logical interpretation urge that the statistical concept of probability presupposes the logical one. For the ascription of a limit (within a certain interval) to a sequence of frequency ratios is itself an hypothesis and must therefore be judged according to the degree of confirmation that such hypotheses possess in the light of their evidence. Contrariwise, the frequency interpretation urges that locutions such as "degree of confirmation" or "weight," whether applied to predictions of single events or to hypotheses of all sorts, are merely *façons de parler*. Basically they all amount to stating frequency ratios which are generalized from statistical findings regarding the events concerned (the occurrence of successful predictions).

We shall not attempt to resolve the issue between these two schools of thought. Our concern is with the justification of induction. And here perhaps the divergence of interpretations makes no fundamental difference. Inductive probability in the sense of a degree of confirmation is a concept whose definition renders analytic every one of its specific applications. If we use this concept as our guide, that is if we believe that it will give us a maximum of successes in inductive guessing, then this could be explicated as an assertion about the (statistical) structure of the world. We are thus led to essentially the same rule of induction which the frequentists propose: Generalize on the basis of the broadest background of available evidence with a minimum of arbitrariness. This principle of straightforward extrapolation in sequences of frequency ratios applies to a world which need not display a deterministic order. Statistical regularity is sufficient. Does the application of the rule guarantee success? Of course not. Do the past successes of procedures according to this rule indicate, at least with probability, further successes? No again. Hume's arguments refuted this question-begging argument before Mill and others fell victims to the fallacy. Only if we utilize the *logical* concept of probability can we achieve a semblance of plausibility in the criticized argument. But for the reasons stated before, mere definitions cannot settle the issue as to whether our world will be good enough to continue to supply patterns of events in the future which will support this definition as a "useful" one.

The presuppositional analysis sketched thus far has merely disclosed one of the ultimate principles of all empirical inference. Any attempts to validate the principle itself involve question-begging arguments. Its ultimate and apparently ineluctable character can be forcefully brought out by considering how we would behave in a world that is so utterly chaotic and unpredictable that any anticipation of the future on the basis of past experience is doomed to failure. Even in such a world after countless efforts at inductive extrapolation had been frustrated we would (if by some miracle we managed to survive) abandon all further attempts to attain foresight. But would not even this yet be another inductive inference, viz., to the effect that the disorder will continue? The inductive principle thus is ultimately presupposed but in turn does not presuppose any further assumptions.

The various attempts (by Keynes, Broad, Nicod, Russell, and others) to deduce or render probable the principle of induction on the basis of some very general assumption concerning the structure of the world seem to me, if not metaphysical and hence irrelevant, merely to beg the question at issue. Assumptions of Permanent Kinds and of Limited Variety, provided they are genuine assertions regarding the constitution of the universe, themselves require inductive validation. To assign to such vast hypotheses a finite initial (or "antecedent") probability makes sense only if "probability" means subjective confidence. But nothing is gained in this manner. Any objective probability (logical or statistical) would presuppose a principle of induction by means of which we could ascertain the probability of such world hypotheses in comparison with the (infinite) range of their alternatives.

We have reached the limit of justification in the sense of validation. Can we then in any fashion provide a "reason" for this acknowledged principle of "reasonability"? Obviously not, if "reason" is meant in the sense of validating grounds. What we mean (at least in part, possibly as the most prominent part) by "reasonability" in practical life as well as in science consists precisely in the conformance of our beliefs with the probabilities assigned to them by a rule of induction or by a definition of degree of confirmation. We call expectations (hopes or fears) irrational if they markedly deviate from the best inductive estimates. The attempt to validate one of the major principles of all validation, it must be amply obvious by now, is bound to fail. We would be trying to lift ourselves by our own bootstraps.

The only further question that can be raised here concerns a justification of the *adoption* of the rule of induction (or rather of one of the various rules of induction, or one of the various definitions of inductive probability or of degree of confirmation). Such a justification must have the character of a vindication, a *justificatio actionis.* Our question then concerns the choice of means for the attainment of an end. Our end here is clearly successful prediction, more generally, true conclusions of nondemonstrative inference. No deductively necessary guarantee can be given for the success of such inferences even if we follow some rule of induction. The probability of success can be proved, but that is trivial because we here utilize the concept of probability which our rule of induction implicitly defines. This probability cannot be construed as an estimate of the limit of frequency. We do not know whether such a limit exists. Only if we grant hypothetically that there is such a limit can we assign weights to its various values (i.e., to intervals into which the limit may fall) on the basis of (always finite) statistical samples. What then justifies our optimistic belief in the convergence of statistical sequences toward a limit? Since any attempt at validation would inevitably be circular, we can only ask for a vindication of the rule according to which we posit the existence of limits. Reichenbach's well known but widely misunderstood justification of induction[3] consists, as I see it, in a vindication of the adoption of that rule. It amounts to the *deductive* proof that no method of attaining foresight could conceivably be successful if every sort of induction were bound to fail. Perhaps there are alternative techniques of foresight that might even be more efficient or reliable than the laborious method of scientific generalization. But such alternative methods (let us none too seriously mention crystal gazing, clairvoyance, premonitions, etc.) would themselves have to be appraised by their success; i.e., they would have to be accepted or rejected on the basis of statistical studies of the frequency ratio of correct predictions achieved by them. And our confidence in such "alternative" techniques of foresight would therefore ultimately be justifiable only on the basis of normal induction. *If* there is an order of nature at all (i.e., at least a statistical regularity), not too complex or deeply hidden, then a consistent application of the rule of induction will reveal it. This statement is of course a tautology. It should not be confused with such bolder and undemonstrable factual assertions as that the inductive procedure is the only reliable one, or that it is our best bet. Reliability and optimal wagering presuppose inductive probabilities and thus cannot be invoked for their justification. We cannot even say that straightforward generalization is a necessary condition (never known to be sufficient) for the success

[3] H. Reichenbach, *Experience and Prediction* (1938). See also his article "On the Justification of Induction," *Jour. of Phil.,* XXXVII (1940), p. 101. Also reprinted in H. Feigl and W. Sellars (eds.), *Readings in Philosophical Analysis* (1949).

of predictions. The air of plausibility that this statement shares with its close relative, the common-sense slogan "Nothing ventured, nothing gained" arises only if we disregard (logically conceivable) alternative routes to predictive success, such as sheer inspiration, capricious guessing, intuition, premonition, etc. The unique character that the inductive procedure possesses in contrast with those alternatives rests exclusively in this: The method of induction is *the only one for which it can be proved* (deductively!) that it leads to successful predictions *if* there is an order of nature, i.e., *if* at least some sequences of frequencies do converge in a manner not too difficult to ascertain for human beings with limited experience, patience, and ingenuity. This is the tautology over again, in expanded form, but just as obvious and trivial as before. In the more ordinary contexts of pragmatic justification the validity of induction is invariably presupposed. If we want to attain a certain end, we must make "sure" (i.e., probable) that the means to be chosen will achieve that end. But if we ask for a vindication of the adoption of the very *principium* of all induction, we deal, so to speak, with a degenerate case of justification. We have no assurance that inductive probabilities will prove a useful guide for our lives beyond the present moment. But equally we have no reason to believe that they will fail us. We know furthermore (as a matter of logical necessity or tautology) that *if* success can be had at all, in any manner whatsoever, it can certainly be attained by the inductive method. For this method is according to its very definition so designed as to disclose whatever order or regularity may be present to be disclosed. Furthermore, since the inductive method is self-corrective, it is the most flexible device conceivable for the adaptation and readaptation of our expectations.

The conclusion reached may seem only infinitesimally removed from Hume's skepticism. Philosophers do not seem grateful for small mercies. In their rationist quest for certainty many still hope for a justification of a principle of uniformity of nature. We could offer merely a deductive (and trivial) vindication of the use of the pragmatic rule of induction. But for anyone who has freed himself from the wishful dreams of rationalism the result may nevertheless be helpful and clarifying. It is the final point which a consistent empiricist must add to his outlook. We refuse to countenance such synthetic a priori postulates as Russell (perhaps not with the best intellectual conscience) lately found necessary to stipulate regarding the structure of the universe. We insist that no matter how general or pervasive the assumptions, as long as they are about the *universe*, they fall under the jurisdiction of the rule of induction. This rule itself is not then a factual assertion but the maxim of a procedure or what is tantamount, a definition of inductive probability. In regard to rules or definitions we cannot raise the sort of doubt that is sensibly applicable to factual assertions. In order to settle doubts of the usual sort we must rely on some principles without which neither doubt nor the settlement of doubt makes sense. The maxim of induction is just such a principle.

10

SELF-SUPPORTING INDUCTIVE ARGUMENTS

Max Black

The use of inductive rules has often led to true conclusions about matters of fact. Common sense regards this as a good reason for trusting inductive rules in the future, if due precautions are taken against error. Yet an argument from success in the past to probable success in the future itself uses an inductive rule, and therefore seems circular. Nothing would be accomplished by any argument that needed to assume the reliability of an inductive rule in order to establish that rule's reliability.

Suppose that inferences governed by some inductive rule have usually resulted in true conclusions; and let an inference from this fact to the probable reliability of the rule in the future be called a *second-order* inference. So long as the rule by which the second-order inference is governed differs from the rule whose reliability is to be affirmed, there will be no appearance of circularity. But if the second-order inference is governed by the very same rule of inference whose reliability is affirmed in the conclusion, the vicious circularity seems blatant.

Must we, then, reject forthwith every second-order inductive argument purporting to support the very rule of inference by which the argument itself is governed? Contrary to general opinion, a plausible case can be made for saying, No.[1] Properly constructed and interpreted, such "self-supporting" inferences, as I shall continue to call them, can satisfy all the conditions for legitimate inductive inference: when an inductive rule has been reliable (has generated true conclusions from true premises more often than not) in the

[1] See Max Black, *Problems of Analysis* (Ithaca, N.Y., 1954), chap. 11, and R. B. Braithwaite, *Scientific Explanation* (Cambridge, 1953), chap. 8.

From *The Journal of Philosophy,* vol. 55 (1958), pp. 718–725. Reprinted by permission of the author and *The Journal of Philosophy.*

past, a second-order inductive inference governed by the same rule can show that the rule deserves to be trusted in its next application.

The reasons I have given for this contention have recently been sharply criticized by Professor Wesley C. Salmon.[2] In trying to answer his precisely worded objections, I hope to make clearer the view I have been defending and to dispel some lingering misapprehensions.

My original example of a legitimate self-supporting inductive argument was the following:[3]

> (a) In most instances of the use of *R* in arguments with true premises examined in a wide variety of conditions, *R* has been successful.
>
> *Hence (probably)*:
>
> In the next instance to be encountered of the use of *R* in an argument with a true premise, *R* will be successful.

The rule of inductive inference mentioned in the premise and the conclusion of the above argument is:

> *R:* To argue from *Most instances of A's examined in a wide variety of conditions have been B* to (probably) *The next A to be encountered will be B.*

Thus the second-order argument (a) uses the rule *R* in showing that the same rule will be "successful" (will generate a true conclusion from a true premise[4]) in the next encountered instance of its use.

The rule, *R*, stated above is not intended to be a "supreme rule" of induction, from which all other inductive rules can be derived; nor is it claimed that *R*, as it stands, is a wholly acceptable rule for inductive inference. The unsolved problem of a satisfactory formulation of canons of inductive inference will arise only incidentally in the present discussion. The rule *R* and the associated argument (a) are to serve merely to illustrate the logical problems that arise in connection with self-supporting arguments: the considerations to be adduced in defense of (a) could be adapted to fit many other self-supporting arguments.

The proposed exculpation of the self-supporting argument (a) from the

[2] See Wesley C. Salmon, "Should We Attempt to Justify Induction?" *Philosophical Studies,* vol. VIII, No. 3 (April 1957), pp. 45–47.

[3] See *Problems,* page 197, where the argument is called '(a_2)' and the rule by which it is governed 'R_2'. At that place, I also presented another self-supporting argument with a more sweeping conclusion of the *general* reliability of the corresponding rule. But since I was unable to accept the premise of that argument, or the reliability of the rule it employed, I shall follow Salmon in discussing only the argument presented above.

[4] Here and throughout this discussion, I assume for simplicity that all the premises of any argument or inference considered have been conjoined into a single statement.

charge of vicious circularity is linked to a feature of the corresponding rule *R* that must be carefully noted. Inductive arguments governed by *R* vary in "strength"[5] according to the number and variety of the favorable instances reported in the premise. So, although *R* permits us to assert a certain conclusion categorically, it is to be understood throughout that the strength of the assertion fluctuates with the character of the evidence. If only a small number of instances have been examined and the relative frequency of favorable instances (*A*'s that are *B*) is little better than a half, the strength of the argument may be close to zero; while a vast predominance of favorable instances in a very large sample of observations justifies a conclusion affirmed with nearly maximal strength. The presence of the word "probably" in the original formulation of *R* indicates the variability of strength of the corresponding argument; in more refined substitutes for *R*, provision might be made for some precise measure of the associated degree of strength.

Variability in strength is an important respect in which inductive arguments differ sharply from deductive ones. If a deductive argument is not valid, it must be *in*valid, no intermediate cases being conceivable; but a legitimate inductive argument, whose conclusion may properly be affirmed on the evidence supplied, may still be very weak. Appraisal of an inductive argument admits of degrees.

Similar remarks apply to inductive rules, as contrasted with deductive ones. A deductive rule is either valid or invalid—*tertium non datur*; but at any time in the history of the employment of an inductive rule, it has what may be called a *degree of reliability* depending upon its ratio of successes in previous applications. A legitimate or correct inductive rule may still be a weak one: appraisal of an inductive rule admits of degrees.

Now in claiming that the second-order argument (a) *supports* the rule *R*, I am claiming that the argument raises the degree of reliability of the rule, and hence the strength of the arguments in which it will be used; I have no intention of claiming that the self-supporting argument can definitively establish or demonstrate that the rule is correct. Indeed, I do not know what an outright demonstration of the correctness or legitimacy of an inductive rule would be like. My attempted rebuttal of Salmon's objections will turn upon the possibility of raising the degree of reliability of an inductive rule, as already explained.

The contribution made by the second-order argument (a) to strengthening the rule *R* by which it is governed can be made plain by a hypothetical illustration. Suppose evidence is available that 4/5 of the *A*'s so far examined

[5] In *Problems,* page 193, I spoke, with the same intention, of the "degree of support" given to the conclusion by the premise. If the latter has the form *m/n A's examined in a wide variety of conditions have been B,* it is natural to suppose that the strength of the argument increases as *m* increases, and also as *m/n* increases. A plausible formula for the "strength" of the argument might be $(1-e^{-m})(2m/n-1)$.

have been B, and it is proposed, by an application of the rule R, to draw the inference that the next A to be encountered will be B. For the sake of simplicity the proposed argument may be taken to have a strength of $4/5$.[6] Before accepting the conclusion about the next A, we may wish to consider the available evidence about past successes of the rule R. Suppose, for the sake of argument, that we know R to have been successful in $9/10$ of the cases in which it has been previously used. If so, the second-order argument affirms with strength $9/10$ that R will be successful in the next instance of its use. But the "next instance" is before us, as the argument whose premise is that $4/5$ of the A's have been B. For R to be "successful" in this instance is for the conclusion of the first-order argument to be true; the strength of the second-order argument is therefore immediately transferred to the first-order argument. Before invoking the second-order argument, we were entitled to affirm the conclusion of the first-order argument with a strength of no better than $4/5$, but we are now able to raise the strength to $9/10$. Conversely, if the second-order argument had shown R to have been unsuccessful in less than $4/5$ of its previous uses, our confidence in the proposed conclusion of the first-order argument would have been diminished.

There is no mystery about the transfer of strength from the second-order argument to the first-order argument: the evidence cited in the former amplifies the evidence immediately relevant to the latter. Evidence concerning the proportion of A's found to have been B permits the direct inference, with strength $4/5$, that the next A to be encountered will be B. It is, however, permissible to view the situation in another aspect as concerned with the extrapolation of an already observed statistical association between true premises of a certain sort and a corresponding conclusion. The evidence takes the form: In 9 cases out of 10, the truth of a statement of the form *m/n X's have been found to be Y's* has been found associated in a wide variety of cases with the truth of the statement *The next X to be encountered was Y*.[7] This is better evidence than that cited in the premise of the original first-order argument: it is therefore to be expected that the strength of the conclusion shall be raised.

It should be noticed that the evidence cited in the second-order argument is not merely greater in amount than the evidence cited in the first-order argument. If R has been successfully used for drawing conclusions about fish, neutrons, planets, etc. (the "wide variety of conditions" mentioned in the premise of the second-order argument), it would be illegitimate to coalesce such heterogeneous kinds of objects into a single class for the sake

[6] This means taking m/n as the measure of strength, rather than some more complicated formula like the one suggested in footnote 5 above. The argument does not depend upon the exact form of the measure of strength.

[7] We might wish to restrict the second-order argument to cases in which the ratio m/n was close to $4/5$. Other refinements readily suggest themselves.

of a more extensive *first-order* argument. Proceeding to 'second-order' considerations allows us to combine the results of previous inductive inquiries in a way which would not otherwise be possible.

Nothing in this conception of inductive method requires us to remain satisfied with the second-order argument. If circumstances warrant, and suitable evidence can be found, we might be led to formulate third- or even higher-order arguments. These might conceivably result in lowering the measures of strength we at present attach to certain arguments in which R is used. But if this were to happen, we would not have been shown to have been mistaken in previously attaching these measures of strength. Nor is it required that a first-order argument be checked against a corresponding second-order argument before the former can properly be used. If we have no reason to think that R is unsuccessful most of the time, or is objectionable on some logical grounds, that is enough to make our employment of it so far reasonable. The function of higher-order arguments in the tangled web of inductive method is to permit us to progress from relatively imprecise and uncritical methods to methods whose degrees of reliability and limits of applicability have themselves been checked by inductive investigations. It is in this way that inductive method becomes self-regulating and, if all goes well, self-supporting.

Salmon's objections to the foregoing conception are summarized by him as follows:

> The so-called self-supporting arguments are ... circular in the following precise sense: the conclusiveness of the argument cannot be established without assuming the truth of the conclusion. It happens, in this case, that the assumption of the truth of the conclusion is required to establish the correctness of the rules of inference used rather than the truth of the premises, but that makes the argument no less viciously circular. The circularity lies in regarding the facts stated in the premises as *evidence* for the conclusion, rather than as evidence against the conclusion or as no evidence either positive or negative. To regard the facts in the premises as evidence for the conclusion is to assume that the rule of inference used in the argument is a correct one. And this is precisely what is to be proved. If the conclusion is denied, then the facts stated in the premises are no longer evidence for the conclusion.[8]

Comments: (1) Salmon's reference to "conclusiveness" smacks too much of the appraisal of deductive argument. An inductive argument is not required to be "conclusive" if that means that its conclusion is entailed or logically implied by its premises; it is, of course, required to be correct or legitimate, but that means only that the rule of inductive inference shall be reliable—shall usually lead from true premises to true conclusions. The correctness of an inductive argument could only depend upon the truth of its conclusion if the latter asserted the reliability of the rule by which the

[8] Salmon, *loc. cit.*, p. 47.

argument was governed. But this was not the case in our argument (a). The conclusion there was that *R* would be successful in the next instance of its use: this might very well prove to be false without impugning the reliability of *R*. Salmon was plainly mistaken if he thought that the falsity of (a)'s conclusion entails the incorrectness of the rule by which (a) is governed.[9]

(2) Can the *correctness* of argument (a) be "established without assuming the truth of the conclusion" of (a)? Well, if "established" means the same as "proved by a deductive argument," the answer must be that the correctness of (a) cannot be established at all. But again, a correct inductive argument in support of the rule governing (a) can certainly be constructed without assuming (a)'s conclusion. We do not have to assume that *R* will be successful in the next instance in order to argue correctly that the available evidence supports the reliability of *R*.

(3) Salmon says: "To regard the facts in the premises as evidence for the conclusion is to assume that the rule of inference used in the argument is a correct one." In using the rule of inference we certainly *treat* it as correct: we would not use it if we had good reasons for suspecting it to be unreliable. If this is what Salmon means, what he says is right, but not damaging to the correctness of (a). But he would be plainly wrong if he maintained that an assertion of the correctness of (a) was an additional premise required by (a), or that an argument to the effect that (a) was correct must precede the legitimate use of (a). For if this last demand were pressed, it would render deductive inference no less than inductive inference logically impossible. If we were never entitled to *use* a correct rule of inference before we had formally argued in support of that rule, the process of inference could never get started.

I shall end by considering an ingenious counter-example provided by Salmon. He asks us to consider the following argument:

> (a′): In most instances of the use of *R′* in arguments with true premises in a wide variety of conditions, *R′* has been *un*successful.
>
> *Hence* (*probably*)
>
> In the next instance to be encountered of the use of *R′* in an argument with a true premise, *R′* will be successful.

The relevant rule is the "counter-inductive" one:

[9] I conjecture that Salmon was led into making this mistake by forgetting the conclusion of the argument that he correctly reproduces at the foot of page 45 of his article. It is a sheer blunder to say "A given inductive rule can be established by a self-supporting argument, according to Black" (p. 45)—if "established" means the same as "proved reliable." The self-supporting argument can *strengthen* the rule, and in this way *"support"* it.

R′: To argue from *Most instances of A's examined in a wide variety of conditions have not been B* to (probably) *The next A to be encountered will be B.*

Salmon says that while (a′) must be regarded as a self-supporting argument by my criteria, the rule here supported, *R′*, is in conflict with *R*. From the same premises the two rules "will almost always produce contrary conclusions."[10] This must be granted. But Salmon apparently overlooks an important respect in which the "counter-inductive" rule *R′* must be regarded as illegitimate.

In calling an inductive rule "correct," so that it meets the canons of legitimacy of *inductive* rules of inference, we claim at least that the rule is reliable, in the sense of usually leading from true premises to true conclusions. That is part of what we *mean* by a "correct inductive rule." It can easily be shown that *R′* must fail to meet this condition.

Suppose we were using *R′* to predict the terms of a series of 1's and 0's, of which the first three terms were known to be 1's. Then our first two predictions might be as follows (shown by underlining):

$$1\ \ 1\ \ 1\ \ \underline{0}\ \ \underline{0}$$

At this point, suppose *R′* has been used successfully in each of the two predictions, so that the series is in fact now observed to be 1 1 1 0 0. Since 1's still predominate, direct application of the rule calls for 0 to be predicted next. On the other hand, the second-order argument shows that *R′* has been successful each time and therefore demands that it not be trusted next time, i.e., calls for the prediction of 1. So the very definition of *R′* renders it impossible for the rule to be successful without being *incoherent.*[11] The suggested second-order argument in support of *R′* could be formulated only if *R′* were known to be unreliable, and would therefore be worthless. So we have an *a priori* reason for preferring *R* to its competitor *R′*. But it is easy to produce any number of alternative rules of inductive inference, none of which suffers from the fatal defect of *R′*. The choice between such rules, I suggest, has to be made in the light of experience of their use. I have tried to show in outline how such experience can properly be invoked without logical circularity.

[10] Salmon, *loc. cit.*, p. 46.

[11] A parallel situation would arise in the use of *R* in predicting the members of the 1–0 series only if *R* were to be predominantly *un*successful. But then we would have the best of reasons for assigning *R* zero strength, and the second-order argument would be pointless.

11

ON VINDICATING INDUCTION

Wesley C. Salmon

Although the process of mathematical idealization applied to empirical reality may not be perfectly understood as yet, there can be no doubt of its utility in science and in the logical analysis of scientific method. The philosophical theory of probability which identifies probabilities with limits of relative frequencies in infinite sequences of events certainly involves such idealization. I wish to take this idealization as a point of departure, setting aside the problems connected with the idealization as such. I shall assume, therefore, that when we are dealing with probabilities we are often dealing with large aggregates of events which can reasonably be treated as if they were infinite ordered sequences. I shall assume further that it would be desirable, if possible, to have a way of inferring the value of the limit of the relative frequency of an attribute in this type of infinite sequence. The problem is an inductive problem. Clearly we are concerned with the type of sequence in which our inferences must be based upon observational data, in particular, observation of the relative frequency of the attribute in some finite initial portion of the sequence. Whether or not there exists, in an abstract sense, a mathematical rule governing the development of the sequence, no such rule is available to us as part of our data. The problem to which I shall address myself is the problem of selecting and justifying a rule to govern the inference from a statement about the relative frequency in a finite initial section of a sequence to a statement of the value of the limit of the relative frequency in that sequence. This is precisely the prob-

---

lem Reichenbach attempted to solve with his "pragmatic justification" of induction.[1]

There have been many criticisms of Reichenbach's argument, but the crucial one, it seems to me, is one of which he was quite aware. By his own admission, the same argument which he uses to justify his *rule of induction* justifies equally well any other rule in an infinite class of "asymptotic rules." He attempted to resolve this difficulty by arguing that, since the results given by all of these asymptotic rules converge in the long run to the same value when they are applied to the same sequence of events, we are justified in selecting his *rule of induction* on grounds of "descriptive simplicity."[2] This argument is faulty. Although the rules do, as he says, converge, they do not converge uniformly. As a matter of fact, the class of asymptotic rules contains rules such that, for any observed relative frequency in an initial section of any finite length and for any arbitrarily selected real number between zero and one inclusive, there exists an asymptotic rule permitting the inference of that real number as the value of the limit on the basis of the given observed frequency.[3] This means that the asymptotic rules are not in any sense empirically equivalent. On Reichenbach's own grounds, then, descriptive simplicity cannot be considered as a basis for selecting one rule from the infinite class of asymptotic rules. As long as this objection to Reichenbach's pragmatic justification stands, any other objections seem to me to be superfluous.

Another set of considerations can be introduced. Let A be any reference class (ordered sequence) of events and let $B_1, \ldots, B_k$ be attributes which are mutually exclusive and exhaustive within A. Let $F^n(A, B_i)$ be the relative frequency with which B_i occurs in A up to and including the nth place in the sequence. Simple arithmetical considerations yield the following conclusions:

$$\sum_{i=1}^{k} F^n(A, B_i) = 1, \text{ for every } n \tag{1}$$

$$0 \leqslant F^n(A, B_i) \leqslant 1, \text{ for every } n \tag{2}$$

$$\sum_{i=1}^{k} \lim_{n \to \infty} F^n(A, B_i) = 1 \tag{3}$$

$$0 \leqslant \lim_{n \to \infty} F^n(A, B_i) \leqslant 1 \tag{4}$$

[1] H. Reichenbach, *The Theory of Probability,* Berkeley, University of California Press, 1959, sec. 87.

[2] *Ibid.,* p. 447.

[3] W. Salmon, "The Predictive Inference," *Philosophy of Science,* vol. 24, 2 (April 1957).

Any rule, asymptotic or otherwise, for inferring from the relative frequency in an initial section to the limit of the relative frequency can be characterized in the following manner:

From

$$F^n(A, B_i) = m_i/n$$

$$\tag{5}$$

to infer

$$\lim_{n\to\infty} F^n(A, B_i) = m_i/n + f(x_1, \ldots, x_r)$$

where the nature of the arguments $x_1, \ldots, x_r$ of the function f is, for the moment, purposely left unspecified. By virtue of (3) and (4), in order to avoid selecting rules which will lead to self-contradictory conclusions, the following *normalizing conditions* must be imposed upon our rules:

$$0 \leqslant m_i/n + f(x_1, \ldots, x_r) \leqslant 1 \tag{6}$$

$$\sum_{i=1}^{k} [m_i/n + f(x_1, \ldots, x_r)] = 1 \tag{7}$$

These conditions must be satisfied for each value of n ($n = 1, 2, 3, \ldots$); otherwise, for some n our rule would permit an inference to a self-contradictory conclusion on the basis of a sample of that size.

Unfortunately the normalizing conditions alone will not solve the problem of finding a unique rule, for there are infinitely many asymptotic rules which satisfy these conditions. This infinite class of rules exhibits the kind of non-uniform convergence mentioned above.[4]

There is, however, still a further condition which may be placed upon rules of the type being considered.[5] In order to state this condition it is necessary to explain what is meant by a *purely linguistic difference* between two expressions. We shall say that two expressions in a given language have a *purely linguistic difference* (or alternatively that they differ by a *purely linguistic transformation*) if their equivalence follows from the syntactical and semantical rules of that language. If the expressions are propositional or sentential expressions, the equivalence of which I speak is material equivalence. If the expressions are class expressions, the relevant equivalence is class identity. If the expressions are predicates, equivalence consists in having the same extension. The notion of a purely linguistic difference can be extended to the case of expressions in different languages in a very natural

[4] *Ibid.*

[5] The discussion of the *criterion of linguistic invariance* follows closely a discussion of W. Salmon, "Vindication of Induction," in Feigl and Maxwell, editors, *Current Issues in Philosophy of Science*, New York, Holt, Rinehart & Winston, 1961. In this article the criterion is applied to confirmation functions as well as inductive rules.

way. Two expressions in two different languages have a purely linguistic difference if their equivalence can be demonstrated in a metalanguage containing the semantics and syntax of both languages. The following two examples illustrate what is meant by a purely linguistic difference:

(1) The statement that a certain bar of iron is thirty-six inches long differs purely linguistically from the statement that the same bar of iron is at the same moment three feet long.

(2) If a new predicate, "rend," is introduced which is equivalent by definition to the predicate "both round and red," then the statement "*x* is rend" differs purely linguistically from the statement "*x* is both round and red."

However, even if it should happen, as matter of fact, that the class of featherless bipeds contains the same members as the class of rational animals, the statement, "Socrates is a featherless biped," would *not* differ purely linguistically from the statement, "Socrates is a rational animal," for the identity of the two classes does not follow from syntactical and semantical considerations alone.

I think it is fundamental to theory of probability and induction that we recognize and adopt a certain *principle of invariance with respect to purely linguistic transformations*. This principle, roughly stated, asserts that probability relations, relations of confirmation, and relations of inductive support are not functions of purely linguistic considerations. The principle in turn gives rise to a criterion of adequacy for inductive methods or rules; I call it the *criterion of linguistic invariance*. It places the following requirement upon inductive rules:

> Whenever two inductive inferences are made according to the same rule, if the premises of the one differ purely linguistically from the premises of the other, then the conclusion of the one must not contradict the conclusion of the other.

This criterion is obviously a consistency requirement; violations of it are cases in which an inductive rule permits contradictory conclusions to be derived from the same evidence. The contradiction in question may, so to speak, bridge two languages, but that does not make it any less objectionable.

The force of this criterion may be illustrated by reference to the preceding examples. If, in a scientific experiment, the result that a certain bar of iron is thirty-six inches long confirms some hypothesis to a certain degree, then the criterion requires that the result that the same bar of iron is three feet long must confirm that same hypothesis to the same degree. Similarly, suppose we have observed that the ratio of *A*'s in a particular sample which are both red and round is m/n. It follows, of course, that the ratio of *A*'s in the same sample which are rend is also m/n. If an inductive rule permitted us to infer from the fact that m/n *A*'s are red and round that the limit of the relative frequency of things being red and round in *A* is p, and if the same inductive rule permitted us to infer from the fact that m/n A's (in the same sample) are rend that the limit of the relative frequency of things being both red and round in A is q, where $p \neq q$, this rule would violate the criterion.

The criterion of linguistic invariance strikes me as being eminently reasonable; indeed, almost trivially so. There is, I think, only one possible alternative to adopting it, and that alternative is to maintain that there exists one particular privileged language of science whose special status can be established a priori. If this were the case, then there would be no need for a criterion of linguistic invariance for there would be no occasion to translate statements from the privileged language into an inferior one. This view has, I think, nothing to recommend it; it involves the most egregious sort of metaphysics. If such metaphysics is to be avoided, the criterion of linguistic invariance must be adopted.

Let us see what can be accomplished by application of the criterion of linguistic invariance, utilizing at the same time the normalizing conditions set out above. Since the normalizing conditions, as stated, must be satisfied for every value of n, we will proceed with the argument for cases in which n has some fixed value. In order to carry out this task we must examine the function f which appears in the general characterization of inductive rules, and we must pay particular attention to the arguments $x_1, \ldots, x_r$ of this function. Two facts are immediately apparent. The quantities i and k cannot be arguments of the function f, for these quantities are subject to change by a purely linguistic transformation. The variable i is the index variable for the predicates; the value of i attached to a given attribute can be changed by a mere reordering of the predicates. If the inferred value of the limit of the relative frequency of an attribute were a function of i, this inferred value would not be linguistically invariant. (One might, I suppose, seek to escape this conclusion by making i the index of the attributes themselves, rather than the predicates. This procedure would amount to mentioning specific attributes in the inductive rules, and this, in turn, is tantamount to adopting a privileged language of science—an alternative already rejected.) A similar consideration applies to k, the number of predicates in our exclusive and exhaustive set. This number can be changed by the trivial linguistic transformation of defining two new predicates which are mutually exclusive and whose disjunction is equivalent to one of the predicates in the original set—that is, define $B_{i'}$ and $B_{i''}$ so that $B_i = B_{i'} \vee B_{i''}$. Again, if the inferred value of the limit of the relative frequency of some attribute were a function of k, this inferred value could be changed by the foregoing sort of purely linguistic transformation.

There is one remaining possibility to be considered. Can f be a function of m_i alone? Since we are dealing with cases in which n is fixed, it is equivalent to consider what happens if f is a function of m_i/n alone. Utilizing the normalizing conditions it is easy to show that $f(m_i/n)$ must be identically zero.

(1) Consider the case in which $k = n$; there are the same number of attributes as there are members of the sample. Suppose further that $m_i = 1$ for each i; that is, each attribute occurs once in the sample.
By (7),

$$\sum_{i=1}^{k} [1/n + f(1/n)] = k[1/n + f(1/n)] = 1 + kf(1/n) = 1$$

Hence $\quad f(1/n) = 0$

(2) It can now be shown that $f(m_i/n) = 0$ for any possible value of m_i. Let m_1 have any value such that $0 \leqslant m_1 \leqslant n$. Let $m_i = 1$ for $i = 2, 3, \ldots, k$; that is, we are considering the case in which B_1 occurs an arbitrary number of times m_1 in our sample and each other attribute occurs just once. We have

$$\sum_{i=1}^{k} [m_i/n + f(m_i/n)] = \sum_{i=1}^{k} m_i/n + f(m_1/n) + \sum_{i=2}^{k} f(1/n) = 1$$

By (1),

$$\sum_{i=1}^{k} m_i/n = 1$$

We have just proved that $f(1/n) = 0$.

Hence, $\quad f(m_i/n) = 0$

There is one inductive rule characterized by the condition that the function f is identically zero; this is the Reichenbach rule of induction. We have just shown that no other rule beside this one can fulfill both the normalizing conditions and the criterion of linguistic invariance. We have considered the possible functional dependency of inferred values of limits upon n, m_i, i, and k; these are the only variables which appear in the statement of the normalizing conditions. The argument thus proves that there is at most one consistent rule for inferring values of limits of relative frequencies from observed initial portions. A violation of either the normalizing conditions or the criterion of linguistic invariance by a rule means, as has been noted, that the rule can give rise to inconsistency.

The natural question at this point is whether the rule of induction is a consistent rule. Quite obviously it cannot violate the normalizing conditions. But does it meet the criterion of linguistic invariance? Unfortunately it does not as things stand.[6] This point is demonstrated by a version of the Good-

[6] This objection is stated by S. Barker in his comments upon the article cited in footnote 5. These comments appear in the same volume.

man paradox, a paradox which may appear rather trivial but which turns out to be profound and difficult.[7]

If we have observed a large number of emeralds and found them all to be green, by the rule of induction we would be entitled to infer that all emeralds are green. Now, let *t* be some specified future time (e.g., the beginning of the twenty-first century) and let the predicate "grue" be defined so that it applies to things examined before *t* which are green and it applies to things examined after *t* which are blue. Similarly, let the predicate "bleen" be defined so that it applies to things examined before *t* which are blue and it applies to things examined after *t* which are green. According to the definition of "grue," the above-mentioned observations may equally well be described by the statement that all observed emeralds are grue. From this evidence we are entitled, by the rule of induction, to conclude that all emeralds are grue. This conclusion contradicts the former conclusion that all emeralds are green. In this paradox we have something very close to a violation of the criterion of linguistic invariance.

The immediate intuitive reaction to this paradox is to say that there is something peculiar about the predicates "grue" and "bleen," in particular that they are time-dependent predicates. As Goodman has carefully pointed out, however, this intuition is not easy to justify. It is true, of course, that "grue" and "bleen" are positional—i.e., involve explicit reference to the time *t*—if "green" and "blue" are taken as basic predicates in terms of which they are defined. However, if "grue" and "bleen" are taken as basic, then "green" and "blue" exhibit this peculiarity. In this case, "green" would be defined so as to apply to things examined before t which are grue and to things examined after *t* which are bleen. "Blue" would be defined analogously. In view of this symmetry it is difficult to see in what sense we could say that "grue" and "bleen" are time-dependent predicates while "blue" and "green" are not.

The foregoing formulation of the paradox is essentially that given by Goodman; for our purposes this formulation has certain shortcomings. First, the Goodman formulation does not actually involve a violation of the criterion of linguistic invariance, for the equivalence of the two premises, "All observed emeralds are green" and "All observed emeralds are grue," does not depend solely upon the definitions of the predicates, but also upon the synthetic statement that the observations occur prior to the specified time *t*. Second, temporal positionality is not the only kind of positionality possible; it is desirable to characterize positionality more generally. Third, there is basic ambiguity in the definitions of "grue" and "bleen." Let us reformulate the paradox in a way which will repair these difficulties.[8]

[7] N. Goodman, *Fact, Fiction, and Forecast,* Cambridge, Harvard University Press, 1955, pp. 74 ff.

[8] The reformulation is suggested in R. Carnap, "On the Application of Inductive Logic," *Philosophy and Phenomenological Research,* vol. 8, (1947), p. 133.

Let the universe be taken as an ordered sequence $x_1, x_2, \ldots$, where it is understood that each x_i represents an object of reasonably small spatial and temporal dimensions. "Green" and "blue" have their customary meanings. We choose some subscript, say 1,000, and give the following definitions:

$$x_i \text{ is grue} =_{df} x_i \text{ is green and } i \leqslant 1{,}000 \text{ or } x_i \text{ is blue and } i > 1{,}000$$

$$x_i \text{ is bleen} =_{df} x_i \text{ is blue and } i \leqslant 1{,}000 \text{ or } x_i \text{ is green and } i > 1{,}000$$

Obviously, "green" and "blue" are again definable in terms of "grue" and "bleen," and when so defined they become positional predicates. Now, if we take the two premises:

> x_1 is an emerald which is green, x_2 is an emerald which is green, . . . , x_{100} is an emerald which is green

and

> x_1 is an emerald which is grue, x_2 is an emerald which is grue, . . . , x_{100} is an emerald which is grue

there is a purely linguistic difference between them. The standard inductive rule permits two conclusions which are mutually contradictory on the assumption that the class of emeralds is not exhausted before the 1,000th place:

> For every i, if x_i is an emerald then x_i is green

and

> For every i, if x_i is an emerald then x_i is grue

Two questions arise from this paradox. First, is there any important sense in which Goodman's predicates are time dependent or positional and the ordinary color predicates are not? If the answer is affirmative, the second question arises. Can we give good reasons for treating these time-dependent predicates in a special way which will prevent violations of the criterion of linguistic invariance? There is danger in ignoring the second question. We might find what appears to be an important asymmetry between the Goodman-type predicates and the normal predicates; it might then be very tempting to exploit this asymmetry just for the sake of escaping the Goodman paradox. For example, the Goodman-type predicates are time dependent with respect to the predicates of ordinary language. This is a genuine asymmetry. However, it would seem to me to be pointless to exclude the Goodman-type predicates from our language on this ground, even though it would avoid the Goodman paradox on the basis of a genuine asymmetry. The question would remain: is there any good reason for according the ordinary predicates such a privileged status? Might not a language with extraordinary basic predicates be equally adequate, for all we know, for the purposes of science?

Before attempting any answer to these questions I should like to present the problem in still different terms. As I conceive the problem of induction, it is the problem of finding grounds for selecting a unique rule of inductive inference from the infinity of possible rules. This problem cannot, however, be treated without paying close attention to the predicates involved. The selection of an inductive rule implies, of course, the exclusion of many alternative rules. The difficulty is that there is a degree of symmetry between the selection of an inductive rule and the selection of predicates to which this rule is to be applied. If we select a particular rule but allow its free application to every type of predicate that can possibly be defined, this is tantamount to allowing the use of other inductive rules besides the one selected. The introduction of the Goodman-type predicates which are positional with respect to the normal predicates into a language which utilizes the standard inductive rule has precisely the effect of introducing a conflicting inductive rule. We must, therefore, pay close attention to the sort of language to be used in conjunction with rules of inductive inference. Two difficulties are to be avoided. First, we shall attempt to exclude violations of the criterion of linguistic invariance which arise by allowing complete latitude with respect to the admission of predicates. Second, we shall attempt to preserve the principle of linguistic invariance by refusing to elevate one language with its particular set of predicates to a privileged position. To avoid these two extremes it is necessary to examine the types of language appropriate to inductive inference. This is best accomplished, I think, by considering the manner in which an inductive logic can be developed.

There are many good reasons for regarding an inductive logic as an extension of a deductive logic. Undoubtedly there are many different ways of providing the extension, but it seems to me that the fundamental feature of the extension is the addition to the apparatus of deductive logic of an inductive rule. A deductive logic may be regarded as a formal system containing logical constants, variables of various types, and non-logical constants among its symbols. In the usual manner it will have formation rules, axioms, and rules of immediate deductive inference. If the system has been constructed in the desired way, the rules of immediate inference have the important characteristic of being truth-preserving in any standard interpretation of the system. An inductive rule will not have the truth-preserving characteristic.

It is not necessary to go into great detail concerning the structure of the deductive system underlying inductive logic. The deductive system should be very generously endowed. It should be rich enough to contain whatever mathematics we need for natural science, including the probability calculus. Furthermore, it should contain ample supplies of individual and functional constants and variables. The individual and functional constants will, of course, be uninterpreted; they will comprise the vocabulary which is to become the basic descriptive vocabulary of the system when interpretation occurs.

When we consider adding an inductive rule, such as the Reichenbach rule of induction, to this system, we immediately face the prospect of making it inconsistent. This is the essential point of the Goodman paradox. Our problem is to introduce the rule and restrict it in such a way as to avoid contradiction. No doubt this can be done in a purely syntactical manner, but if we are to find a solution which is not completely *ad hoc* certain non-syntactical considerations must be taken into account. In particular, we must recognize the purposes for which the system is being developed. There are two primary ones, and they are not independent of one another. First, the language is designed to be a descriptive language; it must be capable of expressing true or false factual assertions. This means that the non-logical constants must be interpreted so as to refer to empirical reality. Second, the language is designed to incorporate inductive inference, so it must be supplemented with an inductive rule.

Let us first consider the matter of interpretation. It is quite clear that it cannot be achieved by interdefining the non-logical constants among themselves. We are all aware of the circularity involved in this procedure. Semantic rules of the following sort could be introduced:

"*a*" denotes Smith

"*F*" denotes yellow

This procedure is efficacious only if the metalanguage has already achieved reference for terms such as "Smith" and "yellow." Utilizing such rules would only push the same problem into the metalanguage. The alternative is to recognize that at some level of language we must provide the reference of terms by means of non-verbal definitions, and we might as well do it at the object language level. By non-verbal definition I mean semantic rules which establish directly some sort of referential relationship between symbols and non-linguistic entities as opposed to rules which establish only relationships among symbols of various languages. It is not necessary, of course, that all non-logical constants be interpreted by non-verbal rules, but only that a basic set be so interpreted.

There is no particular difficulty in providing non-verbal rules of reference for individual constants. One may simply indicate an object presented to the senses and decide to let an individual constant denote it. Ostensive definition is required for the interpretation of functional constants. For simplicity of statement I shall confine attention to singulary functional constants; the extension to *n*-ary functional constants is obvious. An ostensive definition contains three parts:

(1) The indication of a number of positive instances, i.e., individuals which have the property to be denoted by that constant.

(2) The indication of a number of negative instances, i.e., individuals which lack the property to be denoted by that constant.

(3) A similarity clause stating that anything resembling all of the posi-

tive instances in some respect, provided it does not also resemble any of the negative instances in that respect, also has the property to be denoted by that constant.

I shall now define a *purely ostensive predicate* as one which has the following characteristics:

(1) It *can* be defined ostensively. (How it is, in fact, defined is immaterial.)

(2) Its positive and negative instances for ostensive definition *can* be indicated non-verbally.

(3) The respect in which the positive instances resemble each other and differ from the negative instances is open to direct inspection, i.e., the resemblance in question is an observable resemblance.

Purely ostensive predicates have at least two fundamentally important characteristics—they are observation predicates and they are open predicates. The sense in which they are observation predicates is obvious. They are open predicates in the sense that their definition does not limit the number of individuals to which they may be correctly applied. It may be, as a matter of fact, that a purely ostensive predicate applies correctly to, say, seventeen individuals only, but this cannot be a consequence of the definition alone.

The language of empirical science requires, I believe, a basic set of open observation predicates. These are needed in order to make that language capable of expressing the descriptive generalizations which play an indispensable role in explanation and prediction. These predicates must be given meaning in some manner which provides them with empirical reference. Purely ostensive predicates seem to be admirably suited to fulfill this function, and I do not know of any other kind of predicate which is thus well suited. I take it that a primary aim of induction is to furnish a method of establishing universal or statistical generalizations on the basis of observational evidence. For this to be possible we need open observation predicates in our descriptive language. I propose, therefore, that a semantic restriction be placed upon the interpretation of basic predicates; namely, that they be interpreted in such a way as to become purely ostensive predicates.

The next step is to show that the Goodman-type predicates are not purely ostensive predicates.[9] This point is rather obvious. If we examine a number of grue things at about the time *t* (referring to Goodman's formulation of the paradox), some just before and some just after, we will see that they do not look alike. Those examined before *t* will look different from those examined after *t*. Goodman has acknowledged that not all grue things match each other, but he regards this feature of the predicate "grue" as too *ad hoc* to be of any significance.[10] It seems to me that he is mistaken in

[9] I am indebted to Professor S. Körner for pointing this fact out to me.

[10] N. Goodman, "Positionality and Pictures," *Philosophical Review,* vol. 49, 4 (Oct. 1960).

thinking this fact to be lacking in deep significance, for it is this fact which implies that "grue" cannot be ostensively defined. It is clearly impossible to cite a number of positive instances of grue things and stipulate that anything which resembles them in some observational characteristic is grue. To be sure, all grue things resemble each other in being grue, but this is not a resemblance with respect to an observational characteristic. Although grue things resemble each other in being grue, they do not look alike.

If I am correct in thinking that open concepts referring to observational characteristics whose meaning can be specified by non-verbal definition stand in a fundamental position in the development of a descriptive language of science, then I think we can say that the Goodman-type predicates are ruled out from the possibility of playing this fundamental role. This is by no means to say that they must be ruled out of the scientific vocabulary entirely. Rather, their position is secondary and they are time dependent with respect to the fundamental predicates—those which are purely ostensive. Goodman has argued for a complete symmetry with respect to positionality between his predicates and the normal predicates. If his predicates are taken as basic, he says, then the normal predicates become positional when defined in terms of them. The answer I am offering amounts to saying that his predicates cannot be taken as basic in the descriptive language of science because they are not purely ostensive. In this way the symmetry is destroyed.

Reichenbach has placed considerable emphasis upon his distinction between primitive and advanced knowledge, insisting that the problem of justification of induction he was attempting to solve is a problem in primitive knowledge.[11] Although he has drawn this distinction in psychologistic terms and has been criticized for so doing, it seems to me to be an important distinction which can be made in logical terms.[12] A primitive inductive inference is one whose premises are observation statements alone. A primitive inductive rule is one which requires no premises which cannot be observation statements. If the Reichenbach rule of induction is to be taken as a primitive rule, then we can place the qualification upon it that the variables A and B_i which appear in it must have their range restricted to purely ostensive predicates. With this restriction the Goodman paradox is eliminated in connection with the rule of induction.

We must now ask whether any violations of the criterion of linguistic invariance can arise with respect to the rule of induction. I think the answer is demonstrably negative. It is to be noted, first of all, that the inferred value of the limit of the relative frequency is a function of the observed relative frequency in an initial portion of the sequence in question. This quantity is linguistically invariant; regardless of how the reference class and attribute class are described, the numerical value remains unchanged. In order for a

[11] Reichenbach, *op. cit.,* p. 364.

[12] Barker, *op. cit.,* exemplifies this type of objection.

violation of linguistic invariance to occur it would be necessary that there be two classes which, *by definition*, coincide in part and fail to coincide in part. More exactly, there must be two classes F and G which have the following relationship: F and G can be partitioned on the basis of their definitions into two non-empty mutually exclusive subsets, F_1, F_2, G_1, G_2, such that $F_1 = G_1$ while F_2 and G_2 have no members in common. Furthermore, this partitioning must be possible on a positional basis; the predicates characterizing the classes F and G must be relatively positional. But, it is evident that not both F and G can be characterized by purely ostensive predicates. Hence, no violation of linguistic invariance can occur.

If the argument up to this point has been satisfactory we have succeeded in eliminating all inductive rules except one from the competition, and we have shown further that the one remaining rule is not subject to rejection on the grounds of the criterion of linguistic invariance. Does this constitute a vindication of that remaining rule?[13] Not quite, I believe. It remains to be shown that there is any positive reason for adopting the rule of induction, over and above the negative fact that it cannot be excluded as a rule which leads to contradiction. The argument requires two steps. First, it must be shown that it would be useful to be able to infer limits of relative frequencies if it is possible to do so. Keeping in mind the fact that a mathematical idealization is involved in talking about infinite sequences of empirical events, I shall assume that such an aim is worth-while. Second, it is necessary to show that the rule of induction is a rule somehow suited to accomplish this aim. Here, I think, Reichenbach's argument is now effective.[14] If the sequence in question has a limit of the relative frequency, repeated application of the rule of induction will achieve inference to the value of the limit within any desired degree of accuracy. If the sequence has no limit, no rule can provide any correct inference concerning its value.

13 I use the term "vindication" in the sense introduced by H. Feigl, "De Principiis Non Disputandum . . . ?" in M. Black, editor, *Philosophical Analysis*, Ithaca, Cornell University Press, 1950.

14 Reichenbach, *op. cit.*, sec. 91.

12

STUDIES IN THE LOGIC OF CONFIRMATION

Carl G. Hempel

To the memory of my wife,
Eva Ahrends Hempel

1. OBJECTIVE OF THE STUDY[1]

The defining characteristic of an empirical statement is its capability of being tested by a confrontation with experimental finding, i.e., with the results of suitable experiments or "focussed" observations. This feature distinguishes statements which have empirical content both from the statements of the formal sciences, logic and mathematics, which require no experimental test for their validation, and from the formulations of trans-empirical metaphysics, which do not admit of any.

The testability here referred to has to be understood in the comprehensive sense of "testability in principle"; there are many empirical statements which, for practical reasons, cannot be actually tested at present. To call a statement of this kind testable in principle means that it is possible to state

[1] The present analysis of confirmation was to a large extent suggested and stimulated by a co-operative study of certain more general problems which were raised by Dr. Paul Oppenheim, and which I have been investigating with him for several years. These problems concern the form and the function of scientific laws and the comparative methodology of the different branches of empirical science. The discussion with Mr. Oppenheim of these issues suggested to me the central problem of the present essay. The more comprehensive problems just referred to will be dealt with by Mr. Oppenheim in a publication which he is now preparing.

In my occupation with the logical aspects of confirmation, I have benefited greatly by discussions with several students of logic, including Professor R. Carnap, Professor A. Tarski, and particularly Dr. Nelson Goodman, to whom I am indebted for several valuable suggestions which will be indicated subsequently.

A detailed exposition of the more technical aspects of the analysis of confirmation presented in this article is included in my article "A Purely Syntactical Definition of Confirmation," *The Journal of Symbolic Logic,* vol. 8 (1943).

From *Mind,* vol. 54, 1945, pp. 1–21, 102–112. Reprinted by permission of the editor of *Mind* and the author.

just what experiential findings, if they were actually obtained, would constitute favourable evidence for it, and what findings or "data," as we shall say for brevity, would constitute unfavourable evidence; in other words, a statement is called testable in principle, if it is possible to describe the kind of data which would confirm or disconfirm it.

The concepts of confirmation and of disconfirmation as here understood are clearly more comprehensive than those of conclusive verification and falsification. Thus, e.g. no finite amount of experiential evidence can conclusively verify a hypothesis expressing a general law such as the law of gravitation, which covers an infinity of potential instances, many of which belong either to the as yet inaccessible future, or to the irretrievable past; but a finite set of relevant data may well be "in accord with" the hypothesis and thus constitute confirming evidence for it. Similarly, an existential hypothesis, asserting, say, the existence of an as yet unknown chemical element with certain specified characteristics, cannot be conclusively proved false by a finite amount of evidence which fails to "bear out" the hypothesis; but such unfavourable data may, under certain conditions, be considered as weakening the hypothesis in question, or as constituting disconfirming evidence for it.[2]

While, in the practice of scientific research, judgments as to the confirming or disconfirming character of experiential data obtained in the test of a hypothesis are often made without hesitation and with a wide consensus of opinion, it can hardly be said that these judgments are based on an explicit theory providing general criteria of confirmation and of disconfirmation. In this respect, the situation is comparable to the manner in which deductive inferences are carried out in the practice of scientific research: This, too, is often done without reference to an explicitly stated system of rules of logical inference. But while criteria of valid deduction can be and have been supplied by formal logic, no satisfactory theory providing general criteria of confirmation and disconfirmation appears to be available so far.

In the present essay, an attempt will be made to provide the elements of a theory of this kind. After a brief survey of the significance and the present status of the problem, I propose to present a detailed critical analysis of some common conceptions of confirmation and disconfirmation and then to construct explicit definitions for these concepts and to formulate some basic principles of what might be called the logic of confirmation.

2. SIGNIFICANCE AND PRESENT STATUS OF THE PROBLEM

The establishment of a general theory of confirmation may well be regarded as one of the most urgent desiderata of the present methodology

[2] This point as well as the possibility of conclusive verification and conclusive falsification will be discussed in some detail in section 10 of the present paper.

of empirical science.[3] Indeed, it seems that a precise analysis of the concept of confirmation is a necessary condition for an adequate solution of various fundamental problems concerning the logical structure of scientific procedure. Let us briefly survey the most outstanding of these problems.

(*a*) In the discussion of scientific method, the concept of relevant evidence plays an important part. And while certain "inductivist" accounts of scientific procedure seem to assume that relevant evidence, or relevant data, can be collected in the context of an inquiry prior to the formulation of any hypothesis, it should be clear upon brief reflection that relevance is a relative concept; experiential data can be said to be relevant or irrelevant only with respect to a given hypothesis; and it is the hypothesis which determines what kind of data or evidence are relevant for it. Indeed, an empirical finding is relevant for a hypothesis if and only if it constitutes either favourable or unfavourable evidence for it; in other words, if it either confirms or disconfirms the hypothesis. Thus, a precise definition of relevance presupposes an analysis of confirmation and disconfirmation.

(*b*) A closely related concept is that of instance of a hypothesis. The so-called method of inductive inference is usually presented as proceeding from specific cases to a general hypothesis of which each of the special cases is an "instance" in the sense that it "conforms to" the general hypothesis in question, and thus constitutes confirming evidence for it.

Thus, any discussion of induction which refers to the establishment of general hypotheses on the strength of particular instances is fraught with all those logical difficulties—soon to be expounded—which beset the concept of confirmation. A precise analysis of this concept is, therefore, a necessary condition for a clear statement of the issues involved in the problem complex of induction and of the ideas suggested for their solution—no matter what their theoretical merits or demerits may be.

(*c*) Another issue customarily connected with the study of scientific method is the quest for "rules of induction." Generally speaking, such rules would enable us to "infer", from a given set of data, that hypothesis or generalization which accounts best for all the particular data in the given set. Recent logical analyses have made it increasingly clear that this way of conceiving the problem involves a misconception: While the process of invention by which scientific discoveries are made is as a rule *psychologically guided and stimulated* by antecedent knowledge of specific facts, its results are *not logically determined* by them; the way in which scientific hypotheses or theories are discovered cannot be mirrored in a set of general rules of inductive inference.[4] One of the crucial considerations which lead to this

[3] Or of the "logic of science," as understood by R. Carnap; cf. *The Logical Syntax of Language* (New York and London, 1937), sec. 72, and the supplementary remarks in *Introduction to Semantics* (Cambridge, Mass., 1942), p. 250.

[4] See the lucid presentation of this point in Karl Popper's *Logik der Forschung* (Wien, 1935), esp. secs. 1, 2, 3, and 25, 26, 27; cf. also Albert Einstein's remarks in his lecture *On the Method of Theoretical Physics* (Oxford, 1933), pp. 11 and

conclusion is the following: Take a scientific theory such as the atomic theory of matter. The evidence on which it rests may be described in terms referring to directly observable phenomena, namely to certain "macroscopic" aspects of the various experimental and observational data which are relevant to the theory. On the other hand, the theory itself contains a large number of highly abstract, non-observational terms such as "atom," "electron," "nucleus," "dissociation," "valence" and others, none of which figures in the description of the observational data. An adequate rule of induction would therefore have to provide, for this and for every conceivable other case, mechanically applicable criteria determining unambiguously, and without any reliance on the inventiveness or additional scientific knowledge of its user, all those new abstract concepts which need to be created for the formulation of the theory that will account for the given evidence. Clearly, this requirement cannot be satisfied by any set of rules, however ingeniously devised; there can be no general rules of induction in the above sense; the demand for them rests on a confusion of logical and psychological issues. What determines the soundness of a hypothesis is not the way it is arrived at (it may even have been suggested by a dream or a hallucination), but the way it stands up when tested, i.e. when confronted with relevant observational data. Accordingly, the quest for rules of induction in the original sense of canons of scientific discovery has to be replaced, in the logic of science, by the quest for general objective criteria determining (A) whether, and—if possible—even (B) to what degree, a hypothesis *H* may be said to be corroborated by a given body of evidence *E*. This approach differs essentially from the inductivist conception of the problem in that it presupposes not only *E*, but also *H* as given and then seeks to determine a certain logical relationship between them. The two parts of this latter problem can be restated in somewhat more precise terms as follows:

(A) To give precise definitions of the two non-quantitative relational concepts of confirmation and of disconfirmation; i.e. to define the meaning of the phrases "*E* confirms *H*" and "*E* disconfirms *H*." (When *E* neither confirms nor disconfirms *H*, we shall say that *E* is neutral, or irrelevant, with respect to *H*.)

(B) (1) To lay down criteria defining a metrical concept "degree of confirmation of *H* with respect to *E*," whose values are real numbers; or, failing this, (2) To lay down criteria defining two relational concepts, "more highly confirmed than" and "equally well confirmed with," which make possible a non-metrical comparison of hypotheses (each with a body of evidence assigned to it) with respect to the extent of their confirmation.

Interestingly, problem B has received much more attention in method-

---

12. Also of interest in this context is the critical discussion of induction by H. Feigl in "The Logical Character of the Principle of Induction," *Philosophy of Science,* vol. 1 (1934).

ological research than problem A; in particular, the various theories of the "probability of hypotheses" may be regarded as concerning this problem complex; we have here adopted[5] the more neutral term "degree of confirmation" instead of "probability" because the latter is used in science in a definite technical sense involving reference to the relative frequency of the occurrence of a given event in a sequence, and it is at least an open question whether the degree of confirmation of a hypothesis can generally be defined as a probability in this statistical sense.

The theories dealing with the probability of hypotheses fall into two main groups: the "logical" theories construe probability as a logical relation between sentences (or propositions; it is not always clear which is meant);[6] the "statistical" theories interpret the probability of a hypothesis in substance as the limit of the relative frequency of its confirming instances among all relevant cases.[7] Now it is a remarkable fact that none of the theories of the first type which have been developed so far provides an explicit general definition of the probability (or degree of confirmation) of a hypothesis H with respect to a body of evidence E; they all limit themselves essentially to the construction of an uninterpreted postulational system of logical probability. For this reason, these theories fail to provide a complete solution of problem B. The statistical approach, on the other hand, would, if successful, provide an explicit numerical definition of the degree of confirmation of a hypothesis; this definition would be formulated in terms of the numbers of confirming and disconfirming instances for H which constitute the body of evidence E. Thus, a necessary condition for an adequate interpretation of degrees of confirmation as statistical probabilities is the establishment of precise criteria of confirmation and disconfirmation, in other words, the solution of problem A.

However, despite their great ingenuity and suggestiveness, the attempts which have been made so far to formulate a precise statistical definition of the degree of confirmation of a hypothesis seem open to certain objections,[8]

[5] Following R. Carnap's usage in "Testability and Meaning," *Philosophy of Science,* vols. 3 (1936) and 4 (1937); esp. sec. 3 (in vol. 3).

[6] This group includes the work of such writers as Janina Hosiasson-Lindenbaum (cf. for instance, her article "Induction et analogie: Comparaison de leur fondement," *Mind,* vol. L (1941); also see n. 24), H. Jeffreys, J. M. Keynes, B. O. Koopman, J. Nicod (see n. 15), St. Mazurkiewicz, F. Waismann. For a brief discussion of this conception of probability, see Ernest Nagel, *Principles of the Theory of Probability* (Internat. Encyclopedia of Unified Science, vol. 1, no. 6, Chicago, 1939), esp. secs. 6 and 8.

[7] The chief proponent of this view is Hans Reichenbach; cf. especially "Ueber Induktion und Wahrscheinlichkeit," *Erkenntnis,* vol. V (1935), and *Experience and Prediction* (Chicago, 1938), Ch. V.

[8] Cf. Karl Popper, *Logik der Forschung* (Wien, 1935), sec. 80; Ernest Nagel, *l.c.,* sec. 8, and "Probability and the Theory of Knowledge," *Philosophy of Science,* vol. 6 (1939); C. G. Hempel, "Le problème de la vérité," *Theoria*

and several authors[9] have expressed doubts as to the possibility of defining the degree of confirmation of a hypothesis as a metrical magnitude, though some of them consider it as possible, under certain conditions, to solve at least the less exacting problem B (2), i.e. to establish standards of non-metrical comparison between hypotheses with respect to the extent of their confirmation. An adequate comparison of this kind might have to take into account a variety of different factors;[10] but again the numbers of the confirming and of the disconfirming instances which the given evidence includes will be among the most important of those factors.

Thus, of the two problems, A and B, the former appears to be the more basic one, first, because it does not presuppose the possibility of defining numerical degrees of confirmation or of comparing different hypotheses as to the extent of their confirmation; and second because our considerations indicate that any attempt to solve problem B—unless it is to remain in the stage of an axiomatized system without interpretation—is likely to require a precise definition of the concepts of confirming and disconfirming instance of a hypothesis before it can proceed to define numerical degrees of confirmation, or to lay down non-metrical standards of comparison.

(*d*) It is now clear that an analysis of confirmation is of fundamental importance also for the study of the central problem of what is customarily called epistemology; this problem may be characterized as the elaboration of "standards of rational belief" or of criteria of warranted assertibility. In the methodology of empirical science this problem is usually phrased as concerning the rules governing the test and the subsequent acceptance or rejection of empirical hypotheses on the basis of experimental or observational findings, while in its "epistemological" version the issue is often formulated as concerning the validation of beliefs by reference to perceptions, sense data, or the like. But no matter how the final empirical evidence is construed and in what terms it is accordingly expressed, the theoretical problem remains the same: to characterize, in precise and general terms, the conditions under which a body of evidence can be said to confirm, or to disconfirm, a hypothesis of empirical character; and that is again our problem A.

(*e*) The same problem arises when one attempts to give a precise statement of the empiricist and operationalist criteria for the empirical meaningfulness of a sentence; these criteria, as is well known, are formulated by

---

(Göteborg), vol. 3 (1937), sec. 5, and "On the Logical Form of Probability Statements," *Erkenntnis,* vol. 7 (1937–38), esp. sec. 5. Cf. also Morton White, "Probability and Confirmation," *The Journal of Philosophy,* vol. 36 (1939).

[9] See, for example, J. M. Keynes, *A Treatise on Probability* (London, 1929), esp. Ch. III; Ernest Nagel, *Principles of the Theory of Probability* (cf. n. 6 above), esp. p. 70; compare also the somewhat less definitely sceptical statement by Carnap, *l.c.* (see n. 5), sec. 3, p. 427.

[10] See especially the survey of such factors given by Ernest Nagel in *Principles of the Theory of Probability* (cf. note 6), pp. 66–73.

reference to the theoretical testability of the sentence by means of experimental evidence,[11] and the concept of theoretical testability, as was pointed out earlier, is closely related to the concepts of confirmation and disconfirmation.[12]

Considering the great importance of the concept of confirmation, it is surprising that no systematic theory of the non-quantitative relation of confirmation seems to have been developed so far. Perhaps this fact reflects the tacit assumption that the concepts of confirmation and of disconfirmation have a sufficiently clear meaning to make explicit definitions unnecessary or at least comparatively trivial. And indeed, as will be shown below, there are certain features which are rather generally associated with the intuitive notion of confirming evidence, and which, at first, seem well-suited to serve as defining characteristics of confirmation. Closer examination will reveal the definitions thus obtainable to be seriously deficient and will make it clear that an adequate definition of confirmation involves considerable difficulties.

Now the very existence of such difficulties suggests the question whether the problem we are considering does not rest on a false assumption: Perhaps there are no objective criteria of confirmation; perhaps the decision as to whether a given hypothesis is acceptable in the light of a given body of evidence is no more subject to rational, objective rules than is the process of inventing a scientific hypothesis or theory; perhaps, in the last analysis, it is a "sense of evidence", or a feeling of plausibility in view of the relevant data, which ultimately decides whether a hypothesis is scientifically acceptable.[13] This view is comparable to the opinion that the validity of a mathematical proof or of a logical argument has to be judged ultimately by reference to a feeling of soundness or convincingness; and both theses have to be rejected on analogous grounds: They involve a confusion of logical and psychological considerations. Clearly, the occurrence or non-occurrence of a feeling of conviction upon the presentation of grounds for an assertion is a subjective matter which varies from person to person, and with the same person in the course of time; it is often deceptive, and can certainly serve neither as a necessary nor as a sufficient condition for the soundness of the given asser-

[11] Cf. for example, A. J. Ayer, *Language, Truth and Logic* (London and New York, 1936), Ch. I; R. Carnap, "Testability and Meaning" (cf. note 5), secs. 1, 2, 3; H. Feigl, *Logical Empiricism* (in *Twentieth Century Philosophy,* ed. by Dagobert D. Runes, New York, 1943); P. W. Bridgman, *The Logic of Modern Physics* (New York, 1928).

[12] It should be noted, however, that in his essay "Testability and Meaning" (cf. note 5), R. Carnap has constructed definitions of testability and confirmability which avoid reference to the concept of confirming and of disconfirming evidence; in fact, no proposal for the definition of these latter concepts is made in that study.

[13] A view of this kind has been expressed, for example, by M. Mandelbaum in "Causal Analyses in History," *Journal of the History of Ideas,* vol. 3 (1942); cf. esp. pp. 46–47.

tion.[14] A rational reconstruction of the standards of scientific validation cannot, therefore, involve reference to a sense of evidence; it has to be based on objective criteria. In fact, it seems reasonable to require that the criteria of empirical confirmation, besides being objective in character, should contain no reference to the specific subject-matter of the hypothesis or of the evidence in question; it ought to be possible, one feels, to set up purely formal criteria of confirmation in a manner similar to that in which deductive logic provides purely formal criteria for the validity of deductive inferences.

With this goal in mind, we now turn to a study of the non-quantitative concept of confirmation. We shall begin by examining some current conceptions of confirmation and exhibiting their logical and methodological inadequacies; in the course of this analysis, we shall develop a set of conditions for the adequacy of any proposed definition of confirmation; and finally, we shall construct a definition of confirmation which satisfies those general standards of adequacy.

3. NICOD'S CRITERION OF CONFIRMATION AND ITS SHORTCOMINGS

We consider first a conception of confirmation which underlies many recent studies of induction and of scientific method. A very explicit statement of this conception has been given by Jean Nicod in the following passage: "Consider the formula or the law: *A entails B*. How can a particular proposition, or more briefly, a fact, affect its probability? If this fact consists of the presence of *B* in a case of *A*, it is favourable to the law '*A entails B*'; on the contrary, if it consists of the absence of *B* in a case of *A*, it is unfavourable to this law. It is conceivable that we have here the only two direct modes in which a fact can influence the probability of a law. ... Thus, the entire influence of particular truths or facts on the probability of universal propositions or laws would operate by means of these two elementary relations which we shall call *confirmation* and *invalidation*."[15] Note that the applicability of this criterion is restricted to hypotheses of the form "*A entails B*." Any hypothesis *H* of this kind may be expressed in the notation of symbolic logic[16] by means of a universal conditional sentence, such as, in the simplest case,

[14] See Karl Popper's pertinent statement, *l.c.*, sect. 8.

[15] Jean Nicod, *Foundations of Geometry and Induction* (transl. by P. P. Wiener) (London, 1930), p. 219; cf. also R. M. Eaton's discussion of "Confirmation and Infirmation," which is based on Nicod's views; it is included in chap. III of his *General Logic* (New York, 1931).

[16] In this paper, only the most elementary devices of this notation are used; the symbolism is essentially that of *Principia Mathematica,* except that parentheses

$$(x)\,(P(x) \supset Q(x)),$$

i.e. "For any object x: if x is a P, then x is a Q," or also "Occurrence of the quality P entails occurrence of the quality Q." According to the above criterion this hypothesis is confirmed by an object *a,* if *a* is P and Q; and the hypothesis is disconfirmed by *a* if *a* is P, but not Q. In other words, an object confirms a universal conditional hypothesis if and only if it satisfies both the antecedent (here: '$P(x)$') and the consequent (here: '$Q(x)$') of the conditional; it disconfirms the hypothesis if and only if it satisfies the antecedent, but not the consequent of the conditional; and (we add this to Nicod's statement) it is neutral, or irrelevant, with respect to the hypothesis if it does not satisfy the antecedent.

This criterion can readily be extended so as to be applicable also to universal conditionals containing more than one quantifier, such as "Twins always resemble each other", or, in symbolic notation, '$(x)\,(y)\,(Twins(x, y) \supset \mathrm{Rsbl}(x, y))$'. In these cases, a confirming instance consists of an ordered couple, or triple, etc., of objects satisfying the antecedent and the consequent of the conditional. (In the case of the last illustration, any two persons who are twins and who resemble each other would confirm the hypothesis; twins who do not resemble each other would disconfirm it; and any two persons not twins—no matter whether they resemble each other or not—would constitute irrelevant evidence.)

We shall refer to this criterion as Nicod's criterion.[17] It states explicitly what is perhaps the most common tacit interpretation of the concept of confirmation. While seemingly quite adequate, it suffers from serious shortcomings, as will now be shown.

(*a*) First, the applicability of this criterion is restricted to hypotheses of universal conditional form; it provides no standards for existential hypotheses (such as "There exists organic life on other stars," or "Poliomyelitis is caused by some virus") or for hypotheses whose explicit formulation calls for the use of both universal and existential quantifiers (such as "Every human being dies some finite number of years after his birth," or the psychological hypothesis, "You can fool all of the people some of the time and some of the people all of the time, but you cannot fool all of the people all of the time," which may be symbolized by '$(x)\,(Et)\mathrm{Fl}(x,t) \cdot (Ex)\,(t)\mathrm{Fl}(x,t) \cdot \sim (x)\,(t)\mathrm{Fl}(x,t)$', (where '$\mathrm{Fl}(x,t)$' stands for "You can fool (person) x at time t"). We note, therefore, the desideratum of establishing a criterion of confirmation which is applicable to hypotheses of any form.[18]

---

are used instead of dots, and that existential quantification is symbolized by '(*E*)' instead of by the inverted '*E*'.

17 This term is chosen for convenience, and in view of the above explicit formulation given by Nicod; it is not, of course, intended to imply that this conception of confirmation originated with Nicod.

18 For a rigorous formulation of the problem, it is necessary first to lay down

(*b*) We now turn to a second shortcoming of Nicod's criterion. Consider the two sentences

S_1: '$(x)\ (\text{Raven}(x) \supset \text{Black}(x))$';
S_2: '$(x)\ ({\sim}\text{Black}(x) \supset {\sim}\text{Raven}(x))$'

(i.e. "All ravens are black" and "Whatever is not black is not a raven"), and let *a*, *b*, *c*, *d* be four objects such that *a* is a raven and black, *b* a raven but not black, *c* not a raven but black, and *d* neither a raven nor black. Then, according to Nicod's criterion, *a* would confirm S_1, but be neutral with respect to S_2; *b* would disconfirm both S_1 and S_2; *c* would be neutral with respect to both S_1 and S_2, and *d* would confirm S_2, but be neutral with respect to S_1.

But S_1 and S_2 are logically equivalent; they have the same content, they are different formulations of the same hypothesis. And yet, by Nicod's criterion, either of the objects *a* and *d* would be confirming for one of the two sentences, but neutral with respect to the other. This means that Nicod's criterion makes confirmation depend not only on the content of the hypothesis, but also on its formulation.[19]

One remarkable consequence of this situation is that every hypothesis to which the criterion is applicable—i.e. every universal conditional—can be stated in a form for which there cannot possibly exist any confirming instances. Thus, e.g. the sentence

$$(x)[(\text{Raven}(x) \cdot \sim \text{Black}(x)) \supset (\text{Raven}(x) \cdot \sim \text{Raven}(x))]$$

is readily recognized as equivalent to both S_1 and S_2 above; yet no object whatever can confirm this sentence, i.e. satisfy both its antecedent and its consequent; for the consequent is contradictory. An analogous transformation is, of course, applicable to any other sentence of universal conditional form.

4. THE EQUIVALENCE CONDITION

The results just obtained call attention to a condition which an adequately defined concept of confirmation should satisfy, and in the light of which Nicod's criterion has to be rejected as inadequate: *Equivalence con-*

assumptions as to the means of expression and the logical structure of the language in which the hypotheses are supposed to be formulated; the desideratum then calls for a definition of confirmation applicable to any hypotheses which can be expressed in the given language. Generally speaking, the problem becomes increasingly difficult with increasing richness and complexity of the assumed "language of science."

[19] This difficulty was pointed out, in substance, in my article "Le problème de la vérité," *Theoria* (Göteborg), vol. 3 (1937), esp. p. 222.

dition: Whatever confirms (disconfirms) one of two equivalent sentences, also confirms (disconfirms) the other.

Fulfilment of this condition makes the confirmation of a hypothesis independent of the way in which it is formulated; and no doubt it will be conceded that this is a necessary condition for the adequacy of any proposed criterion of confirmation. Otherwise, the question as to whether certain data confirm a given hypothesis would have to be answered by saying: "That depends on which of the different equivalent formulations of the hypothesis is considered"—which appears absurd. Furthermore—and this is a more important point than an appeal to a feeling of absurdity—an adequate definition of confirmation will have to do justice to the way in which empirical hypotheses function in theoretical scientific contexts such as explanations and predictions; but when hypotheses are used for purposes of explanation or prediction,[20] they serve as premisses in a deductive argument whose conclusion is a description of the event to be explained or predicted. The deduction is governed by the principles of formal logic, and according to the latter, a deduction which is valid will remain so if some or all of the premisses are replaced by different, but equivalent statements; and indeed, a scientist will feel free, in any theoretical reasoning involving certain hypotheses, to use the latter in whichever of their equivalent formulations is most convenient for the development of his conclusions. But if we adopted a concept of confirmation which did not satisfy the equivalence condition, then it would be possible, and indeed necessary, to argue in certain cases that it was sound scientific procedure to base a prediction on a given hypothesis if formulated in a sentence S_1, because a good deal of confirming evidence had been found for S_1; but that it was altogether inadmissible to base the prediction (say, for convenience of deduction) on an equivalent formulation S_2, because no confirming evidence for S_2 was available. Thus, the equivalence condition has to be regarded as a necessary condition for the adequacy of any definition of confirmation.

5. THE "PARADOXES" OF CONFIRMATION

Perhaps we seem to have been labouring the obvious in stressing the necessity of satisfying the equivalence condition. This impression is likely to vanish upon consideration of certain consequences which derive from a

[20] For a more detailed account of the logical structure of scientific explanation and prediction, cf. C. G. Hempel, "The Function of General Laws in History," *The Journal of Philosophy,* vol. 39, (1942), esp. secs. 2, 3, 4. The characterization, given in that paper as well as in the above text, of explanations and predictions as arguments of a deductive logical structure, embodies an oversimplification: as will be shown in sect. 7 of the present essay, explanations and predictions often involve "quasi-inductive" steps besides deductive ones. This point, however, does not affect the validity of the above argument.

combination of the equivalence condition with a most natural and plausible assumption concerning a sufficient condition of confirmation.

The essence of the criticism we have levelled so far against Nicod's criterion is that it certainly cannot serve as a necessary condition of confirmation; thus, in the illustration given in the beginning of section 3, the object *a* confirms S_1 and should therefore also be considered as confirming S_2, while according to Nicod's criterion it is not. Satisfaction of the latter is therefore not a necessary condition for confirming evidence.

On the other hand, Nicod's criterion might still be considered as stating a particularly obvious and important sufficient condition of confirmation. And indeed, if we restrict ourselves to universal conditional hypotheses in one variable[21]—such as S_1 and S_2 in the above illustration—then it seems perfectly reasonable to qualify an object as confirming such a hypothesis if it satisfies both its antecedent and its consequent. The plausibility of this view will be further corroborated in the course of our subsequent analyses.

Thus, we shall agree that if *a* is both a raven and black, then *a* certainly confirms S_1: '$(x)\ (\text{Raven}(x) \supset \text{Black}(x))$', and if *d* is neither black nor a raven, *d* certainly confirms S_2:

$$`(x)\ ({\sim}\text{Black}(x) \supset {\sim}\text{Raven}(x))'$$

Let us now combine this simple stipulation with the equivalence condition: Since S_1 and S_2 are equivalent, *d* is confirming also for S_1; and thus, we have to recognize as confirming for S_1 any object which is neither black nor a raven. Consequently, any red pencil, any green leaf, any yellow cow,

[21] This restriction is essential: In its general form, which applies to universal conditionals in any number of variables, Nicod's criterion cannot even be construed as expressing a sufficient condition of confirmation. This is shown by the following rather surprising example: Consider the hypothesis S_1: $(x)(y)$ $[\sim (R(x,y) \cdot R(y,x)) \supset (R(x,y) \cdot \sim R(y,x))]$.

Let a, b be two objects such that $R(a,b)$ and $\sim R(b,a)$. Then clearly, the couple (a,b) satisfies both the antecedent and the consequent of the universal conditional S_1; hence, if Nicod's criterion in its general form is accepted as stating a sufficient condition of confirmation, (a,b) constitutes confirming evidence for S_1. However, S_1 can be shown to be equivalent to

$$S_2\colon (x)(y)R(x,y)$$

Now, by hypothesis, we have $\sim R(b,a)$; and this flatly contradicts S_2 and thus S_1. Thus, the couple (a,b), although satisfying both the antecedent and the consequent of the universal conditional S_1 actually constitutes disconfirming evidence of the strongest kind (conclusively disconfirming evidence, as we shall say later) for that sentence. This illustration reveals a striking and—as far as I am aware—hitherto unnoticed weakness of that conception of confirmation which underlies Nicod's criterion. In order to realize the bearing of our illustration upon Nicod's original formulation, let *A* and *B* be $\sim (R(x,y) \cdot R(y,x))$ and $R(x,y) \cdot \sim R(y,x)$ respectively. Then S_1 asserts that *A* entails *B*, and the couple (a,b) is a case of the presence of *B* in the presence of *A*; this should, according to Nicod, be favourable to S_1.

etc., becomes confirming evidence for the hypothesis that all ravens are black. This surprising consequence of two very adequate assumptions (the equivalence condition and the above sufficient condition of confirmation) can be further expanded: The following sentence can readily be shown to be equivalent to S_1: S_3: '$(x)[(\text{Raven}(x) \vee \sim \text{Raven}(x)) \supset (\sim \text{Raven}(x) \vee \text{Black}(x))]$', i.e. "Anything which is or is not a raven is either no raven or black." According to the above sufficient condition, S_3 is certainly confirmed by any object, say *e*, such that (1) *e* is or is not a raven and, in addition, (2) *e* is not a raven or also black. Since (1) is analytic; these conditions reduce to (2). By virtue of the equivalence condition, we have therefore to consider as confirming for S_1 any object which is either no raven or also black (in other words: any object which is no raven at all, or a black raven).

Of the four objects characterized in section 3, *a*, *c* and *d* would therefore constitute confirming evidence for S_1, while *b* would be disconfirming for S_1. This implies that any non-raven represents confirming evidence for the hypothesis that all ravens are black.

We shall refer to these implications of the equivalence criterion and of the above sufficient condition of confirmation as the *paradoxes of confirmation.*

How are these paradoxes to be dealt with? Renouncing the equivalence condition would not represent an acceptable solution, as is shown by the consideration presented in section 4. Nor does it seem possible to dispense with the stipulation that an object satisfying two conditions, C_1 and C_2, should be considered as confirming a general hypothesis to the effect that any object which satisfies C_1, also satisfies C_2.

But the deduction of the above paradoxical results rests on one other assumption which is usually taken for granted, namely, that the meaning of general empirical hypotheses, such as that all ravens are black, or that all sodium salts burn yellow, can be adequately expressed by means of sentences of universal conditional form, such as '(x) $(\text{Raven}(x) \supset \text{Black}(x))$' and '$(x)$ $(\text{Sod. Salt}(x) \supset \text{Burn Yellow}(x))$', etc. Perhaps this customary mode of presentation has to be modified; and perhaps such a modification would automatically remove the paradoxes of confirmation? If this is not so, there seems to be only one alternative left, namely to show that the impression of the paradoxical character of those consequences is due to misunderstanding and can be dispelled, so that no theoretical difficulty remains. We shall now consider these two possibilities in turn: The sub-sections 5.11 and 5.12 are devoted to a discussion of two different proposals for a modified representation of general hypotheses; in subsection 5.2, we shall discuss the second alternative, i.e. the possibility of tracing the impression of paradoxicality back to a misunderstanding.

5.11. It has often been pointed out that while Aristotelian logic, in agreement with prevalent every day usage, confers "existential import" upon sentences of the form "All *P*'s are *Q*'s," a universal conditional sentence, in the sense of modern logic, has no existential import; thus, the sentence

'(x) (Mermaid$(x) \supset$ Green(x))'

does not imply the existence of mermaids; it merely asserts that any object either is not a mermaid at all, or a green mermaid; and it is true simply because of the fact that there are no mermaids. General laws and hypotheses in science, however—so it might be argued—are meant to have existential import; and one might attempt to express the latter by supplementing the customary universal conditional by an existential clause. Thus, the hypothesis that all ravens are black would be expressed by means of the sentence S_1: '(x) (Raven$(x) \supset$ Black(x)) · (Ex)Raven(x)'; and the hypothesis that no non-black things are ravens by S_2: '(x) ($\sim$ Black$(x) \supset \sim$ Raven(x)) · $(Ex) \sim$ Black(x)'. Clearly, these sentences are not equivalent, and of the four objects *a*, *b*, *c*, *d* characterized in section 3, part (*b*), only *a* might reasonably be said to confirm S_1, and only *d* to confirm S_2. Yet this method of avoiding the paradoxes of confirmation is open to serious objections:

(*a*) First of all, the representation of every general hypothesis by a conjunction of a universal conditional and an existential sentence would invalidate many logical inferences which are generally accepted as permissible in a theoretical argument. Thus, for example, the assertions that all sodium salts burn yellow, and that whatever does not burn yellow is no sodium salt are logically equivalent according to customary understanding and usage; and their representation by universal conditionals preserves this equivalence; but if existential clauses are added, the two assertions are no longer equivalent, as is illustrated above by the analogous case of S_1 and S_2.

(*b*) Second, the customary formulation of general hypotheses in empirical science clearly does not contain an existential clause, nor does it, as a rule, even indirectly determine such a clause unambiguously. Thus, consider the hypothesis that if a person after receiving an injection of a certain test substance has a positive skin reaction, he has diphtheria. Should we construe the existential clause here as referring to persons, to persons receiving the injection, or to persons who, upon receiving the injection, show a positive skin reaction? A more or less arbitrary decision has to be made; each of the possible decisions gives a different interpretation to the hypothesis, and none of them seems to be really implied by the latter.

(*c*) Finally, many universal hypotheses cannot be said to imply an existential clause at all. Thus, it may happen that from a certain astrophysical theory a universal hypothesis is deduced concerning the character of the phenomena which would take place under certain specified extreme conditions. A hypothesis of this kind need not (and, as a rule, does not) imply that such extreme conditions ever were or will be realized; it has no existential import. Or consider a biological hypothesis to the effect that whenever man and ape are crossed, the offspring will have such and such characteristics. This is a general hypothesis; it might be contemplated as a mere conjecture, or as a consequence of a broader genetic theory, other implications of which may already have been tested with positive results; but unquestionably the hypothesis does not imply an existential clause asserting that the contemplated kind of cross-breeding referred to will, at some time, actually take place.

While, therefore, the adjunction of an existential clause to the customary symbolization of a general hypothesis cannot be considered as an adequate *general* method of coping with the paradoxes of confirmation, there is a purpose which the use of an existential clause may serve very well, as was pointed out to me by Dr. Paul Oppenheim:[22] if somebody feels that objects of the types *c* and *d* mentioned above are irrelevant rather than confirming for the hypothesis in question, and that qualifying them as confirming evidence does violence to the meaning of the hypothesis, then this may indicate that he is consciously or unconsciously construing the latter as having existential import; and this kind of understanding of general hypothesis is in fact very common. In this case, the "paradox" may be removed by pointing out that an adequate symbolization of the intended meaning requires the adjunction of an existential clause. The formulation thus obtained is more restrictive than the universal conditional alone; and while we have as yet set up no criteria of confirmation applicable to hypotheses of this more complex form, it is clear that according to every acceptable definition of confirmation objects of the types *c* and *d* will fail to qualify as confirming cases. In this manner, the use of an existential clause may prove helpful in distinguishing and rendering explicit different possible interpretations of a given general hypothesis which is stated in non-symbolic terms.

5.12. Perhaps the impression of the paradoxical character of the cases discussed in the beginning of section 5 may be said to grow out of the feeling that the hypothesis that all ravens are black is about ravens, and not about non-black things, nor about all things. The use of an existential clause was one attempt at expressing this presumed peculiarity of the hypothesis. The attempt has failed, and if we wish to reflect the point in question, we shall have to look for a stronger device. The idea suggests itself of representing a general hypothesis by the customary universal conditional, supplemented by the indication of the specific "field of application" of the hypothesis; thus, we might represent the hypothesis that all ravens are black by the sentence '$(x)(\text{Raven}(x) \supset \text{Black}\ (x))$' (or any one of its equivalents), plus the indication "Class of ravens" characterizing the field of application; and we might then require that every confirming instance should belong to the field of application. This procedure would exclude the objects *c* and *d* from those constituting confirming evidence and would thus avoid those undesirable consequences of the existential-clause device which were

[22] This observation is related to Mr. Oppenheim's methodological studies referred to in n. 1.

pointed out in 5.11 (*c*). But apart from this advantage, the second method is open to objections similar to those which apply to the first: (*a*) The way in which general hypotheses are used in science never involves the statement of a field of application; and the choice of the latter in a symbolic formulation of a given hypothesis thus introduces again a considerable measure of arbitrariness. In particular, for a scientific hypothesis to the effect that all *P*'s are *Q*'s, the field of application cannot simply be said to be the class of all *P*'s; for a hypothesis such as that all sodium salts burn yellow finds important applications in tests with negative results; i.e. it may be applied to a substance of which it is not known whether it contains sodium salts, nor whether it burns yellow; and if the flame does not turn yellow, the hypothesis serves to establish the absence of sodium salts. The same is true of all other hypotheses used for tests of this type. (*b*) Again, the consistent use of a domain of application in the formulation of general hypotheses would involve considerable logical complications, and yet would have no counterpart in the theoretical procedure of science, where hypotheses are subjected to various kinds of logical transformation and inference without any consideration that might be regarded as referring to changes in the fields of application. This method of meeting the paradoxes would therefore amount to dodging the problem by means of an *ad hoc* device which cannot be justified by reference to actual scientific procedure.

5.2. We have examined two alternatives to the customary method of representing general hypotheses by means of universal conditionals; neither of them proved an adequate means of precluding the paradoxes of confirmation. We shall now try to show that what is wrong does not lie in the customary way of construing and representing general hypotheses, but rather in our reliance on a misleading intuition in the matter: The impression of a paradoxical situation is not objectively founded; it is a psychological illusion.

(*a*) One source of misunderstanding is the view, referred to before, that a hypothesis of the simple form "Every *P* is a *Q*" such as "All sodium salts burn yellow," asserts something about a certain limited class of objects only, namely, the class of all *P*'s. This idea involves a confusion of logical and practical considerations: Our interest in the hypothesis may be focussed upon its applicability to that particular class of objects, but the hypothesis nevertheless asserts something about, and indeed imposes restrictions upon, *all objects* (within the logical type of the variable occurring in the hypothesis, which in the case of our last illustration might be the class of all physical objects). Indeed, a hypothesis of the form "Every *P* is a *Q*" forbids the occurrence of any objects having the property *P* but lacking the property *Q*; i.e. it restricts all objects whatsoever to the class of those which either lack the property *P* or also have the property *Q*. Now, every object either belongs to this class or falls outside it, and thus, every object—and not only the *P*'s—either conforms to the hypothesis or violates it; there is no object which is not implicitly "referred to" by a hypothesis of this type. In particular, every object which either is no sodium salt or burns yellow conforms to, and thus "bears out" the hypothesis that all sodium salts burn yellow; every other object violates that hypothesis.

The weakness of the idea under consideration is evidenced also by the observation that the class of objects about which a hypothesis is supposed to assert something is in no way clearly determined, and that it changes with the context, as was shown in 5.12 (*a*).

(*b*) A second important source of the appearance of paradoxicality in certain cases of confirmation is exhibited by the following consideration.

Suppose that in support of the assertion "All sodium salts burn yellow" somebody were to adduce an experiment in which a piece of pure ice was held into a colourless flame and did not turn the flame yellow. This result would confirm the assertion, "Whatever does not burn yellow is no sodium salt," and consequently, by virtue of the equivalence condition, it would confirm the original formulation. Why does this impress us as paradoxical? The reason becomes clear when we compare the previous situation with the case of an experiment where an object whose chemical constitution is as yet unknown to us is held into a flame and fails to turn it yellow, and where subsequent analysis reveals it to contain no sodium salt. This outcome, we should no doubt agree, is what was to be expected on the basis of the hypothesis that all sodium salts burn yellow—no matter in which of its various equivalent formulations it may be expressed; thus, the data here obtained constitute confirming evidence for the hypothesis. Now the only difference between the two situations here considered is that in the first case we are told beforehand the test substance is ice, and we happen to "know anyhow" that ice contains no sodium salt; this has the consequence that the outcome of the flame-colour test becomes entirely irrelevant for the confirmation of the hypothesis and thus can yield no new evidence for us. Indeed, if the flame should not turn yellow, the hypothesis requires that the substance contain no sodium salt—and we know beforehand that ice does not—and if the flame should turn yellow, the hypothesis would impose no further restrictions on the substance; hence, either of the possible outcomes of the experiment would be in accord with the hypothesis.

The analysis of this example illustrates a general point: In the seemingly paradoxical cases of confirmation, we are often not actually judging the relation of the given evidence, E alone to the hypothesis H (we fail to observe the "methodological fiction," characteristic of every case of confirmation, that we have no relevant evidence for H other than that included in E); instead, we tacitly introduce a comparison of H with a body of evidence which consists of E in conjunction with an additional amount of information which we happen to have at our disposal; in our illustration, this information includes the knowledge (1) that the substance used in the experiment is ice, and (2) that ice contains no sodium salt. If we assume this additional information as given, then, of course, the outcome of the experiment can add no strength to the hypothesis under consideration. But if we are careful to avoid this tacit reference to additional knowledge (which entirely changes the character of the problem), and if we formulate the question as to the confirming character of the evidence in a manner adequate to the concept of confirmation as used in this paper, we have to ask: Given some object *a* (it happens to be a piece of ice, but this fact is not included in the evidence), and given the fact that *a* does not turn the flame yellow and is no sodium salt—does *a* then constitute confirming evidence for the hypothesis? And now—no matter whether *a* is ice or some other substance—it is clear that the answer has to be in the affirmative; and the paradoxes vanish.

So far, in section (*b*), we have considered mainly that type of paradoxical case which is illustrated by the assertion that any non-black non-raven constitutes confirming evidence for the hypothesis, "All ravens are black." However, the general idea just outlined applies as well to the even more extreme cases exemplified by the assertion that any non-raven as well as any black object confirms the hypothesis in question. Let us illustrate this by reference to the latter case. If the given evidence *E*—i.e. in the sense of the required methodological fiction, all our data relevant for the hypothesis—consists only of one object which, in addition, is black, then *E* may reasonably be said to support even the hypothesis that all objects are black, and *a fortiori E* supports the weaker assertion that all ravens are black. In this case, again, our factual knowledge that not all objects are black tends to create an impression of paradoxicality which is not justified on logical grounds. Other "paradoxical" cases of confirmation may be dealt with analogously, and it thus turns out that the "paradoxes of confirmation," as formulated above, are due to a misguided intuition in the matter rather than to a logical flaw in the two stipulations from which the "paradoxes" were derived.[23, 24, ... 25]

[23] The basic idea of sect. (b) in the above analysis of the "paradoxes of confirmation" is due to Dr. Nelson Goodman, to whom I wish to reiterate my thanks for the help he rendered me, through many discussions, in clarifying my ideas on this point.

[24] The considerations presented in section (*b*) above are also influenced by, though not identical in content with, the very illuminating discussion of the "paradoxes" by the Polish methodologist and logician Janina Hosiasson-Lindenbaum; cf. her article "On Confirmation", *The Journal of Symbolic Logic,* vol. 5 (1940), especially sec. 4. Dr. Hosiasson's attention had been called to the paradoxes by the article referred to in n. 2, and by discussions with the author. To my knowledge, hers has so far been the only publication which presents an explicit attempt to solve the problem. Her solution is based on a theory of degrees of confirmation, which is developed in the form of an uninterpreted axiomatic system (cf. n. 6 and part (b) in sect. 1 of the present article), and most of her arguments presuppose that theoretical framework. I have profited,

8. CONDITIONS OF ADEQUACY FOR ANY DEFINITION OF CONFIRMATION

The two most customary conceptions of confirmation, which were rendered explicit in Nicod's criterion and in the prediction criterion, have thus been found unsuitable for a general definition of confirmation. Besides this negative result, the preceding analysis has also exhibited certain logical characteristics of scientific prediction, explanation, and testing, and it has led to the establishment of certain standards which an adequate definition

---

however, by some of Miss Hosiasson's more general observations which proved relevant for the analysis of the paradoxes of the non-gradated relation of confirmation which forms the object of the present study.

One point in those of Miss Hosiasson's comments which rest on her theory of degrees of confirmation is of particular interest, and I should like to discuss it briefly. Stated in reference to the raven-hypothesis, it consists in the suggestion that the finding of one non-black object which is no raven, while constituting confirming evidence for the hypothesis, would increase the degree of confirmation of the hypothesis by a smaller amount than the finding of one raven which is black. This is said to be so because the class of all ravens is much less numerous than that of all non-black objects, so that—to put the idea in suggestive though somewhat misleading terms—the finding of one black raven confirms a larger portion of the total content of the hypothesis than the finding of one non-black non-raven. In fact, from the basic assumptions of her theory, Miss Hosiasson is able to derive a theorem according to which the above statement about the relative increase in degree of confirmation will hold provided that actually the number of all ravens is small compared with the number of all non-black objects. But is this last numerical assumption actually warranted in the present case and analogously in all other "paradoxical" cases? The answer depends in part upon the logical structure of the language of science. If a "co-ordinate language" is used, in which, say, finite space-time regions figure as individuals, then the raven-hypothesis assumes some such form as "Every space-time region which contains a raven, contains something black"; and even if the total number of ravens ever to exist is finite, the class of space-time regions containing a raven has the power of the continuum, and so does the class of space-time regions containing something non-black; thus, for a co-ordinate language of the type under consideration, the above numerical assumption is not warranted. Now the use of a co-ordinate language may appear quite artificial in this particular illustration; but it will seem very appropriate in many other contexts, such as, e.g., that of physical field theories. On the other hand, Miss Hosiasson's numerical assumption may well be justified on the basis of a "thing language," in which physical objects of finite size function as individuals. Of course, even on this basis, it remains an empirical question, for every hypothesis of the form "All *P*'s are *Q*'s," whether actually the class of non-*Q*'s is much more numerous than the class of *P*'s; and in many cases this question will be very difficult to decide.

[25] [Ed. note: Original footnotes 25–36, which correspond to omitted sections of this article, have been deleted; remaining footnotes have been renumbered for use in this book.]

of confirmation has to satisfy. These standards include the equivalence condition and the requirement that the definition of confirmation be applicable to hypotheses of any degree of logical complexity, rather than to the simplest type of universal conditional only. An adequate definition of confirmation, however, has to satisfy several further logical requirements, to which we now turn.

First of all, it will be agreed that any sentence which is entailed by—i.e. a logical consequence of—a given observation report has to be considered as confirmed by that report: Entailment is a special case of confirmation. Thus, e.g., we want to say that the observation report "*a* is black" confirms the sentence (hypothesis) "*a* is black or grey"; and—to refer to one of the illustrations given in the preceding section—the observation sentence '$R_2(a, b)$' should certainly be confirming evidence for the sentence '$(Ez)R_2(a, z)$'. We are therefore led to the stipulation that any adequate definition of confirmation must insure the fulfilment of the

(8.1) *Entailment Condition:* Any sentence which is entailed by an observation report is confirmed by it.[26]

This condition is suggested by the preceding consideration, but of course not proved by it. To make it a standard of adequacy for the definition of confirmation means to lay down the stipulation that a proposed definition of confirmation will be rejected as logically inadequate if it is not constructed in such a way that (8.1) is unconditionally satisfied. An analogous remark applies to the subsequently proposed further standards of adequacy.

Second, an observation report which confirms certain hypotheses would invariably be qualified as confirming any consequence of those hypotheses. Indeed: any such consequence is but an assertion of all or part of the combined content of the original hypotheses and has therefore to be regarded as confirmed by any evidence which confirms the original hypotheses. This suggests the following condition of adequacy:

(8.2) *Consequence Condition:* If an observation report confirms every one of a class *K* of sentences, then it also confirms any sentence which is a logical consequence of *K*.

If 8.2 is satisfied, then the same is true of the following two more special conditions:

(8.21) *Special Consequence Condition:* If an observation report confirms a hypothesis *H*, then it also confirms every consequence of *H*.

(8.22) *Equivalence Condition:* If an observation report confirms a hypothesis *H*, then it also confirms every hypothesis which is logically equivalent with *H*.

[26] As a consequence of this stipulation, a contradictory observation report, such as {Black(a), $\sim$ Black(a)} confirms every sentence, because it has every sentence as a consequence. Of course, it is possible to exclude the possibility of contradictory observation reports altogether by a slight restriction of the definition of "observation report." There is, however, no important reason to do so.

(This follows from (8.21) in view of the fact that equivalent hypotheses are mutual consequences of each other.) Thus, the satisfaction of the consequence condition entails that of our earlier equivalence condition, and the latter loses its status of an independent requirement.

In view of the apparent obviousness of these conditions, it is interesting to note that the definition of confirmation in terms of successful prediction, while satisfying the equivalence condition, would violate the consequence condition. Consider, for example, the formulation of the prediction-criterion given in the earlier part of the preceding section. Clearly, if the observational findings B_2 can be predicted on the basis of the findings B_1 by means of the hypothesis H, the same prediction is obtainable by means of any equivalent hypothesis, but not generally by means of a weaker one.

On the other hand, any prediction obtainable by means of H can obviously also be established by means of any hypothesis which is stronger than H, i.e. which logically entails H. Thus, while the consequence condition stipulates in effect that whatever confirms a given hypothesis also confirms any weaker hypothesis, the relation of confirmation defined in terms of successful prediction would satisfy the condition that whatever confirms a given hypothesis, also confirms every stronger one.

But is this "converse consequence condition," as it might be called, not reasonable enough, and should it not even be included among our standards of adequacy for the definition of confirmation? The second of these two suggestions can be readily disposed of: The adoption of the new condition, in addition to (8.1) and (8.2), would have the consequence that any observation report B would confirm any hypothesis H whatsoever. Thus, e.g., if B is the report "*a* is a raven" and H is Hooke's law, then, according to (8.1), B confirms the sentence "*a* is a raven," hence B would, according to the converse consequence condition, confirm the stronger sentence "*a* is a raven, and Hooke's law holds"; and finally, by virtue of (8.2), B would confirm H, which is a consequence of the last sentence. Obviously, the same type of argument can be applied in all other cases.

But is it not true, after all, that very often observational data which confirm a hypothesis H are considered also as confirming a stronger hypothesis? Is it not true, for example, that those experimental findings which confirm Galileo's law, or Kepler's laws, are considered also as confirming Newton's law of gravitation?[27] This is indeed the case, but this does not justify the acceptance of the converse entailment condition as a general rule of the logic of confirmation; for in the cases just mentioned, the weaker hypothesis is connected with the stronger one by a logical bond of a particular kind: it is essentially a substitution instance of the stronger one;

[27] Strictly speaking, Galileo's law and Kepler's laws can be deduced from the law of gravitation only if certain additional hypotheses—including the laws of motion—are presupposed; but this does not affect the point under discussion.

thus, e.g., while the law of gravitation refers to the force obtaining between any two bodies, Galileo's law is a specialization referring to the case where one of the bodies is the earth, the other an object near its surface. In the preceding case, however, where Hooke's law was shown to be confirmed by the observation report that *a* is a raven, this situation does not prevail; and here, the rule that whatever confirms a given hypothesis also confirms any stronger one becomes an entirely absurd principle. Thus, the converse consequence condition does not provide a sound general condition of adequacy.[28]

A third condition remains to be stated:[29]

(8.3) *Consistency Condition:* Every logically consistent observation report is logically compatible with the class of all the hypotheses which it confirms.

The two most important implications of this requirement are the following:

(8.31) Unless an observation report is self-contradictory,[30] it does not confirm any hypothesis with which it is not logically compatible.

(8.32) Unless an observation report is self-contradictory, it does not confirm any hypotheses which contradict each other.

The first of these corollaries will readily be accepted; the second, however, —and consequently (8.3) itself—will perhaps be felt to embody a too severe restriction. It might be pointed out, for example, that a finite set of measurements concerning the variation of one physical magnitude, *x*, with another,

[28] William Barrett, in a paper entitled "Discussion on Dewey's Logic" (*The Philosophical Review,* vol. 50, 1941, pp. 305 ff., esp. p. 312) raises some questions closely related to what we have called above the consequence condition and the converse consequence condition. In fact, he invokes the latter (without stating it explicitly) in an argument which is designed to show that "not every observation which confirms a sentence need also confirm all its consequences", in other words, that the special consequence condition (8.21) need not always be satisfied. He supports his point by reference to "the simplest case: the sentence '*C*' is an abbreviation of '*A* · *B*', and the observation 0 confirms '*A*', *and so* '*C*', but is irrelevant to '*B*', which is a consequence of '*C*'." (Italics mine.)

For reasons contained in the above discussion of the consequence condition and the converse consequence condition, the application of the latter in the case under consideration seems to us unjustifiable, so that the illustration does not prove the author's point; and indeed, there seems to be every reason to preserve the unrestricted validity of the consequence condition. As a matter of fact, Mr. Barrett himself argues that "the degree of confirmation for the consequence of a sentence cannot be less than that of the sentence itself"; this is indeed quite sound; but it is hard to see how the recognition of this principle can be reconciled with a renunciation of the special consequence condition, since the latter may be considered simply as the correlate, for the non-gradated relation of confirmation, of the former principle which is adapted to the concept of degree of confirmation.

[29] For a fourth condition, see n. 16.

[30] A contradictory observation report confirms every hypothesis (cf. n. 8) and is, of course, incompatible with every one of the hypotheses it confirms.

y, may conform to, and thus be said to confirm, several different hypotheses as to the particular mathematical function in terms of which the relationship of *x* and *y* can be expressed; but such hypotheses are incompatible because to at least one value of *x*, they will assign different values of *y*.

No doubt it is possible to liberalize the formal standards of adequacy in line with these considerations. This would amount to dropping (8.3) and (8.32) and retaining only (8.31). One of the effects of this measure would be that when a logically consistent observation report *B* confirms each of two hypotheses, it does not necessarily confirm their conjunction; for the hypotheses might be mutually incompatible, hence their conjunction self-contradictory; consequently, by (8.31), *B* could not confirm it.—This consequence is intuitively rather awkward, and one might therefore feel inclined to suggest that while (8.3) should be dropped and (8.31) retained, (8.32) should be replaced by the requirement (8.33): If an observation sentence confirms each of two hypotheses, then it also confirms their conjunction. But it can readily be shown that by virtue of (8.2) this set of conditions entails the fulfilment of (8.32).

If, therefore, the condition (8.3) appears to be too rigorous, the most obvious alternative would seem to lie in replacing (8.3) and its corollaries by the much weaker condition (8.31) alone. [Added in 1971: But as G. J. Massey has pointed out to me, satisfaction of (8.1), (8.2), and (8.31) logically implies satisfaction of (8.3); hence, that alternative fails.] One of the great advantages of a definition which satisfies (8.3) is that it sets a limit, so to speak, to the strength of the hypotheses which can be confirmed by given evidence.[31]

The remainder of the present study, therefore, will be concerned exclusively with the problem of establishing a definition of confirmation which satisfies the more severe formal conditions represented by (8.1), (8.2), and (8.3) together.

The fulfilment of these requirements, which may be regarded as general laws of the logic of confirmation, is of course only a necessary, not a sufficient, condition for the adequacy of any proposed definition of confirmation. Thus, e.g., if "*B* confirms *H*" were defined as meaning "*B* logically entails *H*," then the above three conditions would clearly be satisfied; but the definition would not be adequate because confirmation has to be a more comprehensive relation than entailment (the latter might be referred to as the special case of *conclusive* confirmation). Thus, a definition of confirmation, to be acceptable, also has to to be materially adequate: it has to provide a reasonably close approximation to that conception of confirmation which is implicit in scientific procedure and methodological discussion. That con-

[31] This was pointed out to me by Dr. Nelson Goodman. The definition later to be outlined in this essay, which satisfies conditions (8.1), (8.2) and (8.3), lends itself, however, to certain generalizations which satisfy only the more liberal conditions of adequacy just considered.

ception is vague and to some extent quite unclear, as I have tried to show in earlier parts of this paper; therefore, it would be too much to expect full agreement as to the material adequacy of a proposed definition of confirmation; on the other hand, there will be rather general agreement on certain points; thus, e.g., the identification of confirmation with entailment, or the Nicod criterion of confirmation as analyzed above, or any definition of confirmation by reference to a "sense of evidence," will probably now be admitted not to be adequate approximations to that concept of confirmation which is relevant for the logic of science.

On the other hand, the soundness of the logical analysis (which, in a clear sense, always involves a logical reconstruction) of a theoretical concept cannot be gauged simply by our feelings of satisfaction at a certain proposed analysis; and if there are, say, two alternative proposals for defining a term on the basis of a logical analysis, and if both appear to come fairly close to the intended meaning, then the choice has to be made largely by reference to such features as the logical properties of the two reconstructions, and the comprehensiveness and simplicity of the theories to which they lead.

9. THE SATISFACTION CRITERION OF CONFIRMATION

As has been mentioned before, a precise definition of confirmation requires reference to some definite "language of science," in which all observation reports and all hypotheses under consideration are assumed to be formulated, and whose logical structure is supposed to be precisely determined. The more complex this language, and the richer its logical means of expression, the more difficult it will be, as a rule, to establish an adequate definition of confirmation for it. However, the problem has been solved at least for certain cases: With respect to languages of a comparatively simple logical structure, it has been possible to construct an explicit definition of confirmation which satisfies all of the above logical requirements, and which appears to be intuitively rather adequate. An exposition of the technical details of this definition has been published elsewhere;[32] in the present

[32] In my article referred to in n. 1. The logical structure of the languages to which the definition in question is applicable is that of the lower functional calculus with individual constants, and with predicate constants of any degree. All sentences of the language are assumed to be formed exclusively by means of predicate constants, individual constants, individual variables, universal and existential quantifiers for individual variables, and the connective symbols of denial, conjunction, alternation, and implication. The use of predicate variables or of the identity sign is not permitted.

As to the predicate constants, they are all assumed to belong to the observational vocabulary, i.e. to denote a property or a relation observable by means of the accepted techniques. ("Abstract" predicate terms are supposed to be defined in terms of those of the observational vocabulary and then actually to be

study, which is concerned with the general logical and methodological aspects of the problem of confirmation rather than with technical details, it will be attempted to characterize the definition of confirmation thus obtained as clearly as possible with a minimum of technicalities.

Consider the simple case of the hypothesis H: '$(x)(\text{Raven}(x) \supset \text{Black}(x))$', where 'Raven' and 'Black' are supposed to be terms of our observational vocabulary. Let B be an observation report to the effect that $\text{Raven}(a) \cdot \text{Black}(a) \cdot \sim\text{Raven}(c) \cdot \text{Black}(c) \cdot \sim\text{Raven}(d) \cdot \sim\text{Black}(d)$. Then B may be said to confirm H in the following sense: There are three objects altogether mentioned in B, namely a, c, and d; and as far as these are concerned, B informs us that all those which are ravens (i.e. just the object a) are also black.[33] In other words, from the information contained in B we can infer that the hypothesis H does hold true within the finite class of those objects which are mentioned in B.

Let us apply the same consideration to a hypothesis of a logically more complex structure. Let H be the hypothesis "Everybody like somebody"; in symbols: '$(x)(Ey)\text{Likes}(x, y)$', i.e. for every (person) x, there exists at least one (not necessarily different person) y such that x likes y. (Here again, 'Likes' is supposed to be a relation-term which occurs in our observational vocabulary.) Suppose now that we are given an observation report B in which the names of two persons, say 'e' and 'f', occur. Under what conditions shall we say that B confirms H? The previous illustration suggests the answer: If from B we can infer that H is satisfied within the finite class $\{e, f\}$; i.e. that within $\{e, f\}$ everybody likes somebody. This in turn means that e likes e or f, and f likes e or f. Thus, B would be said to confirm H if B entailed the statement "e likes e or f, and f likes e or f." This latter statement will be called the development of H for the finite class $\{e, f\}$.—

The concept of *development of a hypothesis*, H, *for a finite class of individuals*, C, can be defined in a general fashion; the development of H for C states what H would assert if there existed exclusively those objects which are elements of C.—Thus, e.g., the development of the hypothesis $H_1 =$ '$(x)(P(x) \vee Q(x))$' (i.e. "Every object has the property P or the property

---

replaced by their *definientia*, so that they never occur explicitly.)

As a consequence of these stipulations, an observation report can be characterized simply as a conjunction of sentences of the kind illustrated by '$P(a)$', '$\sim P(b)$', '$R(c, d)$', '$\sim R(e, f)$', etc., where 'P', 'R', etc., belong to the observational vocabulary, and 'a', 'b', 'c', 'd', 'e', 'f', etc., are individual names, denoting specific objects. It is also possible to define an observation report more liberally as any sentence containing no quantifiers, which means that besides conjunctions also alternations and implication sentences formed out of the above kind of components are included among the observation reports.

[33] I am indebted to Dr. Nelson Goodman for having suggested this idea; it initiated all those considerations which finally led to the definition to be outlined below.

Q") for the class $\{a, b\}$ is '$P(a) \vee Q(a)) \cdot (P(b) \vee Q(b))$' (i.e. "$a$ has the property P or the property Q, and b has the property P or the property Q"); the development of the existential hypothesis H_2 that at least one object has the property P, i.e. '$(Ex)P(x)$', for $\{a, b\}$ is '$P(a) \vee P(b)$'; the development of a hypothesis which contains no quantifiers, such as H_3: '$P(c) \vee Q(c)$' is defined as that hypothesis itself, no matter what the reference class of individuals is.

A more detailed formal analysis based on considerations of this type leads to the introduction of a general relation of confirmation in two steps; the first consists in defining a special relation of direct confirmation along the lines just indicated; the second step then defines the general relation of confirmation by reference to direct confirmation.

Omitting minor details, we may summarize the two definitions as follows:

(9.1 Df.) An observation report B directly confirms a hypothesis H if B entails the development of H for the class of those objects which are mentioned in B.

(9.2 Df.) An observation report B confirms a hypothesis H if H is entailed by a class of sentences each of which is directly confirmed by B.

The criterion expressed in these definitions might be called the satisfaction criterion of confirmation because its basic idea consists in construing a hypothesis as confirmed by a given observation report if the hypothesis is satisfied in the finite class of those individuals which are mentioned in the report.—Let us now apply the two definitions to our last examples: The observation report B_1: '$P(a) \cdot Q(b)$' directly confirms (and therefore also confirms) the hypothesis H_1, because it entails the development of H_1 for the class $\{a, b\}$, which was given above.—The hypothesis H_3 is not directly confirmed by B_1, because its development—i.e. H_3 itself—obviously is not entailed by B_1. However, H_3 is entailed by H_1, which is directly confirmed by B_1; hence, by virtue of (9.2), B_1 confirms H_3.

Similarly, it can readily be seen that B_1 directly confirms H_2.

Finally, to refer to the first illustration given in this section: The observation report '$\text{Raven}(a) \cdot \text{Black}(a) \cdot \sim\text{Raven}(c) \cdot \sim\text{Black}(c) \cdot \sim\text{Raven}(d) \cdot \sim \text{Black}(d)$' confirms (even directly) the hypothesis '$(x)(\text{Raven}(x) \supset \text{Black}(x))$', for it entails the development of the latter for the class $\{a, c, d\}$, which can be written as follows: '$(\text{Raven}(a) \supset \text{Black}(a)) \cdot (\text{Raven}(c) \supset \text{Black}(c)) \cdot (\text{Raven}(d) \supset \text{Black}(d))$'.

It is now easy to define disconfirmation and neutrality:

(9.3 Df.) An observation report B disconfirms a hypothesis H if it confirms the denial of H.

(9.4 Df.) An observation report B is neutral with respect to a hypothesis H if B neither confirms nor disconfirms H.

By virtue of the criteria laid down in (9.2), (9.3), (9.4), every consistent observation report, B, divides all possible hypotheses into three mutually exclusive classes: those confirmed by B, those disconfirmed by B, and those with respect to which B is neutral.

The definition of confirmation here proposed can be shown to satisfy all the formal conditions of adequacy embodied in (8.1), (8.2), and (8.3) and their consequences; for the condition (8.2) this is easy to see; for the other conditions the proof is more complicated.[34]

[34] For these proofs, see the article referred to in Part I, n. 1. I should like to take this opportunity to point out and to remedy a certain defect of the definition of confirmation which was developed in that article, and which has been outlined above: this defect was brought to my attention by a discussion with Dr. Olaf Helmer.

It will be agreed that an acceptable definition of confirmation should satisfy the following further condition which might well have been included among the logical standards of adequacy set up in section 8 above: (8.4). If B_1 and B_2 are logically equivalent observation reports and B_1 confirms (disconfirms, is neutral with respect to) a hypothesis H, then B_2, too, confirms (disconfirms, is neutral with respect to) H. This condition is indeed satisfied if observation reports are construed, as they have been in this article, as classes or conjunctions of observation sentences. As was indicated at the end of n. 14, however, this restriction of observation reports to a conjunctive form is not essential; in fact, it has been adopted here only for greater convenience of exposition, and all the preceding results, including especially the definitions and theorems of the present section, remain applicable without change if observation reports are given the more liberal interpretation characterized at the end of n. 14. (In this case, if 'P' and 'Q' belong to the observational vocabulary, such sentences as '$P(a) \vee Q(a)$', '$P(a) \vee \sim Q(b)$', etc., would qualify as observation reports.) This broader conception of observation reports was therefore adopted in the article referred to in Part I, n. 1; but it has turned out that in this case, the definition of confirmation summarized above does not generally satisfy the requirement (8.4). Thus, e.g., the observation reports, $B_1 =$ '$P(a)$' and $B_2 =$ '$P(a) \cdot (Q(b) \vee \sim Q(b))$' are logically equivalent, but while B_1 confirms (and even directly confirms) the hypothesis $H_1 =$ '$(x)P(x)$', the second report does not do so, essentially because it does not entail '$P(a) \cdot P(b)$', which is the development of H_1 for the class of those objects mentioned in B_2. This deficiency can be remedied as follows: The fact that B_2 fails to confirm H_1 is obviously due to the circumstance that B_2 contains the individual constant 'b', without asserting anything about b: The object b is mentioned only in an analytic component of B_2. The atomic constituent '$Q(b)$' will therefore be said to occur (twice) inessentially in B_2. Generally, an atomic constituent A of a molecular sentence S will be said to occur inessentially in S if by virtue of the rules of the sentential calculus S is equivalent to a molecular sentence in which A does not occur at all. Now an object will be said to be mentioned inessentially in an observation report if it is mentioned only in such components of that report as occur inessentially in it. The sentential calculus clearly provides mechanical procedures for deciding whether a given observation report mentions any object inessentially, and for establishing equivalent formulations of the same report in which no object is mentioned inessentially. Finally, let us say that an object is mentioned essentially in an observation report if it is mentioned, but not only mentioned inessentially, in that report. Now we replace 9.1 by the following definition:

Furthermore, the application of the above definition of confirmation is not restricted to hypotheses of universal conditional form (as Nicod's criterion is, for example), nor to universal hypotheses in general; it applies, in fact, to any hypothesis which can be expressed by means of property and relation terms of the observational vocabulary of the given language, individual names, the customary connective symbols for 'not', 'and', 'or', 'if-then', and any number of universal and existential quantifiers.

Finally, as is suggested by the preceding illustrations as well as by the general considerations which underlie the establishment of the above definition, it seems that we have obtained a definition of confirmation which also is materially adequate in the sense of being a reasonable approximation to the intended meaning of confirmation.

(9.1*a*) An observation report *B* directly confirms a hypothesis *H* if *B* entails the development of *H* for the class of those objects which are mentioned essentially in *B*.

The concept of confirmation as defined by (9.1*a*) and (9.2) now satisfies (8.4) in addition to (8.1), (8.2), (8.3) even if observation reports are construed in the broader fashion characterized earlier in this footnote.

13

THE PARADOX OF CONFIRMATION

J. L. Mackie

The paradox of confirmation, also known as the paradox of the ravens, has been much discussed.[1] But the sound points that have been made, I believe, by all contributors are in danger of being lost in the controversy, and particularly in the mutual misunderstandings and charges of misunderstanding. I shall, therefore, try to place in order some of the main points that should emerge from the discussion and also to develop some further considerations.

The paradox is constituted by the three propositions:

(1) Observations of black ravens confirm 'All ravens are black'.

(2) Observations of black shoes, white swans, etc., are neutral to (i.e., do not confirm) 'All ravens are black'.

(3) If observations confirm one formulation of a hypothesis they confirm any logically equivalent formulation.

These three propositions all seem plausible, but they are incompatible;

[1] See, for example: Janina Hosiasson-Lindenbaum, *Journal of Symbolic Logic*, 1940, 5, 133–148; Carl G. Hempel, *Mind*, 1945, 54, 1–26 and 97–121, also *Philosophy*, 1958, 33, 342–348; J. W. N. Watkins, *Philosophy*, 1957, 32, 112–131 and 1958, 33, 349–355, also this *Journal* [*British Journal for the Philosophy of Science*], 1960, 10, 318–320, also *Australasian Journal of Philosophy*, 1959, 37, 240–241 and 1960, 38, 54–58; D. C. Stove, *Australasian Journal of Philosophy*, 1959, 37, 149–151 and 1960, 38, 51–54; H. G. Alexander, this *Journal*, 1958, 9, 227–233 and 1959, 10, 229–234; J. Agassi, this *Journal*, 1959, 9, 311–317; I. J. Good, this *Journal*, 1960, 11, 145–149 and 1961, 12, 63–64.

From *The British Journal for the Philosophy of Science*, vol. 13 (1963), pp. 265–277. Reprinted by permission of the author and Cambridge University Press.

for an observation of a white swan is an observation of a non-black non-raven, this is related to 'All non-black are non-ravens' in the same way that the observation of a black raven is to "All ravens are black," and these two universal propositions are equivalent. Since no-one will reject (3) or deny the equivalence of the two universals, the paradox can be solved only by rejecting or qualifying (1) or (2) or both. Professor Hempel rejects (2); Mr. Watkins qualifies both (1) and (2), and uses the difficulties he sees in Hempel's solution as a support for Professor Popper's theory of confirmation.

For simplicity I shall exclude, for a time, the complications involved in the use of such terms as 'white', 'shoes', 'swans', etc., and take it that (2) asserts that observations of black non-ravens and non-black non-ravens are neutral to the hypothesis (*h*) that all ravens are black.

1

If we consider this problem in the setting in which Hempel proposed it, namely, without reference to any additional knowledge, then we must say that if the observation of a black raven confirms *h* then an observation of a non-black non-raven equally confirms *h*. But in a setting as bare as this there could be no better confirmation of *h* than the observation of black ravens. So if we are to define confirmation in such a way that it can occur in relation to a context of this sort, we must accept (1) and therefore reject (2). And we can do so, because in this context the rejection of (2) is not counter-intuitive; we simply have no intuitions in favour of (2) that are relevant to a setting as bare as this. Thus Hempel's solution of the paradox is adequate for its context; the paradox is not a difficulty for his definition of confirmation in its original setting, and he is justified in using the paradox to reject Nicod's definition according to which a black raven would confirm *h* but a non-black non-raven would not.

But whether we can plausibly speak of confirmation in a setting as bare as this, or what comes to the same thing, whether any definition of confirmation that gives positive results in this setting can be carried over to ordinary contexts or will even throw light on confirmation in ordinary contexts, is a further question.

2

If we take account of the additional information that there are far more non-black things than ravens, then we can solve the paradox in another way. That is, we can still deny (2), but argue that the observation of non-black non-ravens is not nearly so good a confirmation of *h* as the observation of black ravens, and that, being intuitively aware of the relative weakness of the former confirmation, we mistake it for a complete failure to confirm, and so we have, in this context, a mistaken intuition in favour of (2).[2]

This explanation applies generally, for although it is only if there are far more non-ψ's than ϕ's that the observation of something that is both ϕ and ψ is really a much better confirmation of $(x)\ [\phi x \supset \psi x]$ than the observation of something that is both non-ϕ and non-ψ, we are so used to giving affirmative terms to minority classes (if ϕ is an affirmative term, there are nearly always far more non-ϕ's than ϕ's) that we can understand how a mistaken intuition in favour of the generalisation of (2) could arise.

It is important to see, however, that to develop this solution we have to appeal to what I shall call the *Inverse Principle*, that a hypothesis h is confirmed by an observation-report b in relation to background knowledge k if and only if the observation-report is made more probable by the adding of the hypothesis to the background knowledge, that is, if the probability of b in relation to the conjunction of k and h is greater than the probability of b in relation to k alone—in symbols, if $p(b, k.h) > p(b, k)$—and that b confirms h the better the more the adding of h to k raises the probability of b.

For example, in one of Alexander's accounts of this solution, we assume this principle and assign probabilities in accordance with the relative size of the classes 'ravens', 'non-ravens', etc. Thus if we take it as known that the ratio of ravens to non-ravens is x to $1-x$, and that of black things to non-black things y to $1-y$, where $x < y < \frac{1}{2}$, then the probabilities of each of the four sorts of observation-report b_1 ('This is a black raven'), b_2 ('This is a non-black raven'), b_3 ('This is a black non-raven'), and b_4 ('This is a non-black non-raven'), in relation (i) to our background knowledge k alone, and (ii) to the conjunction of k with h, are as set out below:

	b_1	b_2	b_3	b_4
(i)	xy	$x(1-y)$	$y(1-x)$	$(1-x)(1-y)$
(ii)	x	0	$y-x$	$1-y$

Then clearly reports of the form b_1 have their probability raised by the adding of h far more than those of the form b_4, and so b_1 reports confirm h far better than b_4 ones do, while a b_3 report is actually made less probable by the adding of h to the k assumed here, and thus, in relation to *this* background knowledge, disconfirms h, though only slightly.[3]

[2] Hosiasson-Lindenbaum, Alexander, and Good all develop variants of this way of solving the paradox, though they also make other points. [Ed. note: Footnotes have been renumbered for use in this book.]

[3] I have followed one of Alexander's accounts, but his other account, and the more elaborate accounts of Hosiasson-Lindenbaum and Good, all rest upon theories which involve the Inverse Principle and the assigning of probabilities in accordance with the distribution of characteristics.

In this setting also, then, we can solve the paradox by denying (2) and explaining away our intuitions in favour of it. It follows that the paradox cannot be used against an account of confirmation of this sort—an account which does not equate confirming with satisfying a hypothesis, but which bases it on the Inverse Principle and an assignment of probabilities in accordance with a taken-as-known distribution of characteristics. But it is still a further question whether confirmation defined thus, for a context of still very severely limited knowledge, has much to do with confirmation in ordinary contexts. But this qualification works in two directions. If we accept this account in its proper context, we cannot go on and apply it unhesitatingly in other contexts. But equally we should not use difficulties in its application to other contexts to throw doubt on the account itself.

Thus this account of confirmation does not imply that observations of black ravens, however numerous, in ordinary contexts will provide a worthwhile confirmation of *h*, still less that such observations of non-black non-ravens will do so, or even that they will confirm *h* at all, especially if they are actually observations of white swans, etc., which involve further terms which may possibly be related to *h* in other ways.[4] But for this very reason our commonsense doubts whether observations of non-black non-ravens in ordinary contexts could confirm *h* do not reflect adversely upon this account of confirmation or upon this solution of the paradox in its own context.

Thus to deny (2) in this still restricted context does not commit us to any absurd practical policies of investigation, such as trying to confirm *h* by cheaply scoring up enormous numbers of non-black non-ravens, and the absurdity of this policy in no way tells against the denial of (2) in this context.[5] Nor does it commit us to a policy of trying to support currently-accepted hypotheses as opposed to a policy of trying to overthrow them. And even if we wanted to support a certain hypothesis, the fact that a certain kind of observation would, in relation to severely limited background knowledge, confirm that hypothesis, does not imply that a sensible way of supporting it would be to *try* to make observations of that kind; for *trying* to make a certain kind of observation alters our background knowledge, and (as I shall show in section 4 below) may alter it in such a way that observations of that kind now confirm the hypothesis much less, or perhaps not at all.

3

It would seem, then, that Hempel-type solutions of the paradox, denying only proposition (2), can be defended, at any rate within some limited con-

[4] Cf. Good's distinction between 'stoogian' and 'non-stoogian' observations.

[5] Watkins and Agassi repeatedly use this improper extension to discredit the 'instantial theory'. See, e.g., Watkins in *Philosophy*, 1958, 32, 118, and Agassi in this *Journal*, 1959, 9, 315.

texts. But Mr. Watkins has a strong argument in favour of the view that both (1) and (2) should be qualified,[6] which I would restate as follows.

Suppose that there is an observation-procedure which could turn out in just one of only two possible ways. For example, we inspect an object, already known to be black, to see whether it is a raven or not; the outcome must be either that it is a raven or that it is not. Both the alternative outcomes cannot confirm the same hypothesis: if the observation's having turned out one way confirms the hypothesis, its having turned out the other way would not have confirmed it. This, which I shall call the *Alternative Outcome Principle*, is obvious enough in its own right; it is also provable as a consequence of the Inverse Principle. For in relation to a k which includes the fact that a procedure with just these alternative outcomes has been followed, the denial of one outcome is equivalent to the assertion of the other; so if b confirms h, then adding h to k makes b more likely, and therefore makes not-b less likely, and so the alternative outcome, not-b, would not confirm h but disconfirm it.

From this it follows that if we inspect an object already known to be black, its turning out to be a raven and its turning out to be a non-raven cannot both confirm h, though both are compatible with or satisfy the hypothesis; to confirm a hypothesis, then, an observation must do something more than satisfy it.

This seems to show that b_1 and b_3 observations, since they could be such alternative outcomes, cannot both confirm h, and it suggests the following further conclusions. If we inspect an object already known to be a raven and it turns out to be black, this confirms h, for the procedure might have turned out the other way and falsified h; but if we inspect an object already known to be black and it turns out to be a raven, this does not confirm h. Similarly, if we inspect an object already known to be non-black and it turns out to be a non-raven, this confirms h, for the alternative outcome would have falsified h, but if we inspect an object already known to be a non-raven and it turns out to be non-black, this does not confirm h. And on this view there is no order of inspection by which the observation of a black non-raven could confirm h. Thus both (1) and (2) should be qualified.

But these further conclusions are paradoxical. If we adopt in this sense the view that to confirm a hypothesis a procedure must first endanger it, we make confirmation depend not on what we observe but on the order in

[6] *Philosophy*, 1957, 32, 119–120, and this *Journal*, 1960, 10, 319. Watkins has not always made it clear that his conclusion is that both (1) and (2) should be qualified. On p. 116 of the article in *Philosophy* he describes his way out as being to 'deny the instantiation theory of confirmation implicit in proposition (1)', and one might be forgiven for supposing that he was denying or modifying (1) only. But subsequent discussion has brought it out that his account qualifies (2) also, saying that h *may* be confirmed by a b_4 observation if it is made in a *test* of h.

which we notice characteristics. A black raven observed in one way ('Look, a raven—and it's black') confirms *h*, but observed in another way ('Look, something black—why, it's a raven') it does not. And this must be wrong. We might concede that the way in which an observation is made could affect its confirmatory power, but only if the 'way' involved more significant differences than this. The mere order in which characteristics are spotted must be an irrelevant difference. If confirmation is to be an objective relation, and not a psychological one, the total observation of a black raven must have the same confirmatory value in whichever order its components are put together. In other words, the 'further conclusions' suggested by the Alternative Outcome Principle fall foul of a *Principle of Relevance*.

Part of this difficulty is removed if we distinguish carefully the various 'procedures' and 'outcomes'. The procedure of inspecting something already known to be a raven, with the alternative outcomes that it is, or is not, black, is only the second stage of the observation of a black raven; before it there was another procedure, in which a previously uncharacterised object turned out to be a raven, but might have turned out to be a non-raven. The observation of a black raven can be regarded as being made up of two stages, then, in either of the following ways:

1. (1.1) This (previously uncharacterised object) is a raven, and (1.2) This raven is black.
2. (2.1) This (previously uncharacterised object) is black, and (2.2) This black thing is a raven.

Similarly, the observations of a black non-raven and of a non-black non-raven can be divided into stages as follows:

3. (3.1) This is black, and (3.2) This black thing is a non-raven.
4. (4.1) This is a non-raven, and (4.2) This non-raven is black.
5. (5.1) This is a non-raven, and (5.2) This non-raven is non-black.
6. (6.1) This is non-black, and (6.2) This non-black thing is a non-raven.

This analysis makes a solution at least formally possible, for our 'further conclusions' from the Alternative Outcome Principle have stressed differences in the confirming effect of the various second stages, such as 1.2 and 2.2, but the Principle of Relevance requires only that 1 as a whole, the combination of 1.1 and 1.2, should have the same confirming effect as 2 as a whole, the combination of 2.1 and 2.2. Clearly, 2 as a whole could be as good a confirmation as 1 as a whole, even if 1.2 was superior to 2.2, provided that 2.1 was correspondingly superior to 1.1.

We cannot, however, save the extreme view expressed in our 'further conclusions' in this manner. For if 2.2 did not confirm *h* at all, we should have to say that 2.1 alone confirms *h* as well as 1 as a whole does, which is absurd, especially as the reason given for saying that 2.2 does not confirm *h* at all, that the alternative outcome would not falsify *h*, would equally tell against 2.1.

This extreme view, then, must be rejected, and it can be rejected without regret since it is not required by the Alternative Outcome Principle. That principle tells us only that one outcome can confirm only if the other outcome would disconfirm, not that the other outcome would have to falsify.[7] But a more moderate view can be worked out as follows.

1.2 is in general a better confirmation of *h* than 2.2, but 2.1 is in general correspondingly better than 1.1, so that 1 as a whole and 2 as a whole are, as the Principle of Relevance requires, equally good.[8] Similarly, while 2.2 and 3.2, being alternative outcomes of the same procedure, cannot both confirm *h*, it is none the less possible that 2 as a whole and 3 as a whole should both do so, though not equally, if the confirming by 3.1 outweighs the disconfirming by 3.2; and likewise it is possible that 4 as a whole and 5 as a whole should both confirm *h*, though not equally.

In other words, although the finding that something, already known to be black, is a raven, is not in itself as good a confirmation of *h* as the finding that something, already known to be a raven, is black, yet the total observation of a black raven confirms *h* equally well in whichever order the characteristics are noticed. For we have to take account also of the *first* stage in

[7] As Watkins suggests in this *Journal*, 1960, 10, 319: '. . . *could not possibly* lead to its falsification and hence that the results of such an investigation, whatever they might be, could not confirm it according to a testability theory of confirmation.'

[8] An artificial example may be sufficient to illustrate this. Suppose that our background knowledge tells us that 10 per cent of objects are ravens and that 10 per cent of non-ravens are black. Then the probable distributions in relation to (i) this knowledge alone (ii) this knowledge plus *h*, are as follows:

	black ravens	non-black ravens	black non-ravens	non-black non-ravens
(i)	1/100	9/100	9/100	81/100
(ii)	1/10	0	9/100	81/100

Consequently the reports in stages 1.1, 1.2, 2.1, and 2.2. will have the following probabilities relatively to k and $k.h$ respectively

	1.1	1.2	2.1	2.2
(i)	1/10	1/9	1/10	1/9
(ii)	1/10	1	19/100	10/19

Thus the addition of *h* raises the probability of 1.2 in the ratio 9:1, while it raises the probability of 2.2 only in the ratio 90:19, but it raises the probability of 2.1 in the ratio 19:10, while it leaves the probability of 1.1 unchanged. Multiplying the ratios for 1.1 and 1.2, and again for 2.1 and 2.2, we find that the total probabilities of 1 and 2 are raised equally in the ratio 9:1. Of course, when the probabilities are thus based on numerical distributions the answer has to come out right. But the example shows how this can be combined with a marked superiority of 1.2 to 2.2.

each case, the finding that this previously uncharacterised object is black, and the finding that this previously uncharacterised object is a raven, and these first stages will also have some confirmatory effect, and a different effect from one another, in relation to whatever background knowledge we assume. In the same way, it is true that if finding that something, already known to be black, is a raven, confirms *h*, then finding that something, already known to be black, is a non-raven, cannot confirm *h* but must disconfirm it; but it may be true none the less, when we take account of the confirmatory effect of the common first stage, the finding that this previously uncharacterised object is black, that the total observation of a black non-raven confirms *h*, though not so well as the total observation of a black raven.

The Alternative Outcome Principle, then, is itself sound, and its legitimate consequences can be reconciled with the Principle of Relevance. But the extreme view, that only those procedures can confirm a hypothesis which endanger it in the strong sense that one of their possible outcomes is a falsification, must be rejected. It is not required by the Alternative Outcome Principle, and it conflicts with the Principle of Relevance. Indeed we can decisively refute this extreme view by comparing our procedures 1 and 2. Here we have two procedures whose total effects are equivalent; but one of them contains a first stage which does not endanger *h* and a second stage which does endanger *h*, whereas the other contains two stages neither of which endangers *h*. Two procedures, then, which must be equivalent in confirmatory power, may fail to contain the same element of endangering; it follows that endangering, in this strong sense, cannot be essential to confirmation.

4

I have said that the fact that a certain kind of observation would, relatively to limited background knowledge, confirm a certain hypothesis, does not imply that a sensible way of supporting it would be to try to make observations of that kind. I want to develop this hint and to link it with the claim constantly made by Professor Popper and his followers that a hypothesis is confirmed by a (favourable) observation only if that is made in a *test* of the hypothesis, in an attempt to falsify it.

This concept of testing, of attempting to falsify, must not be a psychological one, or this claim would fall foul of the Principle of Relevance. When challenged (by Mr Stove) to state an objective, non-psychological criterion of a *test* Mr Watkins referred to what I have been calling the Inverse Principle, saying that *b* is a genuine test of *h* if $p(b, k \,.\, h) > p(b, k)$; a test is a procedure in which a certain outcome is less probable without the hypothesis than with it. This is, I think, a fair general criterion for what is relevant to confirmation, but it lets in far more than the procedures that anyone would ordinarily count as tests or attempts to falsify in actual scientific practice. As we saw in section 2 above, this criterion would allow casual observations of black ravens—and even (to a slight degree) of non-black non-ravens—to count as tests of h in relation to the kind of background knowledge assumed in Alexander's solution of the paradox. But we can give a narrower (but still objective) account of a *test*. We may start from the concept of *looking for* or *trying to find* something. Ordinarily, no doubt, this concept is partly psychological, but it is not wholly so. It has an objective part, which can be considered on its own; in this respect to look for, say, a lost golf ball is to adopt a procedure which *raises our chances* of coming upon the ball as compared with the procedure of, say, sitting still or wandering aimlessly about. Similarly, to look for black ravens is to adopt a procedure which raises our chances of coming upon black ravens (if there are any) as compared with the procedure of, say, observing anything and everything that happens along. To test a hypothesis, I suggest, in the sense that we require, is analogously to look for a falsification of it, to adopt a procedure which raises our chances of falsifying it, if it is false. Let us see how *tests* and *looking for* various things, thus objectively defined, may affect confirmation.

Suppose, first, that we add, to the background knowledge assumed in Alexander's solution in section 2 above, the fact that we are looking for black ravens. Then this will raise the probability of b_1 reports above the previously assigned value xy. It will also raise the probability of b_1 reports in relation to k plus h above its previously assigned value x, but not in as high a ratio. Therefore the effect of adding to k the fact that we are looking for black ravens is to reduce the degree in which the adding of h increases the probability of b_1 reports, and hence to make such reports confirm h less well. And the more efficiently we look, the more is this so. In other words, as Popper's supporters (among others) have always said, if we try to confirm a hypothesis the favourable observations confirm it less well. This would be nonsense if 'try to confirm' had here a purely psychological meaning, but it makes very good sense if it means 'adopt a procedure which raises the chance of a favourable observation'.

The obverse policy, of not looking for falsifications, has an even more blatant effect. If we very efficiently do not look for non-black ravens, we may reduce to zero our chance of observing a non-black raven, even if there are some. And then the probabilities of reports of the forms b_1, b_2, b_3, b_4, given this policy and Alexander's background knowledge, will be x, 0, $y—x$, and $1—y$ respectively, that is, exactly the same values as when we add the hypothesis h. It follows that, given this policy, h does not raise the probability of any form of observation, and so no observation can confirm h if this policy is pursued with perfect efficiency.

On the other hand, if we (in this sense) try to falsify h, and look for non-black ravens, then we are adopting a policy which in relation to k alone raises our chances of observing non-black ravens, and therefore reduces the probabilities of the three other forms of observation, but which in relation to $k.h$ makes no difference, since if there are no non-black ravens our chances of finding one cannot be raised, however hard we look. Thus, by trying to falsify the hypothesis we create conditions in which the addition of h increases the probability of favourable observation-reports more than it would otherwise; we create conditions in which the observation of a black raven is a better confirmation of h than it would otherwise be.

Although we cannot give numerical measures to these probabilities, it is plausible to say that in ordinary contexts the probability of a certain sort of observation, and hence the degree of confirmation of a hypothesis by an observation of that sort, will be radically affected by the observer's policy, by what he is looking for. Hence, if we restrict the notion of a genuine test more narrowly than Watkins (as quoted above) did, so that we count as tests only procedures which raise considerably the chance of falsifying the hypothesis if it is false, we may well be right in saying that in ordinary contexts worthwhile confirmations come only from genuine tests. On the other hand it will not be true that observations not made in tests do not confirm at all; this is true only if 'test' is given the wider definition that Watkins suggests, when it is simply a re-statement of the Inverse Principle.

This discussion supports Popper's doctrine with regard to actual investigation in ordinary contexts, but at the same time it shows that this does not conflict with something like an instantiation theory of confirmation in restricted contexts. The view that instances of black ravens confirm h in relation to severely limited background knowledge does not imply that we can confirm h by *looking for and finding* such instances (for such looking takes us beyond the limits set), and so it does not imply that it would be sensible to try to confirm h by looking for black ravens. And since this is so, we cannot argue from the fact that this is (obviously) not sensible to the rejection of that view.

5

The paradox of the ravens, then, is to be solved in different ways in different contexts.

(i) Given no additional knowledge, it is to be solved in Hempel's way; in this context, proposition (2) in the set that constitutes the paradox can be denied without qualms.

(ii) Given only the information that ravens are less numerous than non-ravens, and black things than non-black things (and therefore that ravens are less numerous than non-black things), the paradox is to be solved in Alexander's way (or Hosiasson-Lindenbaum's, or Good's); proposition (2) can be denied, but its plausibility is explained by the fact that we mistake a relatively poor confirmation for no confirmation at all.

(iii) Given unrestricted admission of relevant information of kinds that we commonly have, while proposition (1) is in general true and proposition (2) is in general false, they are so only if we take account of all confirmations, including quite weak ones, and it is more to the point to replace both (1) and (2) with (1′) and (2′) as follows:

(1′) Observations of black ravens confirm 'All ravens are black' to a worthwhile degree only if they are made in genuine tests of this hypothesis (as defined in section 4).

(2′) Observations of black non-ravens never confirm 'All ravens are black' to a worthwhile degree, and observations of non-black non-ravens confirm 'All ravens are black' to a worthwhile degree only if they are made in genuine tests of this hypothesis.

It is clear that (1′) and (2′) are together compatible with (3).

This does not mean, however, that we cannot have a single theory of confirmation which applies in contexts of all sorts. In developing solutions of type (ii) we appeal to some theory that involves the Inverse Principle, and this principle is at least the basic idea of Popper's theory of confirmation, which gives us a solution of type (iii). The difference between these types of solution lies not in the fundamental principle that is used, but in the range of information called upon to supply the probabilities from which that principle generates confirmations. Type (ii) solutions draw their probabilities from relative distributions of characteristics, type (iii) solutions draw theirs from relevant information of all sorts, and the probabilities to which they refer can therefore be radically affected by the policies of the investigator himself. If there is a break in continuity, then, it comes not between solutions of types (ii) and (iii), but between those of types (i) and (ii).

But even a type (i) solution can be related to the Inverse Principle. For we might say that in an absolutely bare setting, to which this solution applies, reports of all the four kinds b_1, b_2, b_3, and b_4 are initially equally probable. Adding h in this context merely rules out the possibility of b_2 reports, and therefore increases equally the probabilities of reports of the other three kinds, that is, of all reports which are consistent with or satisfy h. Thus Hempel's satisfaction criterion of confirmation can be understood as the consequence of applying the Inverse Principle to the only probabilities available in a setting as bare as this.

I conclude, then, that a theory of confirmation based on the Inverse Principle can do justice to what the rival schools of thought have said about this paradox. The conflicts between these schools, on issues so far raised, can be resolved. They have arisen partly because the 'rival' theories and 'rival' solutions of the paradox have been interpreted as applying beyond their proper contexts, and partly because the Alternative Outcome Principle has been used over-enthusiastically and without a sufficiently precise specification of procedures and outcomes. The only legitimate dispute, if the Inverse Principle is itself accepted, would be about the probabilities to which it is applied; for clearly a confirmation in terms of this principle can be no more objective than the probabilities from which it is generated.[9]

[9] [Ed. note: The reader is referred to a more recent article by Mackie, "The Relevance Criterion of Confirmation," *British Journal for the Philosophy of Science*, vol. 20 (1969), which takes up several points that were made in discussions provoked partly by the above article.]

14

THAT POSITIVE INSTANCES ARE NO HELP

Hugues Leblanc

Prior to their examining any *A*, both the wise man and the fool could say in their heart: "All examined *A*'s are *B*'s." Only the fool, however, would hark the rule: "If every examined *A* has proved a *B*, trust every unexamined *A* to prove a *B*," for, under the circumstances, "All examined *A*'s are non-*B*'s" would also hold true, and every unexamined *A* could *with equal justice* be trusted to prove a non-*B*.

Suppose, however, one or more *A*'s had been examined and each one of them had proved a *B*. Could the wise man as well as the fool thereupon trust every unexamined *A* to prove a *B*? Up to 1946 handbooks of logic would have sanctioned the inference as chancy, to be sure, but sensible. That year, however, Nelson Goodman argued that the time-honored rule: "If one or more *A*'s have been examined and each one of them has proved a *B*, trust every unexamined *A* to prove a *B*," is downright foolery, since in the name of it every unexamined *A* could also be trusted to prove a non-*B*.[1]

Claims of this magnitude are bound to be challenged, and this one has been, critics countering that the next *A*, trusted though it could be—under the above rule—to prove both a *B* and a non-*B*, could nonetheless not be trusted to be both *with equal justice*, and hence that Goodman's paradox is less damning than he would have it. The objection, to my mind, is ill-taken.

[1] See "A Query on Confirmation," this *Journal* [*The Journal of Philosophy*], 43, 14 (July 4, 1946): 383–385.

From *The Journal of Philosophy*, vol. 60, 1963, pp. 453–462. Reprinted by permission of the author and *The Journal of Philosophy*. [The author wishes] to thank Professors Goodman, Carnap, and Hempel for their helpful criticisms of an earlier version of this paper. [He considers himself] solely accountable, though, for the line of defense and attack adopted here.

Since, however, Goodman's 1946 paper and the later *Fact, Fiction, and Forecast* somewhat court it, I should like to review the whole controversy and restate the paradox in a more telling form.

I

To quote from Goodman's "A Query on Confirmation":

> Induction might roughly be described as the projection of characteristics of the past into the future, or more generally of characteristics of one realm of objects into another. But exact expression of this vague principle is exceedingly difficult. ... Suppose we have drawn a marble from a certain bowl on each one of the ninety-nine days up to and including VE day, and each marble drawn was red. We could expect that the marble drawn on the following day would also be red. So far all is well. Our evidence may be expressed by the conjunction "$R(a_1)$ & $R(a_2)$ & · · · & $R(a_{99})$," which well confirms the prediction "$R(a_{100})$." But increase of credibility, projection, "confirmation" in any intuitive sense, does not occur in the case of every predicate under similar circumstances. Let "S" be the predicate "is drawn by VE day and is red, or is drawn later and is non-red." The evidence of the same drawings above assumed may be expressed by the conjunction "$S(a_1)$ & $S(a_2)$ & · · · & $S(a_{99})$." By the theories of confirmation in question this well confirms the prediction "$S(a_{100})$"; but actually we do not expect that the hundredth marble will be non-red. "$S(a_{100})$" gains no whit of credibility from the evidence offered (383).

Goodman's case thus rests on the following:
first, the conjunction

$$R(a_1) \mathbin{\&} R(a_2) \mathbin{\&} \cdots \mathbin{\&} R(a_{99}), \qquad (1)$$

which, to believe him, sums up all of the evidence on hand;
second, the surmise

$$R(a_{100}), \qquad (2)$$

which he projects on the basis of (1);
third the definition

$$S(x) \rightarrow (D(x) \mathbin{\&} R(x)) \vee ({\sim}D(x) \mathbin{\&} {\sim}R(x)), \qquad (3)$$

where "$D(x)$" is short for "x is drawn by VE day";
fourth, the conjunction

$$S(a_1) \mathbin{\&} S(a_2) \mathbin{\&} \cdots \mathbin{\&} S(a_{99}), \qquad (4)$$

which, to believe him, is tantamount by way of (3) to (1);
and, fifth, the surmise

$$S(a_{100}), \qquad (5)$$

which, to believe him, yields, by way of (3), the negation

$$\sim R(a_{100}) \tag{6}$$

of (2), and hence would discredit the principle of induction, if projected on the basis of (4) as (2) was on the basis of (1).

In "On the Application of Inductive Logic," a paper he published in 1947, Carnap objected that

> Goodman's reasoning with respect to his [two] example[s] is not correct because he violates the principle of total evidence [in both of them]. In the [first] example, "$S(x)$" is meant as "x is a marble drawn before or on VE day and is red, or is drawn later and is non-red." The observational results include the fact that the marble a_1 was drawn ninety-eight days before VE day and was red. Instead of this known fact, Goodman formulates the evidence concerning a_1 simply by "$S(a_1)$." This sentence, however, says merely that a_1 was drawn before or on VE day and was red, or was drawn later and was non-red; it is true that this follows from the fact mentioned, but it is obviously less than is known about a_1."[2, 3]

Carnap definitely scores a point here: (4) rehearses only part of the evidence on hand, and hence (5), projected as it is on the basis of (4), is not projected *with full justice*. He could have scored three others. First, (1) does not rehearse all the evidence on hand either, and hence (2), projected as it is on the basis of (1), is not projected *with full justice* either. Second, (1), far from being equivalent to (4), does not even imply (4); hence (1) could be true without (4) being true. And, third, (5) does not imply (6), and hence could be true without (6) being true.

Goodman's paradox, fortunately, is sturdy enough to survive the onslaught. Suppose, first, that Carnap, obligingly understating the fact that marble a_1 was drawn *ninety-eight days* before VE day, marble a_2 *ninety-seven days* before VE day, and so on, would let the conjunction

$$(D(a_1) \mathbin{\&} R(a_1)) \mathbin{\&} (D(a_2) \mathbin{\&} R(a_2)) \mathbin{\&} \cdots \mathbin{\&} (D(a_{99}) \mathbin{\&} R(a_{99})) \tag{1$'$}$$

sum up all of the evidence on hand, and let (2) or, preferably,

$$\sim D(a_{100}) \supset R(a_{100}) \tag{2$'$}$$

be projected on the basis of (1′). (4), though implied by (1′), does not imply (1′), and hence

$$\sim D(a_{100}) \supset S(a_{100}), \tag{5$'$}$$

if projected on the basis of (4), would indeed be projected *with less justice* than (2′). Consider, however, Carnap's own rewrite of (4), namely,

$$(D(a_1) \mathbin{\&} S(a_1)) \mathbin{\&} (D(a_2) \mathbin{\&} S(a_2)) \mathbin{\&} \cdots \mathbin{\&} (D(a_{99}) \mathbin{\&} S(a_{99})) \tag{4$'$}$$

[2] *Philosophy and Phenomenological Research*, 8, 1: 139–140.

[3] For Goodman's own reply to Carnap, see "On Infirmities of Confirmation-Theory," *ibid.*, 8, 1: 149–151.

(4′), besides being implied by (1′), also implies (1′), as a simple truth-table calculation will aver. Hence (5′), if projected on the basis of (4′) rather than (4), would be projected *with the same justice* as (2′). But (5′), as a simple truth-table calculation will again aver, is equivalent to

$$\sim D(a_{100}) \supset \sim R(a_{100}) \qquad (6')$$

Hence marble a_{100}, if examined after VE day, could be trusted to be both red and non-red, and trusted to be both *with equal justice.*

Suppose, next, that Carnap did press the fact that marble a_1 was drawn *ninety-eight days* before VE day, marble a_2 *ninety-seven days* before VE day, and so on, and hence insisted on (1′) being amended to read:

$$(D_1(a_1) \;\&\; R(a_1)) \;\&\; (D_2(a_2) \;\&\; R(a_2)) \;\&\; \cdots \;\&\; (D_{99}(a_{99}) \;\&\; R(a_{99})), \qquad (1'')$$

where "$D_i(x)$ is short for "x was drawn 100-i days before the morrow of VE day." So long as he would own that "$D_i(x)$" implies "$D(x)$" or "$\sim D(x)$" according as $i \leqslant 99$ or $i > 100$, and he could hardly balk at so doing, (1″) would prove equivalent to

$$(D_1(a_1) \;\&\; S(a_1)) \;\&\; (D_2(a_2) \;\&\; S(a_2)) \;\&\; \cdots \;\&\; (D_{99}(a_{99}) \;\&\; R(a_{99})), \qquad (4'')$$

and

$$D_{100}(a_{100}) \supset S(a_{100}) \qquad (5'')$$

prove equivalent to

$$D_{100}(a_{100}) \supset \sim R(a_{100}) \qquad (6'')$$

So, if

$$D_{100}(a_{100}) \supset R(a_{100}) \qquad (2'')$$

and (5″) were respectively projected on the basis of (1″) and (4″), then (5″) and hence (6″) would be projected *with the same justice* as (2″). Goodman's fateful marble a_{100}, if examined on the morrow of VE day, could therefore be trusted again to be both red and non-red, and be trusted to be both *with equal justice.*

Goodman's paradox reappears—in possibly more familiar garb—in *Fact, Fiction, and Forecast* (1955):

> Suppose that all emeralds examined before a certain time t are green. At time t, then, our observations support the hypothesis that all emeralds are green; and this is in accord with our definition of confirmation. Our evidence statements assert that emerald a is green, that emerald b is green, and so on; and each confirms the general hypothesis that all emeralds are green. So far, so good. Now let me introduce another predicate less familiar than "green." It is the predicate "grue" and it applies to all things examined before t just in case they are green but to other things just in

> case they are blue. Then at time *t* we have, for each evidence statement asserting that a given emerald is green, a parallel evidence statement asserting that that emerald is grue. And the statements that emerald *a* is grue, that emerald *b* is grue, and so on, will each confirm the general hypothesis that all emeralds are grue. Thus according to our definition, the prediction that all emeralds subsequently examined will be green and the prediction that all will be grue are alike confirmed by evidence statements describing the same observations. But if an emerald subsequently examined is grue, it is blue and hence not green. Thus although we are well aware which of the two incompatible predictions is genuinely confirmed, they are equally well confirmed according to our present definition.[4]

Here again Goodman needlessly invites retort, in particular, Hempel's retort in his 1960 paper, "Inductive Inconsistencies":

> In the case of the prediction that the next emerald will be grue, *more* is known than that the emeralds so far observed were all grue, i.e., that they were either examined before *t* and were green or were not examined before *t* and were blue: it is known that they were all examined before *t*. And failure to include this information in the evidence violates the requirement of total evidence."[5]

Suppose, however, Goodman's first batch of evidence statements were strengthened to read: "Emerald *a* has been examined before *t* and is green, emerald *b* has been examined before *t* and is green, and so on," as indeed they must be if they are to convey all of the evidence on hand. Suppose also Goodman's second batch of evidence statements were strengthened to read: "Emerald *a* has been examined before *t* and is grue, emerald *b* has been examined before *t* and is grue, and so on," as they too must be if they are to convey all the evidence on hand and be *truly parallel* to those from the first batch. Then both predictions "All emeralds subsequently examined will be green" and "All emeralds subsequently examined will be grue" could be made *with equal justice*, and the paradox would—by way of "Emerald *v* is examined after *t*," for example—run its inexorable course to both "Emerald *v* is bound to be green" and "Emerald *v* is bound to be blue."

II

To tidy up and generalize our findings, the fool's rule:

> I: To argue at time *t* from "Every *A* examined by time *t* is a *B*" to "Every *A* examined after time *t* is a *B*,"

or, using "$E_t(x)$" as short for "*x* has been examined by time *t*,"

[4] The passage quoted is from pp. 74–75 of *Fact, Fiction, and Forecast* (Harvard University Press, 1955).

[5] See *Synthese*, 12, 4: 461.

I: To argue at time t from "$(\forall x)((A(x) \,\&\, E_t(x)) \supset B(x))$" to "$(\forall x)((A(x) \,\&\, \sim E_t(x)) \supset B(x))$,"

is self-contradictory. Suppose indeed that no marble has been examined by time t, i.e.,

$$\sim(\exists x)(M(x) \,\&\, E_t(x)), \tag{7}$$

where "$M(x)$" is short for "x is a marble." Since (7) implies

$$(\forall x)((M(x) \,\&\, E_t(x)) \supset R(x)), \tag{8}$$

one could at time t argue to

$$(\forall x)((M(x) \,\&\, \sim E_t(x)) \supset R(x)) \tag{9}$$

But (7) also implies

$$(\forall x)((M(x) \,\&\, E_t(x)) \supset \sim R(x)) \tag{10}$$

Hence one could at time t argue *with equal justice* to

$$(\forall x)((M(x) \,\&\, \sim E_t(x)) \supset \sim R(x)) \tag{11}$$

Suppose now object a were a marble that has not been examined by time t, i.e.,

$$M(a) \,\&\, \sim E_t(a) \tag{12}$$

Then from (9), (11), and (12) one could at time t argue *with equal justice* to both

$$R(a)$$

and

$$\sim R(a),$$

a flat contradiction. That no examined A has failed to prove a B is thus no rational ground for trusting every unexamined A to prove a B.

The handbook rule:

II: To argue at time t from "Some A's have been examined by time t" and "Every A examined by time t is a B" to "Every A examined after time t is a B,"

or, using the same abbreviation as above,

II: To argue at time t from "$(\exists x)(A(x) \,\&\, E_t(x))$" and "$(\forall x)((A(x) \,\&\, E_t(x)) \supset B(x))$" to "$(\forall x)((A(x) \,\&\, \sim E_t(x)) \supset B(x))$,"

is equally self-contradictory. Suppose indeed (i) that each one of $n (n \geqslant 1)$ objects $a_1, a_2, \ldots,$ and a_n is a marble, has been examined by time t, and has proved red, and (ii) that no marble other than $a_1, a_2, \ldots,$ and a_n has been examined by time t,[6] i.e.,

$$(M(a_1) \,\&\, E_t(a_1) \,\&\, R(a_1)) \,\&\, (M(a_2) \,\&\, E_t(a_2) \,\&\, R(a_2)) \\ \&\, \cdots \,\&\, (M(a_n) \,\&\, E_t(a_n) \,\&\, R(a_n)) \quad (13)$$

and

$$(\forall x)((M(x) \,\&\, E_t(x)) \supset (x = a_1 \vee x = a_2 \vee \cdots \vee x = a_n)). \quad (14)$$

Suppose also that the predicate "R^*" is defined as follows:

$$R^*(x) \rightarrow (E_t(x) \,\&\, R(x)) \vee (\sim E_t(x) \,\&\, \sim R(x)) \quad (15)$$

Since the conjunction of (13) and (14) implies

$$(\exists x)(M(x) \,\&\, E_t(x)) \,\&\, (\forall x)((M(x) \,\&\, E_t(x)) \supset R(x)), \quad (16)$$

one could at time t argue to

$$(\forall x)((M(x) \,\&\, \sim E_t(x)) \supset R(x)) \quad (17)$$

But, by virtue of (15), the conjunction of (13) and (14) is equivalent to that of

$$(M(a_1) \,\&\, E_t(a_1) \,\&\, R^*(a_1)) \,\&\, (M(a_2) \,\&\, E_t(a_2) \,\&\, R^*(a_2)) \\ \&\, \cdots \,\&\, (M(a_n) \,\&\, E_t(a_n) \,\&\, R^*(a_n)) \quad (18)$$

and (14), a conjunction which implies

$$(\exists x)(M(x) \,\&\, E_t(x)) \,\&\, (\forall x)((M(x) \,\&\, E_t(x)) \supset R^*(x)) \quad (19)$$

Hence one could at time t argue *with equal justice* to

$$(\forall x)((M(x) \,\&\, \sim E_t(x)) \supset R^*(x)) \quad (20)$$

But, by virtue of (15), (20) is equivalent to

$$(\forall x)((M(x) \,\&\, \sim E_t(x)) \supset \sim R(x)) \quad (21)$$

Hence one could at time t argue *with equal justice* to (21). Suppose, finally, object a_{n+1} were a marble that has not been examined by time, t, i.e.,

$$M(a_{n+1}) \,\&\, \sim E_t(a_{n+1}) \quad (22)$$

Then from (17), (21), and (22) one could at time t argue *with equal justice* to both

$$R(a_{n+1}) \quad (23)$$

and

$$\sim R(a_{n+1}), \quad (24)$$

[6] That no marble (emerald) other than $a_1, a_2 \ldots,$ and a_n has been examined by time t, was taken for granted throughout Section I.

a flat contradiction. That some *A*'s have been examined and none of them has failed to prove a *B* is therefore no rational ground either for trusting every unexamined *A* to prove a *B*.

A variant of Rule II which requests a minimum of $n (n \geqslant 1)$ *A*'s to have been examined by time *t*, namely,

> III: To argue at time *t* from "At least *n* *A*'s have been examined by time *t*" and "Every *A* examined by time *t* is a *B*" to "Every *A* examined after time *t* is a *B*,"

is likewise self-contradictory, for from (13), (14), the added bit of evidence:

$$(a_1 \neq a_2 \;\&\; a_1 \neq a_3 \;\&\; \cdots \;\&\; a_1 \neq a_n) \;\&\; (a_2 \neq a_3 \;\&\; a_2 \neq a_4 \;\&\; \cdots \;\&\; a_2 \neq a_n) \;\&\; \cdots \;\&\; a_{n-1} \neq a_n,$$

and (22), one could at time *t* argue *with equal justice* to both (23) and (24). Since Rule III is self-contradictory for large as well as small values of *n*, Goodman has thus given the lie to Mill's dictum:

> Whatever has been found true in *innumerable* instances, and never found to be false after due examination in any, we are safe in acting on as universal provisionally, until an undoubted exception appears; provided the nature of the case be such that a real exception could scarcely have escaped notice.[7]

Statistically-minded handbooks occasionally feature a generalization of Rule III, namely:

> IV: To argue at time *t* from "Exactly *n* *A*'s have been examined by time *t*" and "Of the *A*'s examined by time *t* *m* are *B*'s" to "The proportion of *B*'s among such *A*'s as are examined after time *t* is equal to m/n."

As Goodman himself noted, however, Rule IV fares no better than Rule I, II, or III. Since for any marble examined by time *t* to be an R^* is tantamount to being red, and since for any marble examined after time *t* to be an R^* is tantamount to being non-red, the rule easily proves to be self-contradictory for any ratio m/n other than ½.[8]

Mindful of Goodman's strategy in "A Query on Confirmation," I troubled, as I argued my way to (23) and (24), to call twice on Rule II. And, mindful of Carnap's and Hempel's criticisms of that strategy, I troubled to argue my way to each one of (23) and (24) from the same data, (13), (14),

[7] See *A System of Logic,* book III, chap. XI, sec. 4. The italics are mine.

[8] Rules II and IV are respectively patterned after Black's rules R_1 and R_2 in *Problems of Analysis* (Cornell University Press, 1954), p. 196. Though writing quite a few years after the publication of "A Query on Confirmation," Black, unaccountably enough, pays no heed to Goodman's paradox. His defense of the two rules, brilliant though it may be, is therefore futile.

and (22), and hence—as I put it all along—*with the same justice*. All four of Rules I, II, III, and IV, however, can be shown self-contradictory in a much simpler manner.

Consider, for example, Rule I. (7), besides implying (8) and (10), also implies

$$(\forall x)((M(x) \mathbin{\&} E_t(x)) \supset (R(x) \mathbin{\&} \sim R(x)))$$

Hence in the name of Rule I one could argue at time *t* to

$$(\forall x)((M(x) \mathbin{\&} \sim E_t(x)) \supset (R(x) \mathbin{\&} \sim R(x))),$$

and hence by virtue of (12) to

$$R(a) \mathbin{\&} \sim R(a),$$

a flat contradiction.

Or consider Rule II. By virtue of (15), (13) is equivalent to

$$\begin{aligned}(M(a_1) \mathbin{\&} E_t(a_1) \mathbin{\&} R(a_1) \mathbin{\&} R^*(a_1))\\ \mathbin{\&} (M(a_2) \mathbin{\&} E_t(a_2) \mathbin{\&} R(a_2) \mathbin{\&} R^*(a_2))\\ \mathbin{\&} \cdots \mathbin{\&} (M(a_n) \mathbin{\&} E_t(a_n) \mathbin{\&} R(a_n) \mathbin{\&} R^*(a_n)),\end{aligned}$$

which along with (14) implies

$$(\exists x)(M(x) \mathbin{\&} E_t(x)) \mathbin{\&} (\forall x)((M(x) \mathbin{\&} E_t(x)) \supset (R(x) \mathbin{\&} \mathrm{R}^*(x)))$$

Hence in the name of Rule II one could argue at time *t* to

$$(\forall x)((M(x) \mathbin{\&} \sim E_t(x)) \supset (R(x) \mathbin{\&} R^*(x))),$$

hence by virtue of (15) to

$$(\forall x)((M(x) \mathbin{\&} \sim E_t(x)) \supset (R(x) \mathbin{\&} \sim R(x))),$$

and hence by virtue of (22) to

$$R(a) \mathbin{\&} \sim R(a),$$

a flat contradiction.

That a single recourse to it would convict Rule II (and, by extension, Rules III and IV) of inconsistency may give pause to any future critic of Goodman.[9]

[9] For yet another handling of the matter, see the author's "A Revised Version of Goodman's Paradox on Confirmation," *Philosophical Studies*, 14, 4 (1963): 49–51. Carnap, Pap, Barker and Achinstein, and others have suggested restrictions which, if imposed upon the predicate "*B*" of Rules II, III, and IV, would block Goodman's paradox and might well ensure the consistency of those three rules. The restrictions in question, devised for the most part after the publication of "A Query on Confirmation," should not be considered—as some would have it—a refutation of Goodman's paradox; they are merely means of circumventing it.

III

Goodman aimed his 1946 attack at current theories of confirmation as well as traditional renderings of the principle of induction. What havoc he made of the former remains to be seen.

Consider, in addition to (13), (15), and (18) from Section II, the following four statements:

$$M(b) \mathbin{\&} \sim E_t(b), \tag{25}$$

$$R(b), \tag{26}$$

$$R^*(b), \tag{27}$$

and

$$\sim R(b). \tag{28}$$

Carnap and others would take the conjunction of (13) and (25) to confirm (26) to a degree r, where r is some real number from the interval [0, 1]. Suppose now that the degree to which the conjunction of two statements, one of the form

$$(A(a_1) \mathbin{\&} E_t(a_1) \mathbin{\&} B(a_1)) \mathbin{\&} (A(a_2) \mathbin{\&} E_t(a_2) \mathbin{\&} B(a_2)) \mathbin{\&} \cdots \mathbin{\&} (A(a_n) \mathbin{\&} E_t(a_n) \mathbin{\&} B(a_n)), \tag{29}$$

the other of the form

$$A(b) \mathbin{\&} \sim E_t(b), \tag{30}$$

confirms a surmise of the form

$$B(b), \tag{31}$$

were held to be the same for each and every possible choice of "B." The degree to which the conjunction of (18) and (25) confirms (27) would then have to equal r. But (13) is equivalent, by virtue of (15), to (18). Hence, by a familiar equivalence law of Carnap and others,[10] the degree to which the conjunction of (13) and (25) confirms (27) would also have to equal r. But, given (13) and (25), (27) is equivalent, by virtue of (15), to (28). Hence, by another familiar equivalence law of Carnap and others,[11] the degree to which the conjunction of (13) and (25) confirms (28) would also have to equal r. But, by a familiar addition law of Carnap and others,[12] the degree to which the conjunction of (13) and (25) confirms (26) and

[10] To wit: If e and e' are equivalent, then $c(h,e) = c(h,e')$.

[11] To wit: If e implies $h \equiv h'$, then $c(h,e) = c(h',e)$.

[12] To wit: If e is not logically false, then $c(h,e) + c(\sim h,e) = 1$.

that to which the self-same conjunction confirms (28) must add up to 1. Hence r would have to equal .5.

The result is quite unwelcome. So is an immediate corollary thereof. Since the above r would have to equal .5 whatever the size of n in (13), surmise (26), that is, "Object b is red," where b is a marble that is unexamined by time t, would gain no whit of confirmation as more and more marbles are examined by time t and all of them prove red.[13] It is difficulties of this sort which led Goodman to hint rather broadly at the bankruptcy of current confirmation theories and embark upon a fresh study of the concept of confirmation.

Goodman's bill of indictment was perhaps hastily drawn. Carnap, for whom the degree to which a surmise of form (31) is confirmed by the conjunction of two statements of forms (29) and (30) would increase, and hence vary, with n in all of his languages L but L_∞, explicitly allows that degree to vary from one predicate "B" to another,[14] and so does every follower of his I can think of. It should, however, serve as a constant reminder that predicates in confirmation theory cannot be dealt with on a wholesale basis. Some of them, like Goodman's "grue," demand individual attention, and will be a grief to any confirmation theorist who denies them individual attention.

[13] See, for background material, the author's "The Problem of the Confirmation of Laws," *Philosophical Studies*, 12, 6 (1961): 373–376.

[14] See, for instance, Carnap's *Logical Foundations of Probability* (2nd ed., Chicago University Press, 1962), pp. 567–569; see also his *Continuum of Inductive Methods* (Chicago University Press, 1952).

15

THE PLAUSIBILITY OF THE ENTRENCHMENT CONCEPT

Bernard R. Grunstra

Given the definition of entrenchment employed in Nelson Goodman's prospects for a theory of projection (pp. 87, 94), it is clear that the fear of some is ill-founded that "grue" may come to be much better entrenched than "green" through the literature about it, with resultant disastrous consequences for the theory.[1] Accordingly this paper will fearlessly bring the topic up again. The principal purpose is to inquire how the entrenchment proposal could be considered a reasonable component of a solution to the problem of justifying projection; i.e., to ask how and in what sense predicate entrenchment could plausibly be considered an indispensable and normative factor in the assessment of hypothesis projectibility. As an auxiliary effort the paper will begin with a sketchy and inconclusive comparison of the projection problem raised by the "grue" example with the problem of alternative hypotheses generally, in particular with the curve-plotter's problem. The comparison suggests that perhaps not all hypothesis projection preferences can be correlated with single-predicate entrenchment superiority, but that in any case the study of the curve-plotter's problem is, if promising, also problematic with respect to understanding the role of entrenchment in projection. Since the entrenchment solution of the quest for projection justification was a response to the point of the "grue" example, we next study the features of that example with a view to making the state-

[1] Except where the context indicates otherwise, numbers appearing alone between parentheses are page citations of the second edition of *Fact, Fiction, and Forecast* [4]. Numbers in brackets identify references.

From the *American Philosophical Quarterly Monograph Series*, no. 3 (1969), pp. 100–127. Reprinted by permission of the editor of the *American Philosophical Quarterly* and the author.

ment of its point more plausible than many misunderstandings have permitted. Next we consider other proposed solutions to the projection problem, some which do accept the point of the "grue" example and some which do not. In the light of their weaknesses the entrenchment proposal gains strength by the comparison. Finally we attempt to allay discontent with the entrenchment solution itself, first by noting the "anticipatory" significence of our "descriptive" predicates, secondly by assimilating the proposed entrenchment normativeness to that of a methodological principle in common use in everyday life and the sciences.

I. THE RELEVANCE OF THE CURVE-PLOTTER'S PROBLEM

1. The cutting edge of Goodman's critique of prior confirmation theory appears in his apparent demonstration that any projection whatsoever can be justified on the basis of any evidence whatsoever if we have only the existing logico-syntactical induction rules or confirmation criteria to help us (pp. 74, 75). Both the strength of his attack and its far-reaching implications for inductive theory have apparently often been missed by commentators because of their preoccupation with the idiosyncracies of his illustrative examples. These examples employ unfamiliar predicates of an abnormal breed, such as the now famous pair, "grue" and "bleen." Furthermore there is a kind of informality about both the specification of these predicates and the comparison of their effect on the confirmation criteria with that of normal predicates which amounts to an underspecification of the illustration in matters of detail. Commentators who have in one way or another supplied a fuller specification have frequently been misled either to believe that inductive theory had not been successfully challenged, or else to propose inadequate remedies. Many of the consequent misapprehensions have appeared in the literature and many of these have been answered by Goodman himself or by still other commentators. I shall not attempt to review them here.[2]

What is of more interest here is the possibility that Goodman's case, or one that seems much like it, can be made quite tellingly with the use of predicates that do not seem particularly abnormal and do not allow the glib dismissal of his argument as depending on predicates that no one, particularly practicing scientists, would consider seriously in inductive practice anyhow. This possibility is suggested by similarities with the curve-plotter's problem, most recently remarked by Skyrms ([13], pp. 52–69) and by Hullett and Schwartz [10].

If we consider Goodman's charge apart from his examples, it seems to

[2] How strong Goodman's case remains under appropriate fuller specification of detail can be seen in a clear article by Leblanc [11].

come to this: Since different evidence statements can describe the same empirical data, we can, via criteria governing evidence statements, use this data to support widely different (and incompatible) projections. But it is a familiar feature of data-plotting situations that through a given set of data points an indefinite number of distinct curves may be drawn, each of which leads to different predictions for unexamined points. Is not the essential thrust of the "grue" example, therefore, somehow the same as that made long ago by examples of alternative hypotheses? If so, our study of the role of predicate entrenchment in the assessment of hypothesis projectibility could perhaps be furthered with the use of more familiar, realistic, and acceptable examples than the "grue" example amounts to.

This advantage may not lightly be claimed, however. It depends on establishing how the "grue" problem relates to the alternative hypothesis problem generally, and how the entrenchment solution relates to this general problem. There is reason to think that establishing what these two relationships are calls for considerable discussion. It is not my purpose here to provide such a discussion, but I do want to say enough to indicate why it seems needed.

2. First there is the fact that for a projection problem drawn in terms of competing predicates one can provide a correlated projection problem in which the competition seems to lie elsewhere. Goodman introduced the projection problem in terms of competing predicates because to do so showed the deficiency of existing confirmation criteria even when strengthened by a suggestion of his own (p. 71). But to project (1) "All emeralds are grue" is, on the assumption that nothing is both blue and green, equivalent to projecting (2) "All emeralds examined before *t* are green and all emeralds not examined before *t* are blue." On Hempel's Satisfaction Criterion [9] the latter hypothesis is as well supported by evidence statements about all the emeralds examined before *t* and found green as is the hypothesis (3) "All emeralds examined before *t* are green and all emeralds not examined before *t* are green," which is equivalent to (4) "All emeralds are green." How is (2) to be eliminated in favor of (3)?

If we appeal to the equivalence of (1) and (2), and of (3) and (4), and the superior entrenchment-based projectibility of (4) relative to (1), then we provoke the following questions. (A) How do we know equivalences are all right to use, since by Goodman's own modification of Hempel's Satisfaction Criterion, equivalent hypotheses are not to be expected to be equally well-confirmed by the same evidence? (p. 71).[3] (B) If equivalences are all

[3] See also Scheffler [12], pp. 286–291, and Hanen [8]. Hypothesis (2) must be distinguished from "All emeralds are examined before *t* and green or not examined before *t* and blue." The latter is disfavored in competition with (4) on the same grounds as (1), namely the inferior entrenchment of the extension of the consequent predicate (pp. 95, 97).

right to use, why can't we use the equivalences in the opposite direction, justifying the selection of (4) over (1) by the selection of (3) over (2)— and then try to establish the warrant for the latter selection in some non-entrenchment-based way?

If instead we attempt to establish the superiority of (3) over (2) by some non-entrenchment-based criterion, i.e., by a syntactic criterion obtained by some suitable augmentation of the Satisfaction Criterion, these questions arise. (C) Assuming we try to limit ourselves to the Selective Confirmation Criterion ([12], pp. 286–291), what is the proper form of a contrary of (2) which will help us to eliminate (2) as not selectively confirmed, but to retain (3) as selectively confirmed—supposing that is desirable? (D) On the same assumption, whether or not we can retain the selective confirmation of (3), what equivalences are we allowed to use, since on the Selective Confirmation Criterion not all will do? (E) Why is it that a difference should be drawn by such diverse methods between hypothesis pairs as equivalent as (4) and (1), or (3) and (2) are in many ways? The possibility of the questions just raised suggests that it may be no shortcut in understanding entrenchment to appeal to the alternative hypothesis problem in its general form.

3. A further consideration of the curve-plotter's problem suggests the same thing. In the form in which Hullett and Schwartz [10] cite this problem it becomes clearer that the point of the "grue" example is independent of the use of "time t" in its characterization.[4] But it remains the case that one is still appealing to abnormal predicates (unless one has switched to a non-predicate-competition example, such as that just discussed). Suppose we have examined nine samples of gas at constant pressure P_0 and constant temperature T_0, nine at higher pressure P_1 and temperature $T_0, \ldots$ and nine at still higher pressure P_{10} and temperature T_0. After allowing for errors we agree that each of the nine investigated at a given pressure have the same volume, and that where P represents any of the indicated pressures and V the volume found to correspond to it, V is discovered to vary inversely with the pressure according to the rule (i) $PV = k_0T_0$, where k_0 is a fixed constant. But then the 99 examined samples also obey the following rules (among many others), none of which is projectible:

(ii) $(P \leqslant P_{10} \,\&\, PV = k_0T_0) \vee (P > P_{10} \,\&\, PV = 2k_0T_0)$

(iii) $((P = P_0 \vee P = P_1 \vee \ldots \vee P = P_{10}) \,\&\, PV = k_0T_0) \vee$
$(\sim (P = P_0 \vee P = P_1 \vee \ldots \vee P = P_{10}) \,\&\, PV = 2k_0T_0)$

(iv) (x is any of the 99 examined (k_0, T_0) samples, suitably identi-

[4] This was clear already from these other predicates Goodman uses: "in this room and English-speaking" (p. 37), "in this room and a third son" (p. 37), "in zig A" (applies to things in some "helter-skelter selection" of marbles, p. 104), and "bagleet" (applies to naval fleets and to a particular bagful of marbles, p. 111).

fied & $P(x)V(x) = k_0(x)T_0(x)) \vee (x$ is a (k_0T_0) sample but not one of the 99 examined samples & $P(x)V(x) = 2k_0(x)T_0(x))$.

However, Goodman has made clear that the trouble he was pointing to was not to be attributed to unfamiliar predicates but to unentrenched extensions (pp. 95, 97). "Examined before *t* and green or not so examined and blue" is just as bad as "grue," and "examined before *t* and grue or not so examined and bleen (or, green)" is as good as "green." The predicate "$(P \leqslant P_{10}$ & $V = k_0T_0/P) \vee (P > P_{10}$ & $V = 2k_0T_0/P)$," especially in competition with "$V = k_0T_0/P$," is as abnormal as "grue."[5]

4. But then one wonders if one can make the point of the "grue" example in the context of the curve-plotter's problem without using abnormal predicates. While it may not be sufficient to the creation of a predicate that will make the point "grue" is used to make that the predicate's extension be poorly entrenched, it is perhaps necessary. It is not clear, therefore, that we can study the role of predicate entrenchment in projectibility assessments with the help of a curve-plotting problem using competing normal predicates of reasonable entrenchment.

This seems borne out if we investigate such a problem. Consider a set of five data points, taken for some measurable property *R* at different values of some measurable property *Q*, such that when *R* values are plotted against *Q* values, the points appear to fall at irregular intervals along a straight line parallel to the *Q*-axis. One is inclined to say that *R* is constant with *Q*, and to proceed to draw a straight line through all five points parallel to the *Q*-axis. There are however, an indefinite number of other graphing possibilities. Let curve *A* be a sine wave of some frequency *f* such that it goes through zero displacement from its average value just at the data points; let curve *B* be a square wave that meets the same condition, but at some other frequency; let curve *C* be an erratic curve for which there is no name, but which goes through all five points. No doubt the extension corresponding to curve C is ill-entrenched, but this cannot be said for the extensions corresponding to curves *A* and *B*. We are in the area where we are dealing with competitions between what Goodman has called "presumptively projectible hypotheses"; hypotheses all of which are supported, unviolated, and unexhausted, and which have survived elimination by both of his two rules

[5] In terms of this context, the analogue of the "grue" example would be a simplified version of the situation described, in which we have examined, say nine samples of (P_0, T_0, k_0) gas and found volume V_0 and "abvolume" A_0, where a thing has abvolume A_0 if either it is one of the first nine cases of (P_0, T_0, k_0) samples examined and has volume V_0 or if it is some other case of (P_0, T_0, k_0) and has volume $2V_0$. The Hullett-Schwartz analysis does not enable us to circumvent the problem of abnormal predicates or extensions, but suggests that it was all along a special case of the curve-plotter's problem.

designed to eliminate radically unprojectible hypotheses (p. 108). In this area, Goodman tells us, we must not even expect all conflicts to be resolved. Each of "two conflicting hypotheses may be equally valid, with the choice between them depending solely upon decisive further evidence" (p. 108). It is true that he proposes to assess the projectibility of presumptively projectible hypotheses in terms of comparative predicate entrenchment, earned and inherited (pp. 108–117). Presumably on this basis the extensions corresponding to curves *A* and *B* will not come off well compared to that corresponding to the extension of "straight line." But to decide whether it is believable that this has anything to do with the superior projectibility of the straight line hypothesis surely requires some discussion.[6] This does not mean that entrenchment considerations never bear on graph competitions, for they do (as the abvolume example shows), especially when inherited entrenchment (pp. 108–117) is taken into account. The present point is that such competitions do not offer strong promise of illuminating the relation

[6] The curve-plotting problem presents a variety of interesting and relevant features which can perhaps be considered further on another occasion. Here are some: (1) Any intuition that a straight line hypothesis should be favored over a sine wave hypothesis is in the first place an intuition that the set of data points we have is a straight-line-set of data points. That is, we have here an analogue of the judgment that a thing is an emerald and will continue to present an emerald-pattern-of-appearances, or that a thing is green and will continue to present a green-pattern-of-appearances. It is less like the claim that all emeralds will be green because such-and-such emeralds have been green than like the claim that this is an emerald. (2) In practical cases our intuition that a straight line hypothesis should be favored is formed in part by a total empirical context. Given a difference in context, then even despite our expectation that a sine wave property should within five points have revealed itself in some departure from the parallel to the Q-axis, we might expect a sine wave far more than a straight line. Thus if all examined situation *S* cases have been sine wave cases, then we shall not be content to overthrow the indicated hypothesis with this one case of five "in-line" points. Similarly, if all examined emeralds have been green, one blue-looking emerald is going to be given a longer look before being accepted as such. Goodman's discussion of comparative entrenchment is surely not in defiance of this point, but in abstraction from it. (3) In this connection, where "This is an emerald" or "This is green" appears itself as an induction from non-demonstrative evidence, we might be prepared to entertain the conjecture that not just isolated predicates become entrenched by actual projections, but predicate pairs or complexes. (4) In the discussion above, complications introduced by "inaccuracies" of measurement, or of judgment generally, have been neglected. Where we must allow for a range of error in the determination of our five data points, they may be so placed that a slightly convex curve, never named (and correspondent, therefore, to an unentrenched extension), may be seriously competitive with the straight line hypothesis and much more favored than the sine or square wave hypotheses (despite the far superior entrenchment of corresponding extensions). This is the analog of the situation where we are not sure whether we have an emerald or some other crystalline mineral form. In actual inductive practice our projections reflect the uncertainty consistent with the evidence.

between projectibility and entrenchment unless they also involve abnormal predicates. For the study of this relation, therefore, the "simple" "grue" kind of abnormal predicate example may prove to have been a clearer case.

5. We have tried to state reasons why we cannot glibly study the relevance and role of predicate entrenchment in hypothesis projectibility through the use of less abnormal predicates than Goodman has used. Now we must combine with these reasons the consideration that in order to be sure we could, with normal-predicate examples account for the contribution of the abnormal-predicate examples to the analysis of projectibility we apparently would have to make sure what the latter contribution is. This suggests that we must study examples like the "grue" example in any case. For this reason we now turn our attention to the "grue" example itself.

II. THE RELEVANCE OF THE "GRUE" EXAMPLE

1. To state the point of the "grue" example does not require much effort or space, but if the statement is to appear plausible we must first consider at some length the characterizing features of the example relevant to its point. There are three or four features in particular which seem to need mention, because misunderstandings have arisen in connection with them.

(*a*) The first has to do with the reference to time t in the initial characterizing sentence for "grue" (p. 74). This sentence itself does not make clear whether t, as metavariable of the characterization, is instantiable with names or constants designating particular times, or with variables any of which is itself instantiable with constants designating particular times. Nor does it make clear that any such particular time is to be that of some particular inductive occasion. The context, both immediate and remote, definitely establishes that t is to be instantiated with the name of a particular time, a time which is that of a particular inductive occasion.[7]

[7] Support for this is found in Goodman's earlier illustration about the inspection of marbles prior to VE day [5]. It is perfectly possible to mischaracterize "grue" so that the force of this result is blunted. One way is to understand t to be instantiable with any variable ranging over all projection times; i.e., to understand it to apply to anything examined before any projection time just in case it is green but to anything else just in case it is blue. In the characterization of "grue" the predicate "green" is akin to a variable in the sense that it can connote a multitude of distinct greens. It is not vital to the "grue" example, however, that "green" be such a variable in the example itself. We can treat the "green" of the characterization as a metalinguistic variable and "instantiate" it by supposing it to connote some one very sharply specified hue of green. But if t is to be, under instantiation, a variable in the "grue" example itself, then even where "green" and "blue" are thus sharply specified, "grue" connotes a variety of grues. An entity examined between t_k and t_{k+1} and found green is not grue_1, $\text{grue}_2, \ldots$, nor grue_k, but it is grue_{k+1}, $\text{grue}_{k+2}, \ldots$, grue_n, where t_n is the time of the present inductive occasion; and so on. By the time t_n, "grue_j" would

(*b*) The reference to a time t in the characterization of "grue" is relevant to the question of the conditions of applicability of the predicate. Here we have a second feature of the example to discuss. Goodman has framed his example to require that at least by the arrival of the inductive occasion of interest the abnormal predicate may be known to apply whenever the normal predicate is then known to apply. This accounts for Goodman's insistence that we are to compare only hypotheses equally well supported by the evidence on the given occasion (pp. 94, 99–103). The evidence classes for the competing hypotheses must be identical.[8] However, he has also framed the example to allow, though not to require, that the applicability of the abnormal predicate to individual cases may be determined independently of the arrival of the inductive occasion. It is not the case that the point of his example depends on the use of a predicate whose applicability cannot be known until the inductive occasion actually arrives at which it can be seen competing with a normal predicate. As for "grue" itself, if we find an emerald green at a time of examination t' which we know to be earlier than t, then we already know at t' both that "green" applies and also that "grue" applies. And in general there is nothing in what Goodman says to rule out the possibility that the applicability of the abnormal predicate be

---

appear in a violated hypothesis for $1 \leqslant j < n$. Though the hypothesis using "grue$_n$" would be unviolated and as well supported as the competing hypothesis using "green," it could be now maintained that the relation of the "grue$_n$" hypothesis to the failed "grue$_j$" hypothesis counted against its predicate's entrenchment and its projectibility. It is this kind of situation, perhaps, that people have in mind who are persuaded that while "green" has applied to past cases and been successfully projected at past times, "grue" has not applied and would not have been successful if tried in projection. (See Sec. III, 2.) Whatever the merits of considering such a predicate it is not Goodman's "grue"; his characterization is precise enough to make that clear. But even if it were not, we would get his hard problem back simply by specifying his characterization as we have already done in the text. In fact the problem can in a sense be made harder than Goodman has made it. To achieve this, we understand t to refer, not to the time t_n of a particular inductive occasion, but to a particular time far removed in the future from t_n. Then, although one can still specify a crucial experiment to decide between this "grue" (no longer Goodman's "grue") and "green," one cannot specify one that will decide before that remote time t.

[8] A thing is in the evidence class for a hypothesis if it is a thing, of the sort described by the antecedent predicate, which is named in a positive instance of the hypothesis, i.e., in an instantiation already determined to be true. A thing is in the projective class for the hypothesis if it is a thing of the same sort not named in a positive or negative case (p. 90). (See also [6].) Incidentally, "grue" cannot be specified as equivalent to "green before t and blue after," for the latter presumably could not be known applicable before t, and if it could, could not apply to just the same cases as "green before t and green after," whose applicability ought to be equally well-known before t (see ns. 9, 18). Hence, "green before t and blue after," though ill-entrenched compared to "green" would not for that reason lose a competition to it, being rather disqualified at the outset.

known as soon as, or even sooner than, the applicability of the normal, if someone can propose a pair of predicates for which this conjecture can be made tolerable. Certainly the conjecture can be tolerated that we may know the applicability of "grue" as soon as we know that of "green," if we already know whether *t* has passed or not, and otherwise as shortly after as is required to determine whether *t* has passed or not. Furthermore, if we do wish to characterize a certain abnormal predicate in such a way that its applicability may be known independently of the arrival of a given occasion, we need not mention the time. We may designate the inductive occasion *e*, or some subsequent event *e′*, in some other way than by its time, or we may tie *e* (or *e′*) to the number of cases *n* to be examined before *e* (*e′*) occurs, or to a designation (naming) of all the individual cases to be examined before *e* (or *e′*) occurs, or to all the values of a certain property (e.g., pressure) which shall pertain to cases examined on occasions prior to the occurrence of *e* (*e′*), or the like. Hence it is a mistake to suppose that on Goodman's characterization "grue" cannot be known to apply (or fail to apply) until time *t*, or until after time *t*. It is also a mistake to suppose that if this were so the problem of abnormal predicates would be solved, since we should only have to characterize "grue" as Goodman has in fact done, or in one of the other ways just suggested, in order to have the problem we in fact have.

(*c*) A third feature of the "grue" example has to do with the role of the word "examined" which appears in the initial characterizing sentence (p. 74). The extension of "grue" apparently includes things blue before *t*, provided there are any such that have not been examined before *t*, and it does not include all things green before *t*, provided there are any such not examined before *t*. The word "examined," therefore, seems to introduce another distinction between grue things and green things besides that introduced by the reference to time *t* at least if we allow a distinction between being green and being examined and found to be green. Then the extensions of green and grue differ not only in respect to what exists, with what color, after *t*, but also in respect to what exists, with what color, before *t*. Accordingly, if we read the characterizing sentence quite literally, we cannot also follow a characterization (A) which says that "grue" applies "to a thing at a given time if and only if either the thing is then green and the time is prior to *t*, or the thing is then blue and the time not prior to *t*" [2]. On the other hand, the context of Goodman's introduction of "grue" suggests that he is not building anything important on a distinction between examined green things and green things generally, but rather abstracting from such a possible distinction in the interest of what is vital to a certain kind of case of induction. His purpose is to design a predicate that will catch the evidence class (p. 90) for the projection, at time *t*, of "All emeralds are green," while at the same time this predicate will have an extension different from that of "green" in a way that would lead to conflicting predictions for unexamined emeralds. For this purpose "grue" might have been characterized to apply (B) to all emeralds examined before t just in case they are green but to other things just in case they are blue, with some augmentation of abnormality, or as in (A) above, to all things existing before t just in case they are green, etc., with some diminution of abnormality. Each of the variant characterizations offers its own small advantages and disadvantages, which need not be rehearsed here. While Goodman's incorporation of "examined" in the characterization of "grue" is not essential to the point of his example, it is essential that the abnormal predicate be characterized so as to be known applicable before t to the cases examined by t.

(*d*) Since the examination of which we speak is intended to delineate the evidence class for a certain hypothesis on a certain inductive occasion, it is clear that "examination" is to be contrasted with "projection," i.e., it is clear that examination definitely establishes something to be the case, while a projection that something is the case is going beyond what has yet been established. This is a fourth feature of the "grue" example that seems to need mention. To decide in a practical situation that a thing belongs in an evidence class, e.g., is green, no doubt takes some measure of what can only fairly be called projection, even when the decision builds on what is quite properly called an examination. But relative to subsequent projection, projection on the basis of this evidence, what is in the evidence class must be contrasted with what is in the projective class. Within the abstraction of the "grue" example, "examination" of a case for a certain property of interest, e.g., greenness, cannot stop with a projection that that case is a case of greenness. Hence the example must not be understood in such a way that the examination of a thing can be found at odds with a projection for the thing. What has once and for all been established by examination is established; it cannot be in conflict with what is projected, which is what has not yet been established; neither can it be in conflict with the results of a subsequent examination, which surely cannot set aside what has already been established (or else "established" was incorrectly used in the first place).

Any objection to meeting this requirement can at best be a claim that it is unrealistic to suppose that it could be met in empirical practice. But while this is true, it is largely because it is unrealistic to suppose that in empirical practice we could establish the color of a thing (any more than the abcolor) right through t by an examination (completed) at t'. This point is confused by linguistic practice, in accord with which we do not honor with our predicates a sharp distinction between what is established and what is only projected. Our normal and "realistic" use of "is established green" or "is found green" applies such expressions to things like emeralds and not things like emerald-temporal-segments, and is projective.[9] If we try to adhere to

[9] If we speak of "establishing" an emerald to be green, without any mention of the time at which and over which it was established to be green, we almost

this use, we must be careful to avoid begging the question. That is, if we construe the "grue" example as though emeralds could at *t′* prior to *t* be examined and established as to color through examination time *t″* subsequent to *t*, then we must allow that the same can be done for "abcolor." In particular, we are not free to construe the example to allow a second examination of an emerald, or at least a second examination issuing in an assignment of "grue" and "not grue" to "the same emerald." Furthermore, suppose we examine an emerald and establish it to be blue for a time interval that ends before *t*. We nevertheless presumably establish for a time interval that extends indefinitely beyond *t* that this emerald was examined before *t*. This fact is relevant, given Goodman's characterization of "grue" as I have represented it above, to the applicability of abcolor predicates. If we allow a second examination after *t*, we shall face perplexities unless we are careful to construe the examination of a thing for abcolor as the characterization requires. That is, the abcolor examination must have the same effect as if, in addition to examining a thing for color we examine it also for examination-prior-to-*t* (see [15], p. 530, and [10], p. 270). However, it is likely that Goodman's original characterization has in view the simplifying assumption that examining an emerald for color (or abcolor) establishes its color (or abcolor) indefinitely. Goodman's evidence class seems full of emeralds established green, not emeralds established green over some limited temporal interval.[10]

invariably are making a tacit projection over the emerald's past and future as an emerald. That is, we certainly have not in fact established, for example, that the emerald has been and will be green at least for so long as it has been and will be an emerald. What we have established at most is that a relatively small temporal segment of the emerald in question, perhaps inclusive of, or overlapping, the temporal segment which underwent examination, is green. In fact, even this much is doubtful, when we consider that we must allow for the distinction between "is green" and "looks green," and for the fact that the very application of the former probably generally involves projection. (See n. 6, comment 1.) The implications of the fact that our "realistic" use of "is established green" is partially projective are further developed in Sec. IV, 1. In particular, compare n. 18.

[10] Of course, we can avoid examining things for examination-prior-to-*t* by recharacterizing "grue" appropriately ([15], p. 531). Independently of that point, we can decide that it is in fact more realistic to apply "green at time *T*" to emeralds than to apply "green," or to apply "green" to emerald-temporal-segments than to emeralds. Such moves, however, bring along their own unfamiliarities. One can get emerald-temporal-segments for predicate extensions through a rather ordinary mode of speech, by attributing "green" or "grue" to things only at given times, e.g., at times when they are examined and found green (or grue) or at times when they are green (or grue). This approach has been followed by several of Goodman's commentators, including Skyrms ([13], p. 57) and Barker and Achinstein ([2]; see Sec. II, 1c). This device does not necessarily eliminate infringements of ordinary locution, however. If "emerald" in (i) "All emeralds are green" denotes emeralds, then establishing that emerald *a* is green

2. We have discussed four features of the "grue" example that are relevant to its point but have been foci of misunderstanding. We now want to emphasize that what is essential to this example, or any other designed to make the same point, is that the predicates represented as competing at a given projection time *t* must differ in applicability, but only over cases not yet examined at *t* (in the above sense). Thus the extensions of "grue" and "green" must differ, but anything, emerald or something else, already examined at *t* and established to belong to either extension must also belong to the other. Nothing literally established by projection time *t*, as distinct from what may have been conjectured by prior projection to hold beyond *t*, may distinguish a case as satisfying one of the competing predicates but not the other. In terms of their descriptive adequacy to the empirical instances available at *t* we must have no basis to choose between the two predicates, while yet their conditions of applicability guarantee that over some cases as yet unexamined not both can apply.

Having said this, we can now say what the point of the "grue" example is. This is to illustrate that our intuitive projection preferences cannot be explained, nor a justifying principle of comparative predicate projectibility constituted, solely in terms of the comparative descriptive adequacy of predicates to what has been literally established about the evidence cases.

3. If this point is tentatively accepted, then the question arises: What is required of a predicate besides adequacy to the evidence? To this Goodman has made the reply that adequate entrenchment is needed as well. One may refuse to accept this point, of course. If he does so, he presumably must return to the attempt to explain the superior projectibility of "green" over "grue" hypothesis in terms of a difference in their descriptive adequacy to the data available at *t*. One immediately thinks that the example, along with the confirmation theory to which it is oriented, may exhibit too narrow a conception of available data. Should we not seek a more realistic conception, perhaps even take account of how certain predicates have fared in

at a given time *T* is evidently not to establish that emerald *a* is green. Hence we do not have a positive instance of the hypothesis in question, although we do of the hypothesis (ii) "All emeralds are green at time *T*," and even of (iii) "All emerald-segments are green." Hypothesis (ii) would be a far-fetched reading of (i), and to get to (i) from (ii) would incorporate further projecting, the analysis of which would require dealing with a variety of predicates differing in specification of *T*. (We cannot, incidentally, understand "Emerald *x* is green at *T*" to mean "Emerald *x* is a green thing and an exists-at-*T* thing," at least without a convention that "green" applies to *x* so long as "exists-at-*T*" applies, but not necessarily longer. For otherwise "green" in "Emerald *x* is a green thing" would in general be understood to apply longer, perhaps as long as "emerald" applies.) Hypothesis (i) can be regarded as entailed by (iii), and so confirmed by the evidence about *a* at *T*, but then the confirmation is not direct in Hempel's sense [9]. If we want direct or "instance" confirmation, then "emerald" in (i) must be read as denoting emerald-segments.

projection, or would have fared had they been tried, perhaps even consider a theory of predicate generation? There are three things we should remember if we adopt this approach.

(*a*) The appeal to entrenchment is explicitly an appeal to a wider conception of data, in fact to data about the relative success of predicates in projection (pp. 85–87, 92–94). We cannot suppose that every useful predicate has become well-entrenched, nor that those which have could not be dispensed with in favor of others never tried. Neither can we suppose that a given well-entrenched predicate has appeared only in unviolated hypotheses, but we can be sure it has not appeared only in quickly and repeatedly violated hypotheses. The success and failure of prior projections "underdetermine" the selection of our stock of well-entrenched predicates, but in part they do determine it.

(*b*) A second point to remember is that Goodman's analysis explicitly endorses our appreciation that predicate selection is in part constrained by empirical considerations other than those having to do with how many times the predicate has been projected (entrenchment). This is, after all, why only supported, unviolated, and unexhausted hypotheses are considered in the two elimination rules (pp. 94, 100, 108). However well-entrenched a predicate may be, a violated hypothesis using it is not found projectible. However ill-entrenched a predicate may be, a supported, unviolated, unexhausted predicate using it is not found unprojectible if it has no competitor meeting the same criteria and using a better-entrenched predicate (p. 97). Hence to be projectible at all a hypothesis must in the usual way answer to the way the world extralinguistically is, and in certain cases a hypothesis which does so is projectible even if it uses very poorly entrenched predicates.

(*c*) Finally we must remember that by his own claim Goodman's primary interest is in isolating a reliable indicator of projectibility (p. 98). His fundamental hypothesis is that superior entrenchment is a necessary and sufficient condition for a valid projective preference (pp. 98, 108 ff.), and hence may be cited as a justification for a particular projective choice (pp. 64, 65).[11] Hence an entrenchment solution is not necessarily incompatible

[11] "Hypothesis" may not be the very best word, but it will keep us aware that Goodman does not attempt to establish a projectibility-indicator role for entrenchment by arguing that an explicit justifying appeal to superior entrenchment is the characteristic feature we find in what we regard as samples of reliable projection. His suggestion that the "judgment of projectibility has derived from the habitual projection, rather than the habitual projection from the judgment of projectibility" (p. 98) must not be stretched that far. Incidentally, the latter suggestion seems to provide ground for regarding superior entrenchment as necessary for superior projectibility. Goodman speaks as though he would be satisfied to have superior entrenchment provide merely a sufficient condition (p. 98). This may be on the analogy of deductive rules, each of which is sufficient but not necessary for deductive validity. But if we may have superior projec-

with a solution in terms of still other data; perhaps, for example, data founding a particular theory of predicate generation and retention in a language.

The latter point is equally true if one accepts the point of the "grue" example, but seeks a solution in either formal or psychological terms, e.g., in terms of formal or psychological "simplicity," ease of applicability, or the like. However, some of these proposed solutions appear no more promising, under investigation, than do proposals which refuse to accept the point of the "grue" example. In fact, when one considers these alternative explanations of projectibility judgments, of whatever sort, many of them turn out to be inadequate, or else too vaguely programmatic for further evaluation. Both the "grue" example and the entrenchment solution, on the other hand, gain strength in the comparison. For this reason we first consider a sample of these alternative proposals, including a number already discussed by Goodman either in [4] or in a reply to one or another objecting article. After that we shall try to allay discontent with the entrenchment solution itself.

III. ALTERNATIVE SOLUTIONS TO THE PROJECTION PROBLEM

Our interest in alternatives to entrenchment is not in classifying them. We shall not attempt to distinguish evidential, formal, or psychological solutions, nor mixtures of these. Similarly no significance attaches to the order of their consideration. For convenience, and because it will help us to distinguish proposals building on mere idiosyncrasies of the "grue" example, we draw up most proposals in terms of that example.

"All emeralds are green," we may say, is to be preferred to "All emeralds are grue," because:

1. "Green" is a more readily applicable predicate than "grue." The idea here would be that to apply "grue" one must meet every condition needed to apply "green" together with some other.

This proposal seems to require an analysis and formalizing of condition-stating. Even if we grant "grue" were in the stated terms less readily applicable than "green" would we be safe in eliminating it for that reason? Suppose *G*, "looks green under green light," applies to all examined emeralds. There are many things to which this applies that "green" does not; to apply the latter seems therefore to require meeting more conditions. Does it also require meeting more conditions than to apply *R*, "looks grue under

---

tibility without superior entrenchment then we must have a supplemental rule to help us recognize a case to which entrenchment comparisons apply. It is true that Goodman's elimination rules are explicitly circumscribed in their application, but entrenchment comparisons are meant to apply more broadly (pp. 106, 118, 119).

grue light," which would perhaps be said to require meeting one more condition (a check of time or date) than *G*? If so, then should we project *R* in preference to "green," which surely conflicts with it over unexamined emeralds?[12]

2. Predicates are proposed in order to try them against future cases, and "grue," had it been projected at earlier times than the present occasion of inductive choice, would immediately have failed subsequent tests prior to the present occasion, whereas "green" would have been successful.

Unfortunately this line of reasoning has in fact missed an important implication of Goodman's "grue" example. This is that if "grue" ever had come into actual trial in projection instead of "green" it would have been projected as successfully as "green" at all projection times prior to t. The t in the characterization of "grue" refers us to some particular time, let us call it t_n, which is that of a particular inductive occasion at which both "grue" and "green" are seen to apply to all the same things (see Sec. II, 1a above). On any prior inductive occasion, of time $t_k < t_n$ the projection of "grue" would have been found successful in, for example, the next examined case, provided the projection of "green" would have been found successful in that case, since this case would be a case of something examined before t_n and green. Neither the nonentrenchment of a predicate, nor its nonprojectibility (if we distinguish them), tells us that it would not have been successful had it been tried previously.

3. The generation of a predicate is psychologically in the nature of a response, at least in part, to "what is presented." What has not yet been experienced elicits no response, in word or otherwise (although later it may prompt responses now unimaginable). The "distribution" and "intensity" of language-eliciting factors or aspects in what has been experienced is reflected in the predicates we propose. Predicates are not mere words aimlessly created in a vacuum, unshaped by any external empirical influence, the product of a random predicate generator. "Green" reflects the way the world is in its eliciting of predicates. "Grue" should be eliminated because, although it would be successful if tried, its coming to trial requires a causal sequence that has not been realized in our world. It requires that at some time t' prior to t (a more or less contemporary projection time), time t has somehow made its "importance" felt in a way that calls for a covering predicate; and this is almost certainly not the case in fact (and is if anything the more unlikely, the greater the interval between t' and t), though as a matter of sheer logical possibility it is not excluded. Similarly any predicate should be eliminated whose conditions of applicability involve an appeal to

[12] Even *G* competes with "green" over unexamined emeralds in the sense that subsequently examined emeralds might satisfy the former but not the latter.

any factor which cannot have causally provoked, at some time or other, a particularly suited linguistic response.

This criterion offers some promise of application in terms of a "fitting" of words to extralinguistic features. However, it does so at the expense of an unrealistic theory of linguistic behavior. Let it be granted that in order actually to have used and projected "grue" with significance at any time, early or late, we must have known at that time approximately under what circumstances to apply it or not to apply it. Still a predicate is proposed not just to answer past experience, but to answer future experience as well. One of the features of human "creative" thinking is that it involves the synthesizing of what has been analyzed as elements in the naive unity of experience into groupings, linkages, and patterns never actually experienced as such. This feature appears in projective behavior when for example, crucial experiments are proposed to decide between hypotheses. Who is to say that "in the early days" of "the mind's" trial of predicates (p. 87) some future time *t* may not have figured in someone's synthesis, with which he intended to unify future and past experience? It is not inconsistent with even a causal theory of language generation to say this, but only with a narrow and unrealistic causal theory. On the basis of a better theory, it is not at all clear that "grue" might not have been tried.[13]

To be sure, the point of the criterion we are considering may be made immune to the objection of the preceding paragraph by proposing a better theory. We need only ask why we may not seek the recurrent extralinguistically oriented characteristics of any or all predicates we would be caused to try, however we might be caused to try them, and also the characteristics of any that may be mentioned which we would not be caused to try, and certify projectibility in terms of the former. But about this proposal three things may be said. First, it is not yet a criterion, but only a proposal for one, since it depends on the promulgation of a successful causal theory of predicate generation. Would we, or would we not, be caused to try "grue"? Secondly, although it is a proposal for an extralinguistically oriented criterion, we cannot suppose in advance that it is a proposal for a criterion in which predicates are preferred solely because they provide a superior "fit" with extralinguistic phenomena. The predicates preferred are to be those which could be consequent upon empirical states of affairs, but we cannot suppose that in view of this they will be predicates which better "picture"

[13] It may be worth remarking that Goodman's speaking of "the mind as in motion from the start, striking out with spontaneous predictions in dozens of directions" (p. 87) of course does not imply an *a*-causal theory of predicate generation. The common conviction that language has conventional (or happenstance) elements may be understood as reflecting something about the facts of actual language generation for which a sophisticated causal theory must account.

or "answer to" those states of affairs. Thirdly, we cannot suppose in advance that we shall even be able to achieve a criterion drawn wholly in terms of extralinguistically oriented characteristics of predicates. It is logically possible that initial projections would appear as consequent upon extralinguistic phenomena but later projections as caused partly by linguistic phenomena in such a way that only by reference to such linguistic phenomena could we distinguish a recurrent set of characteristics to be correlated with projectibility. Entrenchment, of course, would be a candidate for just such a linguistic phenomenon.[14]

4. The definition of "grue" involves a reference to time, place, or finite number of individuals.

As Goodman has pointed out (p. 80) this is true of "grue" defined in terms of "green" and "blue"; it is also true of "green" defined in terms of "grue" and "bleen."

5. The semantic meaning or significance of "grue" involves a reference to time, place, or finite number of individuals. Insofar as this is a suggestion different from the previous one it must have to do with a distinction that can be made between "grue" or "green" in terms of what we would actually have to do to come to employ the terms, or to understand them, or to apply them satisfactorily. The first of these has been considered under suggestion 3 above. The kind of reply that could be made to the others is indicated by what has been said under 1 above.[15]

[14] In the absence of an adequate scientific theory of predicate generation, our present persuasion that a predicate is not one that would have been tried (and successfully so) could lie in the fact that it is not entrenched. This at least is what Goodman is saying.

[15] Goodman's discussion of "qualitativeness" (pp. 79, 80) is not drawn in terms of a distinction between a syntactic (e.g., proposal 4) and a semantic criterion (e.g., proposal 5) of qualitativeness. I am inclined to suppose he is addressing himself primarily to something like 5 as the stronger of the two, and I take him to be making at least two points. The first is that it is doubtful if an independent criterion of qualitativeness can be achieved which does not beg the projectibility question. (As Hullett and Schwartz observed, [10], pp. 264–266, if we suggest a certain criterion for distinguishing "purely qualitative" predicates, we still have to make plausible the hypothesis that such predicates must always be the ones favored in a projection decision, which seems, in fact, unlikely.) Goodman's second point is that while we can indeed specify a concept of qualitativeness with the help of the particular sort of time-involving characterization he has given for "grue" and "bleen," this concept is not absolute but relative. That is, if "green" and "blue" are temporally qualitative predicates, then "grue" and "bleen" must be nonqualitative. But since "green" and "blue" can be characterized in the same sort of way, it is also true that if "grue" and "bleen" are temporally qualitative, then "green" and "blue" must be nonqualitative. But these facts can ground only a judgment of relative nonqualitativeness, unless supplemented with an independent criterion of absolute qualitativeness. Fain's recent discussion of qualitativeness [3] seems to presume what Goodman doubts, that

6. "Green" is simpler than "grue."

In what sense simpler? If graphical simplicity is intended, presumably what is meant is that a plot against time of emeralds examined for color looks simpler if all the emeralds fall on the line representing the color green (straight line), than if those examined after *t* fall on the line representing blue (step-function). This is true, but similarly a plot of emeralds examined for abcolor looks simpler if all the emeralds fall on the line representing the abcolor grue (straight line), than if those examined before *t* fall on the line representing grue while those examined after *t* fall on the line representing bleen (step-function). The criterion does not help us to decide between examining for color and for abcolor.

But perhaps a criterion of psychological simplicity, e.g., of applicability, or of response to what has reached us in experience, is intended. Such a suggestion would not imply a necessary conflict with a projectibility criterion, as already noted. But the kind of problems that might arise in connection with it have been indicated under proposals 1 and 3 above.

7. The hypothesis using "green" is related to other hypotheses which contribute to its endorsement; not so that using "grue."

This is true as a matter of fact, but again it does not help us to distinguish "grue" from "green" except by the fact that "green" is in use in various hypotheses, some of which mutually support one another, while "grue" is not. Depending on one's understanding of the essential features of plausible

---

such an independent, absolute criterion can be achieved which will not beg the projectibility question. Furthermore he seems to presume that according to some such criterion "green" is obviously qualitative (and of course that "grue" is not). Fain says it is as absurd to call "green" temporally nonqualitative because it can be defined with the help of a "temporal" expression as to call "red" self-contradictory because it can be defined wth the help of a self-contradictory expression. This is true but not at issue. Whatever the merits of arguing analogically from the actuality of the "semantic self-consistency" (my term) of "red" to the possibility of the "semantic qualitativeness" of "green," it is the actuality of the latter that Goodman doubts. Fain's contention that qualitativeness is no more a relative matter than contradictoriness depends on our being able to tell whether a given predicate is semantically qualitative as we can tell whether it is semantically self-consistent (supposing the latter granted). In this connection it is worth noting that Goodman ([7], p. 330) has argued against Thomson ([14]; see also her reply [15]) that "grue" can be applied correctly without meeting the alleged extra condition of applicability, i.e., knowledge of the current time, which would presumably be the sort of thing Fain has in mind as definitely establishing the semantic nonqualitativeness of "grue" in contrast to "green." (Ackermann's interesting discussion in [1], pp. 30–32, is relevant here, and the more evidently so if we modify his example to allow his specification of "grue" to approximate more closely Goodman's original characterization.) Once again, however, the more important question is whether in any case the specification of such an extra condition of applicability can be generalized to capture all and only nonprojectible predicates without question-begging.

endorsement mechanisms (pp. 107–118), one can even make a case for the continuance of all endorsement relations under the uniform substitution of "grue" for "green." However that may be, it seems that for a given hypothesis *h* endorsing the hypothesis using "green" one can devise hypothesis *h′*, competing with *h*, which endorses the use of "grue," the competition being of the sort already at issue (pp. 76, 77).

8. If the characterization of "grue" in terms of "green" is relativized to any inductive occasion prior to the current one whose time is t_n, the time t becomes that of that prior occasion, t_k. Then the projection to the next case is in disagreement with the results of the already past examination of that case. We take as our criterion that no predicate is to be allowed which cannot pass this relativization test.

This criterion appears promising at first, because it seems to distinguish between "green" and "grue" in favor of the former. Even if "green" is characterized in terms of "grue" and "bleen" and t, when this characterization is relativized to a prior inductive occasion of time t_k the projection to the next case agrees with the now past examination of that case; i.e., that case is one examined after t_k and green. But while this is true, it is true only if we relativize the characterization of "grue" and "bleen" under the constraint that "green" and "blue" retain the extensions they have in present usage. On what non-question-begging ground can we justify using a test which thus favors the normal predicates? Suppose instead I consider "green" to be the defined term and insist on a relativization of its characterization (i.e., taking 't_n' into 't_k') under the constraint that "grue" and "bleen" retain the extensions given them by Goodman's actual characterizations (in terms of 't_n') of them. Then the projection of "grue" at t_k would have been shown successful at the next examination and that of "green" (as relativized) a failure.

Furthermore, the proposed criterion has two other deficiencies.[16] The first is that it cannot be satisfactorily generalized in any obvious way. We realize at once that the unprojectibility of unprojectible predicates cannot always be tied to the time of the examination of evidence cases. Suppose we introduce "grupe" to apply to those things examined at place *p* and found green or to other things found blue. Then if all our evidence cases up to the time of the present inductive occasion have been examined at *p*, a relativization in terms of time of inductive occasion will clearly leave "green" and "grupe" on equal footing. Can we then eliminate "grupe" by relativizing in terms of place of examination? All evidence cases have in fact been examined at place *p*. How can we say what would have been found true of one of them

[16] In a preliminary version of this paper these were the only two deficiencies noted. The "asymmetry" between "green" and "grue" established by the proposed criterion I took at face value until a questioning comment by Professor Goodman led me to think further about it.

had it been examined at some other place without begging the question at issue on the present inductive occasion? It is clear that while projection may depend on a time difference between examined and unexamined cases the "grue" example does not get its point by any unfair exploitation of that fact.

In the second place, the present criterion, even in the restricted version we are considering, seems to throw out predicates we might very well want. There are surely situations in which we want to predict, for example, that some alteration in phenomenal aspect of certain entities will take place at such and such a time. In a competition to describe these entities, and project over them, a predicate will surely be preferred that cannot pass the relativization test.

9. Consider two competing hypotheses, *H* and *H′*, where *H* uses a predicate whose conditions of application would not prevent a crucial experiment's having been carried out prior to the time *t* of the present inductive occasion for every hypothesis using the predicate, while *H′* uses a predicate whose conditions of application would prevent such an experiment in the case of every hypothesis using that predicate and formed to compete with some hypothesis using the alternative predicate; *H* is to be projected rather than *H′*.

The idea here is plain enough. For no hypothesis pair using "green" and "grue" respectively does it appear that one could manage a test case to decide between the members of the pair prior to time *t*. What is not so plain is how we can specify which predicate's conditions of applicability are at fault without prejudice in favor of the one that happens to be in use, namely "green." For we must preserve the relationship between "green" and "grue" determined by Goodman's characterization of the latter, or else we have altered the example we are trying to explain. Hence "green" and "grue" must apply equally well over all examined cases of all hypotheses using either before *t*, and disagree over all examined cases after *t*. How is it any more the fault of the conditions of applicability of "grue" than of those of "green" that they should not be able to be found in disagreement before *t*, except that "green" was here first? It is quite true that on the inductive occasion of time *t* we expect future examined cases to be "continuous in greenness" with past examined cases, and to be "discontinuous in grueness." But this agreement on the one hand, and this disagreement on the other, are not at time *t* matters of "descriptive knowledge," but of expectation, and neither can be used at time *t* to count for, or against either hypothesis. Had we been all along thinking of green things not as such but as grue things and been expecting them to be continuous in grueness, then we should be inclined to blame the conditions of applicability of "green" for our not being able to decide between "green" and "grue" before *t*.

Hence it appears that this criterion cannot discriminate projectible from unprojectible hypotheses. But even if it could, and if it succeeded then in rejecting only unprojectible hypotheses, it could not reject all unprojectible hypotheses. For it may be that no cases of green things examined by time *t* happen to be of class *B*, so that we can think to project "All emeralds are green" and "All emeralds are grube," where "grube" applies to anything not in class *B* just in case it is green, and to anything in class *B* just in case it is blue, and where nothing in the conditions of applicability of "green" or of "class *B*" (and therefore of "grube") precludes that something of class *B* has been examined before *t*. (E.g., class *B* might be the class of things located at any of places $P_1, P_2, \ldots, P_i$, or the class of things broken in two by any hydraulic press of a certain sort, and so on.) But "All emeralds are grube" is no more projectible than "All emeralds are grue," by comparison with "All emeralds are green."[17]

IV. COATING THE ENTRENCHMENT PILL

Certainly other proposals might be considered, and perhaps eventually one of them may be made to work, either ruling out the entrenchment proposal or "explaining" it in other terms. But the negative results just noted make it worthwhile to give the entrenchment proposal a sympathetic hearing. This means, in part, giving credence to the possibility that the intended point of the "grue" example is well taken, i.e., that we cannot explain intuitive projection preferences, nor ground a justifying principle of comparative projectibility, in the comparative descriptive adequacy of predicates to what has been literally established in "the evidence." But it also means what may for some be harder to allow, entertaining the possibility that what is needed in a predicate besides descriptive adequacy is a sufficient degree of comparative entrenchment. In the remainder of this paper we try to make it easier to entertain both these possibilities by relating them to other, presumably familiar, features of our epistemological response to experience. We do so first by arguing that we have a familiar way of ascribing properties to things, reflected in the usual significance of our predicates, which goes beyond what has in fact been established in our experience and is therefore in part anticipatory; if we take note of this fact, both possibilities become more believable in a rather obvious way. Secondly we argue that the entrenchment proposal is quite credible as an answer to the need suggested by the point of the "grue" example because it can be seen as a special case of a methodological principle employed in answer to similar needs fairly general both in science

[17] A variation of proposal 9 would rule out any predicate "disjunctive" in conditions of applicability, one of whose disjuncts is empty of examined cases at *t*. Unfortunately, "examined after *t* and bleen" or "examined after *t* and green" alike constitute such disjuncts.

and in our everyday response to our environment. To these two arguments we now turn.

1. Goodman observes that Kant has taught us to be wary of supposing that "the way things are" is independent of our knowing that they are that way. Then he suggests that what we mean by "the way things are" at any given time is in part determined by the words in use in our language, and that it is just to this part that we must have recourse to explain our inductive preferences. This seems hard to take because it seems to require that things turn out to be green rather than grue because we happen to have started applying "green" to things rather than "grue." But no matter how many emeralds may have been found to be grue up to this time, we simply don't believe that those examined after *t* will be found grue. And this, we feel, has nothing to do with the mere fact that "grue" has never been actually projected before, because that has nothing to do with the question whether the emerald next examined after *t* is blue or green.

However, since Goodman explicitly denounces the attempt to decide the validity of projections by reference to unexperienced cases (p. 99), we should seek an understanding of his suggestion without this unacceptable implication. To say that the way things are is partly determined by the words we use would be truistic if it meant merely that what we expect to experience in the future is evidenced in the predicates with which we project over future cases; and it would be misleading (if not worse) if it were only an alternative locution for a recommendation that we continue to use the predicates in successful use. A better understanding of the suggestion is forthcoming, I believe, if we consider what lies behind our utterance when we insist that things just "are" green and just "are not" grue. There is a common and legitimate use of "are" which will tolerate the construction of such an utterance. Indeed when we say, "the way things are," we are almost always making such a use. It is a use which incorporates a projective element. In my judgment, it is of this "are" which anticipates what shall be, and does not just reflect on what has been literally established, that a predicate-in-use can be understood as partly determinative.

To revert to our example, if we insist that despite the applicability of "grue" to all and only those tried cases to which "green" applies nevertheless things just aren't grue but they are green, we must be using "are" in a way that anticipates untried cases. If we were talking only about tried cases we would be flatly contradicting ourselves, since the characterization of "grue" guarantees its application to the same tried cases as "green."[18] In this

[18] Exactly the same can be true even where the past tense is used, as in "Emeralds have been green rather than grue," and even where there seems to be an explicit restriction to tried cases, as in, "All tried cases of emeralds have been green rather than grue." In the respect in which an emerald would normally be considered a

sense of "are" where we are already talking about cases in general, and about what exhaustive examination alone can decide, and not just about what can be affirmed on the basis of cases actually tried, the way things "are" depends in part, as Goodman says, on "how the world . . . has been described and anticipated in words" (p. 119). Not with the careful, worried "are" of epistemological discourse, in which we distinguish what has actually been established from what may perhaps be found to be, do we say that the way things are depends on the predicates we use, but with the generous, trusting everyday "are" that reports our persuaded anticipation about things as a facet of a single insight into their observed nature. What we are willing to concede to hold for all examined emeralds, namely that they are both grue and green, does not depend on the words we use, but on what has been presented. What we are persuaded holds both for examined and unexamined emeralds, namely that they are green, depends not just on what is presented, but on the words that have been used, and used successfully, in "organizing" the presented (p. 96). The usage of words is partly determinative, not of truth, but of projective validity.

2. But we have yet to say how it can be satisfactory to regard superior entrenchment of a predicate as the normative mark of superior projectibility of a hypothesis using that predicate. The only answer that can be given, it seems to me, is that nothing in the way of a superior fit to the data would be provided by an alternative choice of (a less-well-entrenched) predicate at the present projection time, and nothing in the way of superior projective promise. "Green" has done for us, in all examined cases, everything that "grue" could have done for us, both descriptively and predictively. To see things as grue rather than green is admittedly logically possible, but gains us nothing that we do not already have in seeing them as green. Where a newly suggested predicate differs relevantly from the old only in untried cases of every hypothesis in which the old has always appeared, what would

---

tried case, the emerald has not been tried through the time *t* of the present inductive occasion. (See Sec. II, 1d, and n. 9.) The trial, therefore, cannot have distinguished any emerald as green *rather than* grue. To speak in this way is to take the actual trial to have established something about an emerald over what is, by hypothesis of the "grue" example, the temporal span of the projective. It is to anticipate that future scrutiny will show already scrutinized emeralds continuous in greenness through *t*, and is therefore already to make a projection, despite the past tense. Of course we can, for simplicity, assume that trial of an emerald establishes its color through *t*, whence it must, by the hypothesis of this assumption, be wholly in the evidence class of emeralds at *t* and wholly excluded from the projective class (n. 8). But since fixing color behavior through hypothesis entails a concurrent fixing of abcolor behavior, it remains true that pre-*t* tried emeralds cannot be green and not also be grue. The same holds if we take "tried case of emerald" to mean, also unnaturally, what is meant by "exhaustively scrutinized emerald-pre-*t*-temporal-segment."

prompt a switch to the new?[19] To put it another way, we don't consider projecting a new predicate without a good reason, and the wholly satisfactory service of an old predicate is not a good reason. Our reason for preferring "green" to "grue" at the present projection time is that we have no good reason to prefer "grue" to "green."

From a logical standpoint there is no doubt that this amounts to a kind of conventionalism. However it is the same sort of more or less virtuous conventionalism that operates in much of scientific methodology, particularly in our reliance on hypotheses that have proved useful. We do not set one hypothesis aside in favor of a competitor that does nothing for any existing evidence whatsoever that the first does not do (i.e., without a reason in the empirical situation).[20] May we not become accustomed to the thought that we do not set one predicate aside in favor of a competitor that does nothing for any existing evidence whatsoever that it does not do? To choose a better entrenched predicate in a projection seems to be merely the following of an extension of the same methodological principle of insufficient reason used in the selection of a better established hypothesis over alternative hypotheses. We have seen this principle employed in the latter way long before Goodman showed us new ways of generating such alternative hypotheses through predicate tailoring.

What we are saying, after all, amounts to this, that a predicate is a part of what is being tried against the world when we try the hypothesis in which it appears against the world. It is being tried for its success in projecting as the hypothesis itself is being tried for its success as a projection. Indeed, in a certain sense a predicate is a kind of hypothesis. For as I tried to argue above, even to say that a particular thing of a certain kind is green is quite commonly to mean that it is not, among other things, grue. But it makes no sense to regard such an "excluding" assertion merely as a statement of fact. It is a tacit hypothesis about what shall be found to hold about other things of the same kind. It is a statement whose function has gone halfway towards that of the generic particular form, more readily recognizable as hypothesis, "The emerald is green" (compare "The tiger is a mammal").

19 "New" here refers us to predicates with a poor record of actual projective use. There is no question there may be reason to try in a certain area predicates new to that area, but otherwise old. For example, cases of *A*, always found to be *B*, may come to be seen as special cases of *A′*, which have always been found to be *C*. This could motivate the projection that every *A* is a *C* also, though only in as-yet-unexamined cases of *A* could the difference between *B* and *C* be determined, and though '*C*' has never been applied to an *A* before.

20 Even in the case of broad, integrative theories, attractive on non-empirical grounds (e.g., scope, simplicity), a competing older theory is not regarded as empirically inferior (e.g., "merely approximate") unless an empirical basis for this devaluation is found.

Of course, there is a difference in the estimation of the success of a given predicate and of a given hypothesis employing it, because the two are not held to the same criterion of utility. This is only to repeat what has been said earlier. The failure of the given hypothesis does not by itself annul the predicate's success-in-projecting character. This is true if the predicate has been often successfully projected, perhaps even in this hypothesis. It is true also if the predicate has never been tried before, or tried but never successfully, for that matter.[21] Similarly a hypothesis may succeed in one or more trials, while the predicate it uses is one that has been remarkably unsuccessful in other hypotheses and may shortly turn out the same in the present one also.

Nevertheless, in a general way, predicates are tried for success-in-projecting as the hypotheses in which they appear are tried for success as projections. As an unviolated, confirmed hypothesis is only one of a great number of alternatives which are equally well supported by the same positive instances, so its entrenched predicate *P* is only one of a great number of alternatives which equally well apply to all the same entities described by all positive instances of all the hypotheses in which *P* has appeared. To rely on predicate entrenchment in the normative assessment of projectibility seems neither more nor less harmful than to rely on previously satisfactory hypotheses generally.

The present discussion brings us full circle to the problem of Sec. I. May we reduce the problem of alternative predicates entirely to that of alternative hypotheses, or vice versa? As promised in Sec. I, this is a question we consider no further here, along with such as the following: How does the firm violation of a hypothesis, which certainly necessitates a new hypothesis, stand in respect to the necessitation of new predicates? Can we count on relative predicate entrenchment to mark relative psychological simplicity in the reactions of "the human race" to the presented? Do we invoke new predicates (new extensions) not just because of hypothesis violation but also to serve alleged hypothesis broadening to cover more evidence from diverse areas? The discussion of these questions may be allowed to wait another opportunity.[22]

21 Incidentally, were not entrenchment a relatively coarse measure of projectibility anyhow, one would perhaps want to define degree of entrenchment in terms of number of successful projections, rather than in terms of mere projections, and perhaps even introduce positive and negative entrenchment. Ignoring the distinction is warranted, of course, by the consideration that for a well-entrenched predicate there has presumably been until now an average temporal convergence of number of successful projections toward total number of projections.

22 Work on which this paper is based was supported in part by stipend of a Mellon Postdoctoral Fellowship at the University of Pittsburgh, 1966–67. I should like to thank Robert Schwartz, Robert Ackermann, and Professor Nelson Goodman for reading this paper evaluatively. As a consequence of their com-

REFERENCES

[1] Robert Ackermann, *Nondeductive Inference* (New York, Dover, 1966).

[2] S. F. Barker and Peter Achinstein, "On the New Riddle of Induction," *The Philosophical Review*, vol. 69 (1960), pp. 511–522.

[3] Haskell Fain, "The Very Thought of Grue," *The Philosophical Review*, vol. 76 (1967), pp. 61–73.

[4] Nelson Goodman, *Fact, Fiction, and Forecast* (Cambridge, Massachusetts, Harvard University Press, 1955; 2d ed.; New York, Bobbs-Merrill, 1965). (Page numbers appearing in parentheses in the text are references to the second edition of this work.)

[5] ——— "A Query on Confirmation," *The Journal of Philosophy*, vol. 43 (1946), pp. 383–385.

[6] ——— "Faulty Formalization," *The Journal of Philosophy*, vol. 60 (1963), pp. 578–579.

[7] ——— "Comments," *The Journal of Philosophy*, vol. 63 (1966), pp. 328–331.

[8] Marsha Hanen, "Goodman, Wallace, and the Equivalence Condition," *The Journal of Philosophy*, vol. 64 (1967), pp. 271–280.

[9] Carl Hempel, "Studies in the Logic of Confirmation," *Mind*, vol. 54 (1945), pp. 1–26, 97–121; reprinted in Hempel, *Aspects of Scientific Explanation* (New York, Free Press, 1965), pp. 3–46, with a postscript written in 1964, pp. 47–51.

[10] James Hullett and Robert Schwartz, "Grue: Some Remarks," *The Journal of Philosophy*, vol. 64 (1967), pp. 259–271.

[11] Hughes Leblanc, "That Positive Instances Are No Help," *The Journal of Philosophy*, vol. 60 (1963), pp. 453–462.

[12] Israel Scheffler, *Anatomy of Inquiry* (New York, A. A. Knopf, 1963).

[13] Brian Skyrms, *Choice and Chance* (Belmont, California, Dickenson, 1966).

[14] Judith Jarvis Thomson, "Grue," *The Journal of Philosophy*, vol. 63 (1966), pp. 289–309.

[15] ——— "More Grue," *The Journal of Philosophy*, vol. 63 (1966), pp. 528–534.

---

ments and questions a number of modifications have been incorporated into the text in the interest of a clearer and more correct statement of "grue" problem complexities.

Postscript

ARE THERE NON-DEDUCTIVE LOGICS?

Wilfrid Sellars

I

(*1*) Without attempting to define what is meant by a 'logic', it seems reasonable to say that, however many 'logics' there are, they are 'logics' by virtue of their concern with what makes an argument sound. In the case of 'deductive logic', the concept of a sound argument is that of an argument which is such that if its premisses are true, its conclusion *must* be true. A *good* deductive argument is one which is not only sound (valid) but has true premisses, and hence a true conclusion.

(*2*) An argument purports to be 'deductive' if its conclusion is qualified by the parenthetical adverb 'necessarily', thus:

All men are mortal.
Socrates is a man.
So (necessarily) Socrates is mortal.

All Texans are rich.
Getty is rich.
So (necessarily)
Getty is a Texan.

Although both these arguments *purport* to be deductive, the former is correct (valid) and indeed sound, while the second is incorrect (invalid).

(*3*) Notice that the parenthetical comment 'necessarily' does not mean that the conclusion by itself is a necessary truth. The necessity indicated is *relative* necessity; the conclusion (if the argument is sound) is necessary relatively to the truth of the premisses. Of course, the conclusion of an argument may be a *statement of necessity*, thus

From Nicholas Rescher *et al.* (eds.), *Essays in Honor of Carl G. Hempel.* Copyright 1969, D. Reidel Publishing Company, Dordrecht, Holland. Reprinted by permission of the publisher and the author.

All propositions the denial of which is self-contradictory are necessary.
The denial of the proposition 'Either snow is white or snow is not white' is self-contradictory.
So (necessarily) the proposition 'Either snow is white or snow is not white' is necessary.

(Note the occurrence of *both* the parenthetical adverb 'necessarily' and the modal adjective 'necessary'.)

(4) In the conclusion of an argument which purports to be deductive

So (necessarily) *p*,

the conclusion *asserts* '*p*' and *signifies* by the parenthetical comment that '*p*' is the conclusion of a deductively good argument, i.e., one which is valid and has true premisses.

(5) An argument which purports to be deductive may be incomplete (an 'enthymeme') in the sense that one or more premisses are left to be understood. Thus,

All men are mortal.
So (necessarily) Socrates is mortal.

leaves the premiss 'Socrates is a man' to be understood. If it is supplied, one gets a *complete* argument which not only purports to be deductive, but is actually both valid and good. It is notoriously difficult, in many cases in which one is offered an argument which purports to be deductive, to decide whether the person who offers it is leaving premisses to be understood which, if supplied, would turn it into a valid deductive argument, or whether he thinks, mistakenly, that the argument as it stands is deductively valid. Thus it is often not clear whether a bad argument is intended as complete and should therefore be criticized as invalid, or whether, when understood premisses are supplied, it turns out to be deductively valid, but bad, because these premisses are false.

(6) Now, *prima facie*, there are good arguments which, if taken to be complete, are not only *not* deductively valid, but do not even *purport* to be so. Consider, for example

Black clouds are gathering.
So (probably) it will shortly rain.

This argument does not even purport to be deductively valid, as is indicated by the use of the parenthetical comment 'probably' (as contrasted with the 'necessarily' of the arguments we have previously considered). Let us say that it purports to be a good probability argument. Even as a probability argument, however, it is clearly not complete. It is plausible to suggest that we are to supply some such premiss as

When black clouds gather, it usually rains.

When this premiss is made explicit, the argument becomes

When black clouds gather, it usually rains.
Black clouds are gathering.
So (probably) it will shortly rain.

It now strikes us as sound, even though it neither is, nor purports to be, deductively valid. If we suppose that the premisses are true, we are tempted to say that it is not only sound, but good; a good probability argument.

(7) Notice that the conclusion does not say

It is probable that it will shortly rain

anymore than the conclusion of 'old mortality' said

It is necessary that Socrates is mortal.

Just as the deductive conclusion *asserted* that Socrates is mortal, and *implied* (or *signified*) that the conclusion was necessary relatively to true premisses; so the conclusion of the present argument *asserts* that it will shortly rain, and *implies* that this conclusion is *probable* relatively to premisses which are true and which satisfy a further condition which can be expressed by saying that the premisses formulate all relevant knowledge. I shall henceforth abbreviate this condition by the letter '*Q*'.

(8) In the case of 'old mortality', we pointed out that the adverbial qualifier '(necessarily)' implies that the conclusion is necessary relatively to the premisses (and that the premisses are true). By parity of reasoning we would expect that if our probability argument is good, then the following statement is true.

'It will shortly rain' is probable relatively to the premisses, 'When black clouds gather, it usually rains' and 'Black clouds are gathering', which are true and contain all that is known to be relevant.

(9) A (complete) deductive argument is good, if its conclusion stands in a certain logical relation to its premisses, and its premisses are true. By parity of reasoning a (complete) probabilistic argument is *good* if its conclusion stands in a certain relation (which it is presumably proper to call a logical relation) to its premisses, and its premisses are true (and satisfy *Q*).

II

(*10*) There are interestingly different varieties of argument having the general form

g & *Q*
So (probably) *p*.

In the first place, there is the distinction between those cases in which we say simply

g & Q
So (more probably than not) p,

and those in which we say something like

g & Q
So (probably to degree .9) p.

Abstractly put, we distinguish between a purely *comparative* and a *metrical* concept of probability. If we distinguish, as we are wont, between such species of probability as

The probability of theories
The probability of laws (law-like statements)
The probability of statements of proportion
The probability of singular statements

(a list not intended to be exhaustive), we have in mind such probabilistic arguments as

g & Q
So (probably) T.

where 'T' represents a specific theory;

g & Q
So (probably) LL,

where 'LL' represents a law-like statement (i.e., a statement which purports to formulate a scientific law);

g & Q
So (probably) n/NC_A is B,

where 'n/NC_A is B' says that n/N of a finite class of items of kind A are B; and, to complete our list,

g & Q
So (probably) x_1 is B.

In the last two cases we are often in a position to offer the stronger arguments

g & Q
So (probably to degree .75) n/NC_A is B,

and

g & Q
So (probably to degree .75) x_1 is B.

(*11*) To be confident that we understand the generic concept of probability, i.e., that we understand the nature of probability arguments, and what it is which makes them sound (even though they are not deductively valid), we must examine a variety of these specific forms of argument. To concentrate on one, to take one of them as the 'paradigm', is to run the danger of being misled by its specific character into misconstruing the generic notion of probability, and, therefore, of misconstruing even the 'paradigm'. (For to misconstrue a genus is *ipso facto* to misconstrue all its species.)

III

(*12*) I have argued that if the argument

g & *Q*
So (probably) *p*

is to be a good argument, there must be a relation (which can in a suitably broad sense be called 'logical' between '*g* & *Q*' and '*p*', such that the truth of the former makes it reasonable to accept the latter, even though the purported deductive argument

g & *Q*
So (necessarily) *p*

would be invalid. Now since the relation is to be 'logical' and to be of a kind which makes it reasonable to accept the conclusion if the premisses are true, it seems proper to speak of it as an implication relation, and to characterize it as 'probabilistic implication', since it is not the kind of implication which makes deductive arguments valid.

(*13*) This suggests that

g & *Q*
So (probably) *p*

is a good probabilistic argument if '*g* & *Q*' is true and '*g* & *Q*' probabilistically implies '*p*'.

(*14*) Can we say something more specific about this probabilistic implication? Well what sorts of things can we say about deductive implication which might be suggestive? One familiar line of thought is the following: Consider implication statements of the form

'*p* and (if *p*, then *q*)' (deductively) implies '*q*'

and consider the deductively valid arguments it authorizes, schematically,

p and (if *p*, then *q*)
So (necessarily) *q*.

All of these arguments have the property of being truth-preserving, i.e., they are such that if the premiss is true, the conclusion must be true. Now sound probabilistic arguments do not have this property. But perhaps they have *something like* truth-preservingness. Consider,

$(\frac{3}{4}\ C_A$ is $B)$ & Q
x_1 is C_A
So (probably) x_1 is B.

Arguments of this form are not truth-preserving. It is possible for an argument of this form to be sound, yet have true premisses and a false conclusion.

(*15*) However, let us represent the members of the class C_A by 'a_1', ... 'a_n' (C_A being a finite class of A's). Consider the class of specific arguments

$(\frac{3}{4}\ C_A$ is $B)$ & Q ... $(\frac{3}{4}\ C_A$ is $B)$ & Q
So (probably) A_1 is B ... So (probably) A_n is B.

(This class of arguments has n members—for we are concerned with *species* of arguments and not with argument occasions.) We see that $\frac{3}{4}$ of these arguments have true conclusions, if the premiss '$(\frac{3}{4}\ C_A$ is $B)$ & Q' is true. It seems, therefore, appropriate to say that this class of arguments is 'truth-preserving to degree .75'. By contrast, the class of arguments represented by

All C_A is B
x_i is C_A
So (necessarily) x_i is B

will *all* have true conclusions, if 'All C_A is B' is true. This class of arguments could be said to be 'truth-preserving to degree 1.00'.

(*16*) In contrast to the above class of probability arguments we could, indeed, have considered the stronger class

$(\frac{3}{4}\ C_A$ is $B)$ & Q ... $(\frac{3}{4}\ C_A$ is $B)$ & Q
So (probably to degree .75) a_1 is B ... So (probably to degree .75) a_n is B

thus making use of the quantitative concept of probability. But the arguments we have actually been considering are sufficient for our present purposes.

(*17*) Now, we might be tempted to generalize this line of thought, by construing the soundness of all probabilistic arguments

g & Q
So (probably) p

in terms of the idea that each such argument, if sound, belongs to a class of argument types of which it can be shown that the majority have true conclusions if their premisses are true.

(*18*) If we characterize an argument as 'successful' if its premisses and conclusion are both true, then to take this approach is to tie the concept of 'probabilistic soundness' to *success* in the sense that to know that such an argument is *sound* and its premisses are *true*, is to know that it belongs to a class of arguments, a majority of which are, in this sense, successful.

IV

(*19*) Now in the case of the probabilistic arguments we have been considering, there is an independent check on the success of such arguments. Thus, whereas we *could* use another argument of the same general form to establish that a particular argument (say α) is successful, e.g.,

$\frac{3}{4}$ arguments of the form
 ($\frac{3}{4}$ C_A is B) & Q
 So (probably) a_i is B
  have true conclusions, if the premiss is true.
α is an argument of this form.
The premiss of α is true.
So (probably) α is successful.

But we can also, given the appropriate observations, argue

The conclusion of α is '(probably) a_1 is B'
The premiss of α is '$\frac{3}{4}$ C_A is B'
This premiss is true (for we have examined C_A)
a_1 is B (observation), so 'a_1 is B' is true
So α is successful.

V

(*20*) Before we turn our attention to probability arguments which do *not* appear to be amenable to this treatment, let us note that it can, with considerable plausibility, be extended to arguments of the form (where '*OA*' stands for *A*'s which have been observed)

($\frac{3}{4}$ OA are B) & Q $OA = \{a_1, \ldots, a_{100}\}$
So (probably) a_{101} is B.

(*21*) For, let *PA* be the population consisting of $OA + a_{101}$. Clearly, if *OA* has, as specified, 100 members, the proportion of *B*'s in *PA* cannot be less than 75/101. Thus, at least 75/101 of the arguments

At least (75/101 PA is B) & Q
So (probably) a_i is B $a_i \in \{a_1, \ldots, a_{101}\}$

will have true conclusions if the premiss is true. Therefore, derivatively, at least 75/101 of the arguments

(75/100 *OA* is *B*) & *Q* $a_i \in \{a_1, \ldots, a_{101}\}$
So (probably) a_i is *B*

will also have true conclusions if the premiss is true. Consequently, the following would be a derivatively sound probabilistic argument,

(75/100 *OA* is *B*) & *Q*
So (probably to degree at least 75/101) a_{101} is *B*.

(*22*) The account we have been giving of certain probability arguments can, therefore, it would seem, be extended to arguments from a finite number of *observed* cases to an unobserved case. Indeed, by the use of a somewhat more complicated apparatus, it can be extended to arguments of the form

(75/100 *OA* are *B*) & *Q*
So (probably) approximately 75/100 C_A is *B*,

where C_A is a finite class of unobserved *A*'s. For, given a suitable relationship in size between the observed and the unobserved classes of *A*'s, the 'probably' will amount to 'probably to a considerable degree' and the degree of approximation of the proportion of *B*'s in C_A to 75/100 will be high.

(*23*) This extension would account for the soundness (within specifiable limits of the inference from the *B*-composition of a class of *observed A*'s to the composition of a finite class of *unobserved A*'s.

VI

(*24*) On the other hand, the preceding account would throw no light on the soundness of probabilistic arguments from statements about the composition of finite observed samples to *statistical law-like statements.* If we suppose that a statistical law-like statement has the form

The limit frequency of *B*'s in an infinite series of *A*'s is *n/N*,

it is clear that the preceding account throws no light on the soundness, if it be sound, of

(*n/N OA* is *B*) & *Q*
So (probably) the limit frequency of *B* in A_∞ is *n/N*.

And even if we reject this account of the form of a statistical law-like statement, the 'open-endedness' of law-like statements rules out the possibility that their form is akin to those we have been considering. To put it generally, our recent discussion throws no light on arguments of the form

(*n/N OA* is *B*) & *Q*
So (probably) $LL_{A,\, B,\, n/N}$,

where '$LL_{A,\,B,\,n/N}$' represents a law-like statement, whatever its proper form, which relates the proportion of *B*'s among *A*'s to the ratio n/N.

(*25*) It does not seem possible to construe the soundness of probabilistic arguments of the latter form as a matter of their being members of classes of arguments of which it can be shown that a majority have true conclusions, if their premisses are true, i.e., are successful, if *their* premisses are true.

(*26*) The possibility arises, therefore, that the account we have been giving of the soundness of arguments of the form

$(\frac{3}{4}\ C_A$ is $B)$ & Q
So (probably) a_i is B

contains a fundamental error, though enough truth to keep the error from bursting out into the open. To gain perspective, it will be useful to consider an ostensible type of probabilistic argument with respect to which—if it be recognized at all—quite a different account *must* be given. If we take seriously the idea that there is such a thing as the probability of theories, where by 'theory' I mean the kind of theory which, as we say, postulates entities, and processes which are not independently observable,[1] we commit ourselves to the idea that there are good and sound arguments of the form

g & Q
So (probably) T.

(*27*) What might the grounds represented by 'g' be? Intuitively, we feel that the probability of a theory is a function of its explaining established laws, suggesting new law-like statements capable of empirical tests, and, perhaps, being the simplest (whatever exactly *that* means) of the set of alternatives which are equally effective in the previous respects.

(*28*) Now it might be said that considerations of this kind have *nothing* to do with 'probability' of a theory, but rather with the *reasonableness of accepting it*. Well, they certainly have a lot to do with the latter, but they would *also* have a lot to do with the former, if to say of a theory that it is *probable* were the same thing as to say that *it is* (all things considered) *reasonable to accept it*.[2]

(*29*) Notice that if an argument of the form

g & Q
So, it is reasonable to accept T,

[1] I say 'not independently observable' because there is a dependent sense in which, e.g., the motion of an electron is observable—cf. gas and bubble chambers—but observable in a sense which involves a tacit appeal to the theory.

[2] This use of 'accept' must be distinguished from the ordinary sense in which one 'accepts', for example, the statements (testimony) of others. I use 'accept', in the first instance, as roughly equivalent to 'come to believe'. Believing and coming to believe are things that one does only in that broad sense in which anything expressed by a verb not in the passive voice is a 'doing'.

were sound and good (though it might be incomplete in the sense that certain additional premisses were 'understood'), then, since it would be reasonable to accept *T* (with the qualifier 'probably') and hence, for example, to inscribe the principles of *T* (with the qualifier 'probably') on a blackboard, it would then be reasonable to inscribe the *pair* of propositions '*g* & *Q*', '(probably) *T*' on the blackboard, thus

> *g* & *Q*
> (Probably) *T*.

Whether it would be appropriate to add 'so', thus

> *g* & *Q*
> *So* (probably) *T*

hinges on whether the '*g* & *Q*' is to be construed as contained among the premisses of an argument of which the conclusion is '(probably) *T*'. For if the only arguments which establish the reasonableness of inscribing '(probably) *T*' are arguments which have as their conclusion *not* '(probably) *T*', but rather

> So, it is reasonable to accept *T*.

then the reasonableness of inscribing the sequence

> *g* & *Q*
> (Probably) *T*

would *not* entitle us to write it as

> *g* & *Q*
> *So* (probably) *T*.

(*30*) As a remote parallel let us note that it is reasonable to accept '2 plus 2 = 4' *and* reasonable to accept 'The moon is round', and hence to inscribe

2 plus 2 = 4 The moon is round.	but *not*	2 plus 2 = 4 *So* the moon is round.

(*31*) If this line of thought is correct, then, even though the sequence

> *g* & *Q*
> (Probably) *T*

is correct and proper, there is no such thing as a probability *argument* of which the conclusion is

> So (probably) *T*.

(*32*) This suggests the possibility that in *no* case is there a probability *argument* of the form

> *g* & *Q*
> So (probably) *p*

i.e., that the concept of such probability arguments is an illusion; the division of argument into 'deductive' and 'probability' arguments a mistake. Notice that by a probability argument I mean an argument of which the conclusion is

So (probably) *p*

which *asserts p*, though in a qualified way. I do not mean to say that there are no probability arguments, if by this is meant an argument which has as its conclusion

So it is probable that *p*.

The latter conclusion does not assert '*p*'; it asserts a higher order proposition about '*p*'—perhaps the higher order proposition that it is reasonable to assert that -*p*.

VII

(*33*) Notice that correlated with every good deductively valid argument, for example,

(If *p*, then *q*) & *p*
So (necessarily) *q*

is a certain kind of second order argument, thus (roughly)

'(If *p*, then *q*) & *p*' implies '*q*'
'(If *p*, then *q*) & *p*' is true
So it is (epistemically) reasonable to accept '*q*'.

This second-order argument has a 'deductive' flavor which is worth savoring. Indeed, we notice that it would be deductively valid, if we add the premiss

It is ε-reasonable to accept propositions which are
implied by true propositions.

What about this premiss? It might be thought that it is a tautology; in particular that

'*q*' is implied by a true proposition

has the same sense as

It is ε-reasonable to accept '*q*'.

Yet it does not seem so. Implication is essentially truth-preservingness, whereas the latter concerns the reasonableness of doing something, accepting a proposition.

(*34*) Notice that if

> '*q*' is true

meant the same as

> It is ε-reasonable to accept '*q*',

then truth-preservation would amount to preserving 'ε-reasonableness of acceptance', and

> '*q*' is implied by a true proposition

would entail

> It is ε-reasonable to accept '*q*'.

(*35*) But

> '*q*' is true

does not mean the same as 'It is ε-reasonable to accept "*q*"', for it can be ε-reasonable for Jones to accept '*q*' even though '*q*' is, in point of fact, false.

(*36*) On the other hand, if we are prepared to say something like

> '*q*' is true = it is *ideally* ε-reasonable to accept '*q*'

and to distinguish *ideal* ε-reasonableness from ε-reasonableness *for Jones at time t*, then we could put the previous point by saying that while it cannot be *ideally* epistemically reasonable to accept a false proposition, it can be ε-reasonable (for Jones, at *t*) to accept it.

(*37*) It might be pointed out that while we are able to distinguish between the *concept* of a true proposition and the *concept* of a proposition which it is ε-reasonable (for me, now) to accept, statements of the form

> Although '*p*' is true, it is not ε-reasonable (for me, now)
> to accept '*p*'

are *pragmatically* absurd. Thus, in a certain sense of 'imply',

> '"*p*" is true' *implies* 'It is ε-reasonable (for me, now) to accept "*p*"'.

(*38*) From this point of view

> '*q*' is implied by a true proposition

implies (though it is not synonymous with)

> It is ε-reasonable (for me, now) to accept '*q*'

i.e., is, in this particular case a *ground* for the latter.

(*39*) If, following this line of thought, we distinguish between the '*ground*' of its being ε-reasonable (for me, now) to accept '*q*' and the *mean-*

ing of 'It is ε-reasonable (for me, now) to accept "*q*"' (as we distinguish between the ground of an obligation, e.g.,

I promise to do *A*

and the meaning of 'It is morally reasonable for me to do *A*', then we see that there might be *other* grounds for its being ε-reasonable (for me, now) to accept a proposition, than the fact that it is implied by a true proposition.

(*40*) Indeed we have already considered such another ground in our discussion of the proportional syllogism

($\frac{3}{4}$ C_A is *B*) & *Q*
So (probably) a_i is *B*.

A moment ago we paralleled the deductive argument

(If *p*, then *q*) & *p*
So (necessarily) *q*

with

'(If *p*, then *q*) & *p*' implies '*q*'
'(If *p*, then *q*) & *p*' is true.
So it is ε-reasonable (for me, now) to accept '*q*'.

Let us therefore construct the higher order counterpart of what we have been construing as a probability argument with the premiss '($\frac{3}{4}$ C_A is *B*) & *Q*'. It would be something like

'a_i is *B*' stands in R_L to '($\frac{3}{4}$ C_A is *B*) & *Q*'
'($\frac{3}{4}$ C_A is *B*) & *Q*' is true
So it is ε-reasonable (for me, now) to accept 'a_i is *B*'.

(*41*) In our previous discussion we construed R_L as the relation of

Being the conclusion of an argument belonging to a family of arguments $\frac{3}{4}$ of which have true conclusions if their premisses are true.

But in the light of our discussion of the probability of theories, it occurs to us that this might be a mistake, not because something *like* this relation does not obtain, but because (a) a more basic relation obtains, and (b) this more basic relation insures that $\frac{3}{4}$ of the *pairs of statements*—not necessarily *arguments*, though they were previously construed as arguments—

($\frac{3}{4}$ C_A is *B*) & *Q*
(Probably) a_1 is *B*

.
.
.

($\frac{3}{4}$ C_A is *B*) & *Q*
(Probably) a_n is *B*

have true second members, if the first member is true.

(*42*) The relation in question between

'a_i is B' and '($\frac{3}{4}$ C_A is B) & Q'

consists in the fact that it is a logical truth that

If '$\frac{3}{4}$ C_A is B' is true, then $\frac{3}{4}$ {'a_1 is B', 'a_2 is B', . . . , 'a_n is B'} are true.

(*43*) This suggests that the rationale of accepting the statement-pair

($\frac{3}{4}$ C_A is B) & Q
(Probably) a_i is B

rests on a rationale for accepting 'a_i is B' *which is not a matter of* '*(probably)* a_i *is B' being the conclusion of an argument.* The suggestion works out neatly as follows

It is ε-reasonable (for me, now) to accept all members of a set of propositions $\frac{3}{4}$ of which are true, if Q
$\frac{3}{4}$ {'a_1 is B', 'a_2 is B', . . . , 'a_n is B'} are true
Q
So it is ε-reasonable (for me, now) to accept all members of 'a_1 is B', 'a_2 is B', . . . , 'a_n is B'.

Both the reasonableness expressed by the major premiss and the reasonableness of accepting the deductive implications of true premisses would seem to concern the reasonableness of possessing a stock of propositions which are either true or within which a rationally controlled proportion are false.

(*44*) If this interpretation is correct, there is no such thing as an argument

($\frac{3}{4}$ C_A is B)
So (probably) a_i is B

i.e., no such thing as a non-deductive probability argument. The argument by virtue of which it is reasonable to accept '(Probably) a_i is B' has as its conclusion not '(Probably) a_i is B', but rather

It is ε-reasonable (for me, now) to accept 'a_i is B'

and the argument of which *this* is the conclusion is a *deductively* valid argument.[3]

VIII

(*45*) Suppose all this is correct, what is the larger framework into which it fits? What are the broader implications of the concept of epistemic rea-

[3] For a more detailed account of probability statements and arguments which explores the inter-relationships between the various modes of probability see my 'Induction as Vindication', *Philosophy of Science* vol. 31 (1964), pp. 197–231.

sonableness (for s, at t)? In the first place, the epistemic reasonableness in question is, as we saw, that of accepting a proposition. If, for the moment, we construe 'accepting a proposition' as 'bringing it about that one believes the proposition', we tie in with the concept of the reasonableness of an action.[4] To say that an action is (all things considered) reasonable is (in first approximation) to say that one has, all things considered, a good argument for doing it. This argument will be of the kind traditionally called practical. Such arguments have a variety of forms of which the following is an illustration

(All things considered) I shall bring about *E*.
Bringing about *E* implies doing *A* at *t*, if in circumstances *C* at *t*.
I am in *C* now.
So I shall do *A*.

The phrase 'all things considered' is essential, *for it is the root of the principle of total evidence.*

(46) The above argument contains a 'practical' premiss ('I shall bring about *E*'), two matter-of-factual premisses of quite different kinds ('bringing about *E* implies doing A_i at t, if in C_j at t' and 'I am in C_j now, at t' and a practical conclusion ('I shall do *A*'). It is clear that the skeleton of the argument is deductive, as can be seen by considering the argument

Jones brought about *E*.
Bringing about *E* implies doing A_i at t, if in C_j at t.
Jones was in C_j at t.
So Jones did A_i.

But just how this deductive skeleton is involved in the specific nature of practical reasoning is a complicated topic which I have explored on other occasions.[5]

(47) We saw that a *deductive* argument is good, if it is (a) valid, (b) its premisses (and hence its conclusion) are true. The argument about Jones would be a *good* one if its premisses were true. What about the original *practical* argument?

(48) Insofar as its factual premisses are true, it can be said to be 'factually' good. But its practical premiss, 'I shall bring about *E*', is, on the desired interpretation, neither true nor false. It has a force akin to 'Would that I

[4] A finer grained analysis of the context 'It is reasonable to accept that-*p*' would require careful use of the distinction between *rules of action* and *rules of criticism* developed, among other places, in 'Language as Thought and as Communication', *Philosophy and Phenomenological Research* vol. 29 (1969), pp. 528–535.

[5] Most recently in Chapter VII of *Science and Metaphysics,* Routledge and Kegan-Paul, London, 1968.

brought about E' or 'Let me bring about E'. The same is true of the conclusion.

(*49*) The argument could be said to be *good without qualification* if 'statements of intention', though neither true nor false, had a character analogous to 'truth'—'practical objectivity' we might call it, or even 'practical truth'. Thus if

(All things considered) I shall promote E

had 'practical truth', then the argument

All things considered I shall promote E.
Promoting E implies doing A_i at t, if in C_j at t.
I am in C_j now.
So I shall do A_i.

would be both *practically* 'good' as well as *factually* 'good'.

(*50*) If a practical argument is *practically* good, its conclusion will be said to be *categorically reasonable*. If it is merely *factually* good, its conclusion will be said to be *hypothetically reasonable* or reasonable relatively to the purpose expressed by the major premiss.

(*51*) If the argument is merely factually good, thus:

(All things considered) I shall poison my aunt today.
The only available poison is this prussic acid.

. . . . . .

So I shall give her this prussic acid.

it establishes what has been called a 'hypothetical imperative' i.e.,

If (all things considered) I want to poison my aunt today,
then I ought to give her this prussic acid.

But a practical argument which is merely *factually* good can never generate an unhypothetical or categorical 'ought'.

(*52*) Otherwise put, a practical argument which is merely factually good can generate:

If (all things considered) I want to bring about E, it is reasonable (for me, now) to do A,

but it can never lead to

It is categorically reasonable (for me, now) to do A

even if it is true that all things considered I want to bring about E, i.e., by the use of *modus ponens*.

(*53*) Now I have claimed that the argument which establishes the ε-reasonableness of accepting (i.e., which establishes the probability of) law-like statements, theories, singular statements, etc., are deductive arguments which have as their conclusion

> (All things considered) it is ε-reasonable (for me, now) to accept '*p*'; that is (All things considered) it is probable that *p*.

(*54*) The phrase 'ε-reasonable' carries with it the implication that the end in terms of which the accepting of '*p*' is justified by such arguments is an epistemic one, possibly, let us say, the possession by some person or persons of a maximum of truth.

(*55*) Given that the following of certain policies can be shown to be a necessary condition of achieving this end, there would be *available* factually good practical arguments of the form:

> (All things considered) I shall promote the possession by such-and-such a person or persons of a maximum of truth (E).
> Promoting E implies doing A_i at t, if in C_j at t.
> I am in C_j.
> So I shall do A_i (e.g., accept a certain theory, etc.).

(*56*) If I actually reason this way, my conclusion can be said to be reasonable relatively to the epistemic end in question.

(*57*) But can we remain satisfied with the idea that the reasonableness of accepting law-like statements, theories, singular statements, etc., is simply a function of an end one *happens* to have? Thus, suppose I simply *like* promoting the truth. I could say:

> If I want to promote the truth, I ought to accept '*p*'.

But the mere fact that I *want* to promote the truth—enjoy the idea of promoting it—could never generate the categorical

> I ought to accept '*p*'.

(*58*) There are two considerations which lead me to suppose that *wanting* to promote truth cannot be the end of the story. In the first place, probability statements are *intersubjective*. If Jones says to Smith, 'It is probable that *p*' and Smith agrees, they are agreeing about the same thing. This suggests that

> It is probable that *p*

has something like the sense of

> It is ε-reasonable (for *us*, now) to accept '*p*'.

In the second place whenever a person acts on a probability, he regards the kind of action he decides to do as *reasonable* because he thinks that a substantial series of such actions in the kind of circumstance in which the contingency may or may not occur would probably maximize relevant values or utilities. But, as Peirce pointed out, we regard such actions as reasonable even when we know that we *as individuals* will not be in that kind of circumstance often enough to make this consideration relevant and, to put an extreme case, we regard such actions as reasonable *even when we know that we are about to die*. Peirce concludes, correctly, I think, that in thinking and acting in terms of probability, we are, in a certain sense, identifying ourselves with a continuing community.

(*59*) Both of these considerations suggest that the prime mover of the practical reasoning involved in probabilistic thinking, indeed all logically oriented thinking, 'deductive' as well as 'inductive', 'practical' as well as 'theoretical', is not an idiosyncratic *wish* to promote truth, but the intention *as a member of a cummunity* to promote the total welfare of that community. This intention is implicit, and, when it becomes explicit, can be overcome by impulse and self-interest. But it is because truth is a necessary condition of securing the common good that the search for it presents itself to us, on reflection, as categorically reasonable—in the truest sense a moral obligation.

Bibliography

This Bibliography contains a selected list of texts, anthologies and "classic" books, each of which is highly recommended for those interested in pursuing the study of probability and inductive logic. The volumes listed below contain ample references to the general literature in this field; by far the most comprehensive bibliography in print is that provided in Kyburg [1].

INTRODUCTORY TEXTS

[1] Kyburg, Henry E., Jr. *Probability and Inductive Logic*. London: Macmillan & Co., 1970.

[2] Michalos, Alex C. *Principles of Logic*. Englewood Cliffs, N.J.: Prentice-Hall, 1969.

[3] Salmon, Wesley C. *The Foundations of Scientific Inference*. Pittsburgh: University of Pittsburgh Press, 1966.

[4] Skyrms, Brian. *Choice and Chance*. Belmont, Calif.: Dickenson Publis' ing Co., 1966.

ANTHOLOGIES

[5] Feyerabend, Paul K., and Maxwell, Grover, eds. *Mind, Matter, Method*. Minneapolis: University of Minnesota Press, 1966.

[6] Foster, Marguerite H., and Martin, Michael L., eds. *Probability, (firmation, and Simplicity*. New York: The Odyssey Press, 1966.

[7] Hintikka, Jaakko, and Suppes, Patrick, eds. *Aspects of Inductive Lo* Amsterdam: North-Holland Publishing Co., 1966.

[8] Kyburg, Henry E., Jr., and Nagel, Ernest, eds. *Induction: Some Curr Issues*. Middletown, Conn.: Wesleyan University Press, 1963.

[9] Kyburg, Henry E., Jr., and Smokler, Howard, eds. *Studies in Subjective Probability*. New York: John Wiley & Sons, 1964.

[10] Lakatos, Imre, ed. *The Problem of Inductive Logic*. Amsterdam: North-Holland Publishing Co., 1968.

[11] Morgenbesser, Sidney, ed. *Philosophy of Science Today*. New York: Basic Books, 1967.

[12] Schilpp, Paul Arthur, ed. *The Philosophy of Rudolf Carnap*. La Salle, Ill.: Open Court Publishing Co., 1963.

[13] Swain, Marshall, ed. *Induction, Acceptance, and Rational Belief*. Dordrecht, Holland: D. Reidel Publishing Co., 1969.

"CLASSIC" BOOKS

[14] Barker, S. F. *Induction and Hypothesis*. Ithaca, N.Y.: Cornell University Press, 1957.

[15] Braithwaite, Richard B. *Scientific Explanation*. London: Cambridge University Press, 1953.

[16] Bross, Irwin D. *Design for Decision*. New York: Macmillan Co., 1953.

[17] Carnap, Rudolf. *The Continuum of Inductive Methods*. Chicago: University of Chicago Press, 1952.

[18] Carnap, Rudolf. *The Logical Foundations of Probability*. Chicago: University of Chicago Press, 1962.

[19] Chernoff, Herman, and Moses, Lincoln. *Elementary Decision Theory*. New York: John Wiley & Sons, 1959.

[20] Cramer, Harold. *The Elements of Probability Theory*. New York: John Wiley & Sons, 1955.

[21] Fisher, R. A. *Statistical Methods and Scientific Inference*. New York: Hafner Publishing Co., 1956.

[22] Goodman, Nelson. *Fact, Fiction, and Forecast*. Cambridge, Mass.: Harvard University Press, 1955.

23] Hacking, Ian. *Logic of Statistical Inference*. London: Cambridge University Press, 1965.

] Hanson, Norwood R. *Patterns of Discovery*. London: Cambridge University Press, 1958.

Hempel, Carl G. *Aspects of Scientific Explanation*. New York: The Free Press, 1965.

] Jeffrey, Richard C. *The Logic of Decision*. New York: McGraw-Hill, 1965.

] Jeffreys, Harold. *Scientific Inference*. London: Cambridge University Press, 1957.

8] Keynes, John Maynard. *A Treatise on Probability*. London: Macmillan & Co., 1952.

9] Kneale, William. *Probability and Induction*. London: Oxford University Press, 1949.

[30] Kyburg, Henry E. *Probability and the Logic of Rational Belief.* Middletown, Conn.: Wesleyan University Press, 1961.

[31] Laplace, Pierre Simon marquis de. *A Philosophical Essay on Probabilities.* New York: Dover Publications, 1951.

[32] Levi, Isaac. *Gambling With Truth.* New York: Alfred A. Knopf, 1967.

[33] Luce, Duncan, and Raiffa, Howard. *Games and Decisions.* New York: John Wiley & Sons, 1964.

[34] Mill, John S. *A System of Logic.* New York: Harper & Brothers, 1895.

[35] von Mises, Richard. *Probability, Statistics, and Truth.* New York: Macmillan Co., 1957.

[36] Pap, Arthur. *An Introduction to the Philosophy of Science.* New York: The Free Press, 1962.

[37] Popper, Karl R. *Conjectures and Refutations.* New York: Basic Books, 1965.

[38] Popper, Karl R. *The Logic of Scientific Discovery.* New York: Harper & Row, 1968.

[39] Ramsey, Frank P. *The Foundations of Mathematics.* London: Routledge & Kegan Paul, 1950.

[40] Reichenbach, Hans. *The Theory of Probability.* Berkeley: University of California Press, 1949.

[41] Savage, Leonard J. *Foundations of Statistics.* New York: John Wiley & Sons, 1954.

[42] Scheffler, Israel. *The Anatomy of Inquiry.* New York: Alfred A. Knopf, 1963.

[43] Strawson, Peter F. *Introduction to Logical Theory.* London: Methuen & Co., 1952.

[44] Toulmin, Stephen. *The Uses of Argument.* London: Cambridge University Press, 1958.

[45] von Wright, G. H. *A Treatise on Induction and Probability.* New York Harcourt, Brace & World, 1951.